T0190040

Advances in Information Security

Volume 77

Series editor
Sushil Jajodia, George Mason University, Fairfax, VA, USA

More information about this series at http://www.springer.com/series/5576

Amos R. Omondi

Cryptography Arithmetic

Algorithms and Hardware Architectures

 Springer

Amos R. Omondi
State University of New York – Korea
Songdo, South Korea

ISSN 1568-2633   ISSN 2512-2193   (electronic)
Advances in Information Security
ISBN 978-3-030-34144-2   ISBN 978-3-030-34142-8   (eBook)
https://doi.org/10.1007/978-3-030-34142-8

This Springer imprint is published by the registered company Springer Nature Switzerland AG.
The registered company address is: Gewerbestrasse 11, 6330 Cham, Switzerland

To Yokoyama Masami

Preface

This book has been developed from notes that I wrote for a course on algorithms and hardware architectures for computer arithmetic. The course was a broad one—covering the usual areas of fixed-point arithmetic, floating-point arithmetic, elementary functions, and so forth—with a "tail end" on applications, one of which was cryptography. Parts II and III of the book are from that tail end. Adding Part I, on basic integer arithmetic—the part of computer arithmetic that is relevant for cryptography— makes for a self-contained book on the main subject.

The students who took the aforementioned course were final-year undergraduate and first-year graduate students in computer science and computer engineering. The book is intended to serve as an introduction to such students and others with an interest in the subject. The required background consists of an understanding of digital logic, at the level of a good first course, and the ability to follow basic mathematical reasoning. No knowledge of cryptography is necessary; brief discussions of some helpful basics are included in the book.

Part I is on algorithms and hardware architectures for the basic arithmetic operations: addition, subtraction, multiplication, and division.

Much of the arithmetic of modern cryptography is the arithmetic of finite fields and of two types of field in particular: prime fields (for which the arithmetic is just modular arithmetic with prime moduli) and binary fields. Part II covers the former and Part III the latter. Each part includes a chapter on mathematical fundamentals, a short chapter on well-known cryptosystems (to provide context), and two or more chapters on the arithmetic. Binary-field arithmetic is used in elliptic-curve cryptography (which also uses prime-field arithmetic); an introductory chapter is included on such cryptography.

Cryptography involves numbers of high precision, but there is no more understanding to be gained with examples of such precision, in binary or hexadecimal, than with those of low precision in decimal. Therefore, for the reader's "visual ease" and to ensure that he or she can easily work through the examples, all examples are of small numbers, with most in decimal. It is, however, to be understood that in practice the numbers will be large and the radix will almost always be a power of two; the hardware-related discussions are for powers of two.

A note on writing style: For brevity, and provided no confusion is possible, I have in some places been "sloppy" with the language. As an example, writing "number" instead of the more precise "representation of . . . number." Another example is the use of "speed" and "cost" in the discussion of an architecture; the terms refer to the realization of the architecture.

A note on notation: I hope meaning will be clear from usage, but the following examples should be noted. x denotes a number; x_i denotes bit or digit i in the representation of x; $\mathbf{x}_h$ denotes the number represented by several bits or digits in the representation of x; X_i denotes the value of X in iteration i; and $X_{i,j}$ denotes bit or digit i in the representation of X_i; in algorithms, "$=$" denotes assignment and x_P denotes the x-coordinate of a point P.

A final note is on the repetition of some text (a few algorithms). This is an aspect I have retained from the original lecture notes, as I imagine the reader will find it convenient to not have to go back over numerous pages for a particular algorithm.

This work was supported by the Ministry of Science and ICT (MSIT), Korea, under the ICT Consilience Creative (IITP-2019-H8601-15-1011), supervised by the Institute for Information & Communications Technology Planning & Evaluation (IITP).

Songdo, South Korea   Amos R. Omondi
July 2019

Acknowledgements

All Thanks and Praise to Almighty God

Contents

Part I

Chapter 1
Basic Computer Arithmetic

Abstract This chapter consists of a brief review or introduction, depending on the reader's background, of the basics of computer arithmetic. The first two sections are on algorithms and designs of hardware units for addition and multiplication. (Subtraction is another fundamental operation, but it is almost always realized as the addition of the negation of the subtrahend.) For each of the two operations, a few architectures for hardware implementation are sketched that are sufficiently exemplary of the variety of possibilities. The third section of the chapter is on division, an operation that in its direct form is (in this book) not as significant as addition and multiplication but which may nevertheless be useful in certain cases. The discussions on algorithms and architectures for division are therefore limited.

As appropriate, some remarks are made on high-precision arithmetic, which is what distinguishes "ordinary" arithmetic from cryptography arithmetic: precisions in the latter are typically hundreds of bits versus, say, thirty-two or sixty-four bits in the former. The representational radix in all cases is two; the operational radix too is two, except for a few cases in which it is a larger power of two. For all three operations—addition, multiplication, and division—we shall initially assume unsigned numbers for the operands and subsequently make some remarks on signed numbers.

We shall in various places make broad comments on cost and performance. A crude estimate of cost may be made in terms of number of logic gates in an implementation, and a similar estimate of performance may be made in terms of the number of gate delays through the longest path. In current technology, however, cost is best measured in terms of chip area, and interconnections also contribute to both that (especially by their number) and to operational time (especially by their length). We shall therefore assume a consideration of the number of gates, the number of gate delays in a critical path, and the number and lengths of interconnections, all of which are greatly influenced by the regularity of a structure. The reader who is really keen on better estimates of cost and performance can get some by working from the "data books" semiconductor manufacturers.

© Springer Nature Switzerland AG 2020

A. R. Omondi, *Cryptography Arithmetic*, Advances in Information Security 77,
https://doi.org/10.1007/978-3-030-34142-8_1

1.1 Addition

We describe four types of adder: the *serial adder, the carry-ripple adder*, the *parallel-prefix adder*, and the *carry-select adder*. The serial adder is the simplest possible adder. The core of the adder consists of the most fundamental unit in addition and, indeed, all computer arithmetic—the *full adder*, a unit for the addition of two bits. The serial adder exemplifies the repeated use of the same unit on different parts of the same (multi-bit) operands. (The description of this adder is also useful as a vehicle for the introduction of certain terminology.) The carry-ripple adder shows the straightforward use of replication: a basic unit is replicated as many times as necessary to obtain a unit for the required precision. The parallel-prefix adder shows the techniques used in the design of high-performance adders. And the carry-select adder shows how techniques used in the design of different types of adder can be combined in a single high-speed, high-precision adder.

The four types of adder are also representative of the design space, in terms of cost and operational time. For n-bit operands, the serial adder has constant cost and an operational time proportional to n. The carry-ripple adder too has both cost and operational time proportional to n, but with the constant of proportionality in the operational time smaller than for the serial adder. And the parallel-prefix adder has cost proportional to $n \log_2 n$ and operational time proportional to $\log_2 n$. The measures for the carry-select adder are slightly more complex because the adder is based on a hybrid of techniques used in the other adders. Generally, the cost and performance of the carry-select adder will be somewhere between those for a carry-ripple adder and those for parallel-prefix adder, and we will see that a time proportional to $\sqrt{n}$ is reasonable. Numerous other adder designs will be found in the standard literature on computer arithmetic [1–4].

In the ordinary paper-and-pencil addition of two multi-digit numbers,[1] the outputs of the addition at a particular digit position are a sum digit for that position and a carry to the next position, and the inputs are the two operand digits and a carry from the preceding position. All that can be reflected, in a reasonably straightforward manner, in the design of a digital arithmetic unit; that is the essence of a carry-ripple adder, which consists of a full adder for each bit-position. A serial adder, on the other hand, consists of a single full adder, so it is necessary to separate (in time) the carry from one pair of operand bits and the carry from an adjacent pair. For both the serial adder and the carry-ripple adder the time to add two multi-bit operands is proportional to the number of operand bit-pairs. In the carry-ripple adder, the time corresponds to the worst-case time to *propagate* carries, as in the binary addition of $111\cdots 11$ and $000\cdots 01$; and in the serial adder it corresponds to the number of cycles to add n bit-pairs. The *carry-propagation* delay is the most

---

[1]For brevity, we shall make a distinction between a number and its representation only if confusion is possible.

critical aspect in the performance of an adder, and a key objective in all adder designs is to keep it small. The best possible delay is proportional to $\log_2 n$ for n-bit operands.

In what follows $x_{n-1}x_{n-2}\cdots x_0$ and $y_{n-1}y_{n-2}\cdots y_0$ will denote the two n-bit operands to be added, $s_{n-1}s_{n-2}\cdots s_0$ will denote the result, and c_i will denote the carry from the addition at bit-position i. In ordinary paper-and-pencil addition there is no equivalent of c_{-1}, an initial carry into the addition. In computer arithmetic, however, such a carry is useful—for subtraction, addition with signed numbers, and modular addition.[2]

1.1.1 Serial

The core of the binary serial adder consists of logic to add two operand bits and a carry bit (from the addition of the preceding pair of operand bits) and some temporary storage to separate the carry input for the addition of a bit pair from the carry output of that addition. That logic is used repeatedly, in n cycles to add two n-bit operands. The details of the logic design are as follows.

Consider the addition of the bit pairs at position i of the two operands, x and y. The inputs are x_i, y_i, a carry-in, c_{i-1}, that is the carry-out from position $i-1$; and the outputs are a sum-bit s_i and a carry-out c_i. The corresponding truth table is shown in Table 1.1, whence the logic equations

$$s_i = (x_i \oplus y_i) \oplus c_{i-1} \tag{1.1}$$

$$c_i = x_i y_i + (x_i \oplus y_i)c_{i-1} \tag{1.2}$$

The one-bit addition unit obtained from these equations is a *full adder*.

The term $x_i y_i$ in Eq. 1.2 corresponds to a carry-out that is produced when both operand bits are 1s; this carry-out is independent of the carry-in and may therefore

Table 1.1 Logic table for 1-bit addition

x_i	y_i	c_{i-1}	s_i	c_i
0	0	0	0	0
0	0	1	1	0
0	1	0	1	0
0	1	1	0	1
1	0	0	1	0
1	0	1	0	1
1	1	0	0	1
1	1	1	1	1

---

[2]Modular arithmetic is discussed in subsequent chapters.

be regarded as a carry that is *generated* at that position. And the term $(x_i \oplus y_i)c_{i-1}$ corresponds to carry-out that is produced when the operand bits are such that the carry-in, c_{i-1}, is in essence passed through from the carry input to the carry output at position i; this is therefore known as a *propagated* carry. The distinction between a generated carry and a propagated carry is useful in the design of many types of adder, and in discussing those we shall accordingly make use of the two functions

$$g_i = x_i y_i \qquad \text{carry generate}$$

$$p_i = x_i \oplus y_i \qquad \text{carry propagate}$$

and, therefore, of this form of Eq. 1.2:

$$c_i = g_i + p_i c_{i-1} \tag{1.3}$$

g_i and p_i on their own also correspond to the addition of two bits without an incoming carry included: g_i is the carry output, and p_i is the sum output. The hardware unit for such an addition is a *half adder*.

Equations 1.1 and 1.2 are the most common expressions for s_i and c_i, and in their use advantage may be taken of the common term $x_i \oplus y_i$ to share the logic for s_i and c_i, although such sharing can increase the circuit delay. There are, however, other functions that can be used to propagate carries and which, depending on the realization, may allow s_i and c_i to be produced slightly quicker but at little or no extra cost. As an example of an alternative signal, direct algebraic expansion and simplification of Eq. 1.2 yield

$$c_i = g_i + t_i c_{i-1} \tag{1.4}$$

where t_i, which is known as the *carry transfer* function, is $x_i + y_i$. That t_i will work is evident from the fact that it includes p_i.

The inverse of t_i is known as the *carry kill* function and denoted k_i, and it too can also be used to propagate carries.

$$c_i = g_i + \overline{k_i} c_{i-1} \tag{1.5}$$

where $k_i = \overline{x_i} \cdot \overline{y_i}$. (The signal is named "kill" because if $k_i = 1$, then any incoming carry will not propagate past bit-position i; it will be "killed" there.)

The transfer function t_i is often used in a type of adder known as *carry-skip adder*, and the kill function k_i finds much use in *parallel-prefix adders* (Sect. 1.1.3). Other, less obvious, functions that can be used to propagate carries are discussed in [4, 6].

Figure 1.1 shows one possible gate-level design of a full adder; variations are possible, depending on the logic equations derived from Table 1.1. A serial adder is obtained from the full adder by including two shift registers to hold the operands, one shift register to hold the sum, and a one-bit register to separate carries. The arrangement is shown in Fig. 1.2. Note that delay through the carry-register time will be a significant part of the cycle time for the adder.

Fig. 1.1 Full adder

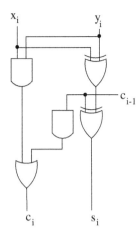

Fig. 1.2 Serial adder

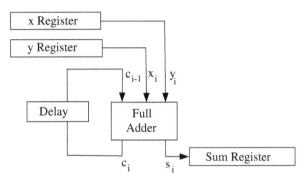

   The addition of two n-bit operands in a serial adder takes n cycles, each of which consists of adding a (new) pair of least-significant operand bits, shifting the output sum bit into the result register, shifting the operands by one bit-position to the right (thus "dropping" one bit pair and forming a new "least-significant pair), and "looping" back the output carry bit. S, the serial adder will be quite slow for anything but low-precision operands, although it has the advantage of very low cost. The nominal low speed notwithstanding, the serial adder can be used very effectively where there are numerous multi-bit additions that can be carried out in parallel. In such a case many serial adders can be employed concurrently, and, therefore, the *average* delay for a single multiple-bit addition will be small.

   The basic principle in the design of the serial adder is the serial processing of a sequence of bit pairs. That principle may be used as the basis of a faster adder, in the following way. Instead of radix-2 digits (binary), imagine digits in a larger radix (the equivalent of multiple bits per digit). And for the digit-pair additions consider one of the fast adders described below, but with the digit-addition still done serially. That is, if a large-radix digit consists of k bits, then the serial addition of every k bits gets replaced with faster, non-serial addition; and the overall addition is therefore faster. A few more details on this are given in Sect. 1.1.5.

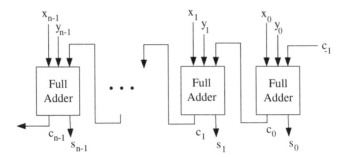

Fig. 1.3 Carry-ripple adder

1.1.2   Carry-Ripple

The design and operation of the carry-ripple adder closely reflects ordinary paper-and-pencil addition. For the addition of n-bit operands, a carry-ripple adder consists of n full adders, with the carry-output point of one full adder connected to the carry-input point of the next full adder (Fig. 1.3). The operational time of the adder is determined by the worst-case of carry propagation, from the least significant bit-position to the most significant—as in the addition of, say, $111\cdots11$ and $000\cdots01$—and is therefore proportional to n. This proportionality is the same as in the serial adder, but the constant factor will be smaller because the carry-ripple adder's time does not include the carry-register delay that is incurred in every cycle of the serial adder. The key challenge in the design of fast adders is to reduce the delay through the *carry-propagation path*, which in Fig. 1.1 is two gate delays per bit-stage.

Compared with several other types of adder, the carry-ripple adder is especially favorable for implementation in current technology, for two main reasons. First, a small basic unit (the full adder) and short interconnections (between and within full adders) mean a compact size and higher actual speed than is immediately apparent from the gate-level design. Second, in realization the carry-propagation path can be made much faster than is apparent from the gate-level structure of Fig. 1.1. One well-known way in which that can be done is by employing, for the carry propagation, what is known as a *Manchester carry chain*,[3] of which the reader will descriptions in standard texts on computer arithmetic and VLSI digital logic (e.g., [1–4, 17, 18]). Other[4] recent work that show how realization technology can be the basis of fast carry-ripple adders will be found in [19].

---

[3]The chain may be viewed as consisting of "switches" that are much faster than "standard" logic gates.

[4]All these provide good examples of the limitations in measuring operational time by simply counting gate delays.

The carry-ripple design can be used easily with any precision; all that is required is an appropriate number of connected full adders. With current technology, the design is excellent for adders of low or moderate precisions, but a carry-ripple adder will be slow for large precisions, for which different designs are therefore required.[5] The *parallel-prefix adder* and the *carry-select adder*, which we describe next, are examples of such designs.

As with the serial adder, the basic idea of rippling carries can be employed for faster addition with high-precision operands: wide operands are split into smaller "pieces," pairs of "pieces" are each added in a small high-speed adder, and carries are rippled between "piece" adders (Sect. 1.1.5).

1.1.3   Parallel-Prefix

In ordinary computer arithmetic, the *parallel-prefix adders* are the best for high performance in medium-sized and large adders, and they can be realized at a reasonable cost.[6] The essence of such an adder is based on two principles: *carry lookahead* and *parallel-prefix computation*. We shall start with an explanation of these.

We have seen above that the performance of an adder is largely determined by the worst-case carry-propagation delay. Therefore, if all carries could be determined before they were actually required, addition would be as fast as possible. That is the underlying idea in carry lookahead, the details of which are as follows.

The basic carry equation is (Eq. 1.3)

$$c_i = g_i + p_i c_{i-1} \tag{1.6}$$

Now consider the design of a 4-bit adder. We can obtain independent equations for all the carries by unwinding Eq. 1.6:

$$c_0 = g_0 + p_0 c_{-1}$$

$$c_1 = g_1 + p_1 g_0 + p_1 p_0 c_{-1}$$

$$c_2 = g_2 + p_2 g_1 + p_2 p_1 g_0 + p_2 p_1 p_0 c_{-1}$$

$$c_3 = g_3 + p_3 g_2 + p_3 p_2 g_1 + p_3 p_2 p_1 g_0 + p_3 p_2 p_1 p_0 c_{-1} \tag{1.7}$$

And these equations can be implemented to produce all the carries in parallel.

---

[5]It should be noted that recent work has shown that for moderate precision, with proper realization a carry-ripple adder can be competitive with adders that are nominally much faster (i.e., with performance is measured in terms of gate delays) [19].

[6]The very high precisions of cryptography arithmetic require the combination of several techniques.

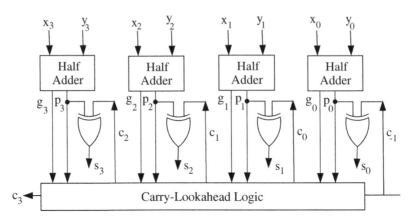

Fig. 1.4 4-bit carry-lookahead adder

The interpretation of the equation

$$c_i = g_i + p_i g_{i-1} + p_i p_{i-1} g_{i-2} + \cdots + p_i p_{i-1} p_{i-2} \cdots p_2 p_1 p_0 c_{-1} \qquad (1.8)$$

is that there is a carry out of bit-position i if a carry is generated at position i, or if a carry is generated at position $i - 1$ and propagated through position i, or if ..., or if, c_{-1}, the carry into the least significant bit-position of the adder is propagated through positions $0, 1, \ldots, i$.

The equation for the sum bits is unchanged:

$$s_i = p_i \oplus c_{i-1}$$

The complete gate-level design of a 4-bit carry-lookahead adder is shown in Fig. 1.4. A half adder consists of an AND gate (for g_i) and an XOR gate (for p_i). The carry-lookahead logic implements Eq. 1.7.

Equations 1.7 and 1.8 (and, partially, Fig. 1.4) show the parallelism inherent in the computation of the carries—which parallelism can be exploited in a various ways—but they also show fundamental problems with a direct application of the carry-lookahead technique: in general, very high fan-in and fan-out requirements and numerous lengthy interconnections. Dealing these problems leads to a large variety of carry-lookahead adders [1], of which the parallel-prefix adder is one good design, because of its high structural regularity. We next state the general *prefix problem*, explain how it is relevant, and then describe the designs of some parallel-prefix adders.

Let **A** be a set $\{a_0, a_1, a_2, \ldots a_{m-1}\}$ that is closed under an associative operator $\bullet$. That is:

- For all a_i and a_j in **A**, $a_i \bullet a_j$ is in **A**. (Closure)
- For all $a_i, a_j,$ and a_k in $(a_i \bullet a_j) \bullet a_k = a_i \bullet (a_j \bullet a_k)$. (Associativity)

(Since the brackets have no "real" effect, either expression in the associativity clause may be written as $a_i \bullet a_j \bullet a_k$.)

The *prefixes* $\mathbf{a}_0, \mathbf{a}_1, \ldots, \mathbf{a}_{m-1}$ are defined as

$$\mathbf{a}_i = a_i \bullet a_{i-1} \bullet \cdots \bullet a_0 \tag{1.9}$$

and their computation is the *prefix problem* [7]. The most important aspect of the associativity property is that a prefix can be computed by forming groups of its constituent elements, concurrently evaluating the groups, and then combining the results of those evaluations. That is, several prefixes can be computed in parallel, with advantage taken of their common subterms.

The computation of carries can be expressed as the computation of prefixes. Given c_{i-1}, c_i is determined by p_i and g_i (Eq. 1.6). We may therefore express c_i in terms of a function $[\cdots]$ that involves p_i and g_i:

$$c_i = [g_i, p_i](c_{i-1}) \tag{1.10}$$

We then define $\bullet$ as

$$[g_j, p_j] \bullet [g_i, p_i] = [g_j + p_j g_i, \; p_j p_i] \tag{1.11}$$

A straightforward exercise in Boolean Algebra will show that $\bullet$ is associative, whence this equivalent of Eq. 1.8:

$$c_i = ([g_i, p_i] \bullet [g_{i-1}, p_{i-1}] \bullet \cdots \bullet [g_0, p_0])c_{-1} \tag{1.12}$$

the "inner" part of which corresponds to Eq. 1.9.

Because $\bullet$ is associative, the subterms in Eq. 1.12 may be evaluated by forming different *blocks*, evaluating these in parallel, and then combining the results. A block will consist of adjacent bit-positions, so we may define *block propagate* (P_i^j) and *block generate* (G_i^j) signals that cover the bit-positions i to j:

$$\left[G_i^j, P_i^j\right] = [g_j, p_j] \bullet [g_{j-1}, p_{j-1}] \bullet \cdots \bullet [g_i, p_i] \tag{1.13}$$

(Note that for all i, $G_i^i = g_i$ and $P_i^i = p_i$.)

The interpretation of Eq. 1.13 is easy to see from full expansions:

$$P_i^j = p_j p_{j-1} \cdots p_i \tag{1.14}$$

$$G_i^j = g_j + p_j g_{j-1} + p_j p_{j-1} g_{j-2} + \cdots + p_j p_{j-1} \cdots p_{i+1} g_i \tag{1.15}$$

P_i^j expresses the propagation of a carry in bit-positions i through j, and G_i^j expresses the generation of a carry in any one of bit-positions in i through j and the subsequent propagation of that carry through the bit-positions up to j. Equations 1.8 and 1.12 are therefore equivalent to

$$c_i = \left[G_0^i,\ P_0^i \right] (c_{-1})$$

$$= G_0^i + P_0^i G_{-1}^i \qquad \text{where } g_{-1} = c_{-1} \qquad (1.16)$$

with

$$G_0^i = G_k^i + P_k^i G_0^{k-1} \qquad 1 \le k \le i \qquad (1.17)$$

For unsigned addition, $c_{-1} = 0$, so

$$c_i = G_0^i \qquad (1.18)$$

Equations 1.14 and 1.15 can be evaluated with different degrees of parallelism, by appropriately grouping subterms into blocks. For example, some of the ways in which P_0^3 may be evaluated are

$$(p_3 p_2 p_1) p_0 = P_1^3 P_0^0$$

$$(p_3 p_2)(p_1 p_0) = P_2^3 P_0^1$$

$$p_3 (p_2 p_1 p_0) = P_3^3 P_0^2$$

And G_0^3, for which the defining expression is $g_3 + p_3 g_2 + p_3 p_2 g_1 + p_3 p_2 p_1 g_0$, may be evaluated as

$$g_3 + p_3 (g_2 + p_2 g_1 + p_2 p_1 g_0) = G_3^3 + P_3^3 G_0^2$$

$$(g_3 + p p_3 g_2) + p_3 p_2 (g_1 + p_1 g_0) = G_2^3 + P_2^3 G_0^1$$

$$(g_3 + p_3 g_2 + p_3 p_2 g_1) + p_3 p_2 p_1 g_0 = G_1^3 + P_1^3 G_0^0$$

The definition given in Eq. 1.11 of $\bullet$ can be extended to the block propagate and generate functions. The functions are defined over adjacent bit-positions, and that extends to adjacent blocks, so

$$\left[G_{j+1}^k,\ P_{j+1}^k \right] \bullet \left[G_i^j,\ P_i^j \right] = \left[G_{j+1}^k + P_{j+1}^k G_i^j,\ P_{j+1}^k P_i^j \right] \qquad i \le j < k$$

$$= \left[G_i^k,\ P_i^k \right]$$

and associativity may now be expressed as

$$\left(\left[G_{k+1}^m,\ P_{k+1}^m \right] \bullet \left[G_{j+1}^k,\ P_{j+1}^k \right] \right) \bullet \left[G_i^j,\ P_i^j \right] = \left[G_{k+1}^m,\ P_{k+1}^m \right] \bullet \left(\left[G_{j+1}^k,\ P_{j+1}^k \right] \right.$$

$$\left. \bullet \left[G_i^j,\ P_i^j \right] \right) i \le j \le k < m$$

with both sides equal to $[G_i^m, P_i^m]$.

Associativity is the most useful property of •, but there is also another useful property: *idempotency*, which is that

$$\left[G_i^j, P_i^j\right] \bullet \left[G_i^j, P_i^j\right] = \left[G_i^j, P_i^j\right] \qquad i \leq j$$

since

$$G_i^j + G_i^j P_i^j = G_i^j \left(1 + P_i^j\right) = G_i^j$$

Idempotency means that group may overlap:

$$\left[G_k^j, P_k^j\right] \bullet \left[G_i^m, P_i^m\right] = \left[G_i^j, P_i^j\right] \qquad k \leq m, i \leq j$$

since

$$\left[G_k^j, P_k^j\right] \bullet [G_i^m, P_i^m] = \left(\left[G_{m+1}^j, P_{m+1}^j\right] \bullet [G_k^m, P_k^m]\right) \bullet \left([G_k^m, P_k^m] \bullet \left[G_i^{k-1}, P_i^{k-1}\right]\right)$$

$$= \left[G_{m+1}^j, P_{m+1}^j\right] \bullet ([G_k^m, P_k^m] \bullet [G_k^m, P_k^m]) \bullet \left[G_i^{k-1}, P_i^{k-1}\right]$$

$$= \left[G_{m+1}^j, P_{m+1}^j\right] \bullet [G_k^m, P_k^m] \bullet \left[G_i^{k-1}, P_i^{k-1}\right]$$

$$= \left[G_{m+1}^j, P_{m+1}^j\right] \bullet [G_i^m, P_i^m]$$

$$= [G_i^m, P_i^m]$$

Such overlap is useful because it gives additional options in how parallelism can be exploited.

We have noted that carries can also be propagated by using *kill* or *transfer* functions instead of *propagate* functions (Sect. 1.1.1). The formulations above may therefore be replaced with ones based on k_i or t_i. For example, for k_i we have

$$\overline{K}_i^j = \begin{cases} \overline{k_i} & \text{if } i = j \\ \overline{K}_{l+1}^j \overline{K}_i^j & \text{if } i \leq l < j \end{cases}$$

$$G_i^j = \begin{cases} G_i & \text{if } i = j \\ G_{l+1}^j + \overline{K}_{l+1}^j G_i^k & \text{if } i \leq l < j \end{cases}$$

$$\left[G_{l+1}^j, \overline{K}_{l+1}^j\right] \bullet \left[G_i^l, \overline{K}_i^l\right]$$

$$= \left[G_{l+1}^j + \overline{K}_{l+1}^j G_i^l, \overline{K}_{l+1}^j \overline{K}_i^l\right] \qquad i \leq l \leq j$$

$$= \left[G_i^j, \overline{K}_i^j\right]$$

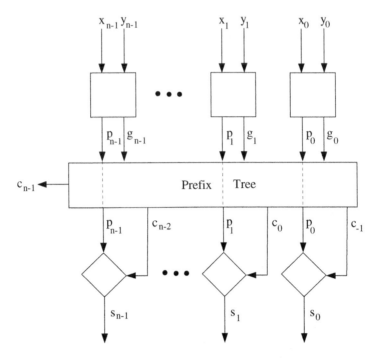

Fig. 1.5 Generic parallel-prefix adder

The use of kill functions is common in realizations of parallel-prefix adders.

There are many types of parallel-prefix adders, all of which have the general form shown in Fig. 1.5. A $\square$ represents logic for $[p_i, g_i]$, and a $\lozenge$ represents sum-formation logic; these parts are the same for all parallel-prefix adders. Parallel-prefix adders differ primarily in the details of the *carry-prefix network*, which is where groupings of the P_i^j and G_i^j signals are formed and then combined—in what we will term *prefix cells*—to produce carries. Different adders are obtained by taking advantage, to different degrees, of the associativity and idempotency of the operator $\bullet$, thus varying the degree of parallelism, the fan-in and fan-out requirements, and the length of the interconnections. Regardless of the details, the operational time of any well-constructed[7] parallel-prefix adder will be proportional to $\log_2 n$, where n is the operand precision, i.e., to the number of levels in the prefix network. The exact number of levels will, however, depend on the particular network design.

In the following examples we will show all prefix operators as similar. It should, however, be noted that the last operator in a chain—i.e., one that produces c_i—can be replaced with a simpler one: the P of the second operand is not required, nor is a P output produced. (The carry is the G output.)

---

[7]The worst case is a linear structure that is essentially a carry-ripple adder.

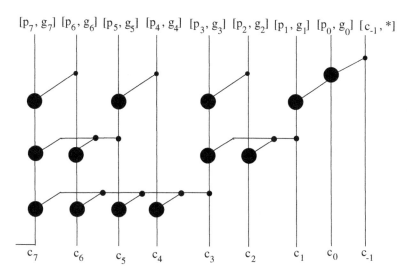

Fig. 1.6 8-bit Ladner–Fischer prefix network

One of the earliest and best known of parallel-prefix adders is the *Ladner–Fischer adder* [9]. The prefix network has the form shown in the example of Fig. 1.6 and is of minimal depth. A problematic aspect of this design is that the lateral fan-out required of the prefix cells doubles at every level—one, two, four, eight, and so forth—and is ultimately substantial for a large adder. The Ladner–Fischer network makes use of the associativity of • but not of its idempotency, which is the case as well for the next two adders.

In the *Kogge–Stone adder* the lateral fan-out of the prefix cells is limited to one [10]. An example of its prefix network is shown in Fig. 1.7. Compared with the Ladner–Fischer network, there are more interconnections here, and they are longer, but the network is of minimal depth; another difference is that more prefix cells are used. Kogge–Stone adders tend to be faster than Ladner–Fischer adders but are more costly.

In the *Kogge–Stone adder* the lateral fan-out of the prefix cells is limited to one. An example of its prefix network is shown in Fig. 1.7. Compared with the Ladner–Fischer network, there are more interconnections here, and they are longer, but the network is of minimal depth; another difference is that more prefix cells are used. Kogge–Stone adders tend to be faster than Ladner–Fischer adders but also more costly (Fig. 1.8).

The *Brent–Kung adder* is another parallel-prefix adder in which all lateral fan-out is limited to one [8]. An example of its prefix-network is shown in Fig. 1.9; this has a minimal number of prefix cells but maximal depth.

The Ladner, Kogge–Stone, and Brent–Kung adders are just three examples in a large design space. Different designs can be obtained by, for example, combining, in a single adder, elements of the preceding three types of adder. An example of such

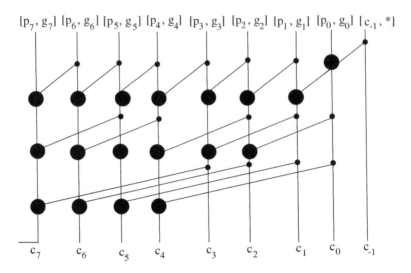

Fig. 1.7 8-bit Kogge–Stone prefix network

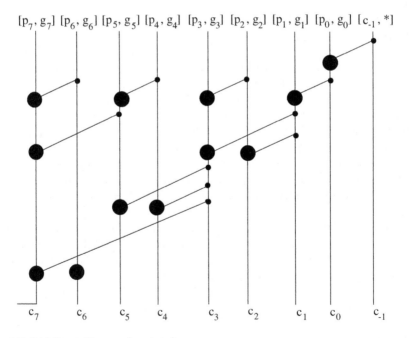

Fig. 1.8 8-bit Brent–Kung prefix network

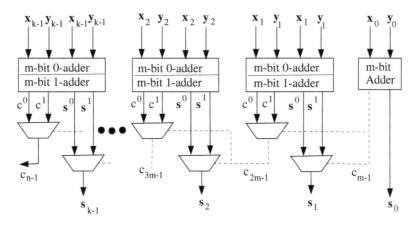

Fig. 1.9 Carry-select adder

a "hybrid" is the *Han–Carlson adder* [12], which combines aspects of the Brent–Kung and Kogge–Stone adders. In general, a variety of prefix networks can be obtained by varying prefix-cell fan-in and fan-out, the number of prefix operators, the length and number of interconnection, and the depth of the operator network, all of which are related [11, 13].

A particular insight in the development of parallel-prefix adders was the categorization by lateral fan-out (by tree level) in the prefix tree [11]. The Kogge–Stone and Ladner–Fischer adders may be considered as the extreme ends of the design space: the lateral-fanout sequence for the former is $\langle 1, 1, 1, 1, 1 \cdots \rangle$ and $\langle 1, 2, 4, 8, 16, \ldots \rangle$ for the latter. Other possibilities exist between those extremes. For example, for an 8-bit adder the possible fanout sequences are $\langle 1, 1, 1 \rangle$, $\langle 1, 1, 2 \rangle$, $\langle 1, 1, 4 \rangle$, $\langle 1, 2, 2 \rangle$, and $\langle 1, 2, 4 \rangle$. We leave it to the reader to confirm that the $\langle 1, 1, 2 \rangle$ tree makes use of idempotency.

1.1.4 Carry-Select

In a carry-ripple adder, the time taken to produce the sum bit at a given bit-position depends on how "far" that position is from the least significant end of the adder, i.e., on how long it takes a carry to propagate between those bit-positions. To reduce such delays, the essential idea in the carry-select adder is to generate the different possible sum bits at the high-order positions before the low-order carries into those positions are known and then immediately select the correct sum bits once the carries from the low-order positions are known. The basic unit of a carry-select adder therefore consists of two "conditional adders" and multiplexers. One conditional adder generates sum bits under the assumption that the carry into the adder is 0, and the other generates sum bits under the assumption that it is 1. Choices are then made in multiplexers when the carry is known. We shall initially assume that the two conditional adders are carry-ripple adders and then make some remarks on other types of adder. Some detailed descriptions of carry-select adders will be found in [14, 15, 18].

Suppose an n-bit adder is divided into $k = n/m$ blocks of m bits each.[8] Let s_j^q denote bit j of the sum computed under the assumption that the carry into the block containing that bit is q. And similarly let c_j^q be the carry from stage j under the assumption that the carry into the block is i. Then, on the basis of Eqs. 1.1 and 1.2, the logic equations for the conditional sum and carry bits in block k are

$$s_i^0 = (x_i \oplus y_i) \oplus c_{i-1}^0 \quad i = jm, jm+1, \ldots, (j+1)m-1, \ j = 1, 2, \ldots, k-1$$

$$s_i^1 = (x_i \oplus y_i) \oplus c_{i-1}^1$$

$$c_i^0 = x_i y_i + (x_i \oplus y_i) c_{i-1}^0$$

$$c_i^1 = x_i y_i + (x_i \oplus y_i) c_{i-1}^1$$

and the multiplexers effect the equations

$$s_i = c_{km-1} s_i^1 + \overline{c_{km-1}} s_i^0$$

$$c_{(j+1)m-1} = c_{jm-1} c_{(j+1)m-1}^1 + \overline{c_{jm-1}} c_{(j-1)m-1}^0$$

The architecture for a carry-select adder is therefore as shown in Fig. 1.9. $\mathbf{x}_i$, $\mathbf{y}_i$, and $\mathbf{s}_i$ denote the m bits of x, y, and s in block i. c^q denote a carry bit produced by a q-adder, and $\mathbf{s}^q$ denotes a block of m sum bits produced by a q-adder.

Assuming the "base" adders are ripple adders, the operational time of the adder in Fig. 1.9 is determined primarily by the delay through the first block of full adders and the delay through the chain of multiplexers. If we assume that the construction of full adders and multiplexer are the straightforward ones at the gate level and that the delay through a gate is τ, then we have 2τ through the carry path of a full adder and 2τ through a multiplexer.[9] The delay through block 0 is $(2m + 1)\tau$, and that through the multiplexers is $2(n/m - 1)\tau$, for a total of $[2(m + n/m) - 1]\tau$. Thus, for example, with two $(n/2)$-bit blocks the operational time would be $(n + 3)\tau$, in contrast with $(2n + 1)\tau$ for an ordinary n-bit carry-ripple adder.

The block size is critical in the performance of the adder: increasing block size increases the delay through the full adders but reduces the delay through the multiplexer chain, and reducing the block size does the opposite. The total number of gate delays in the critical path is

---

[8]For convenience we assume that n is exactly divisible by m; if not, then one block may be made smaller or larger than m.

[9]It is reasonable to exclude inverter delay. It is also worth noting that in current technology a multiplexer can be realized with much greater efficiency (cost and performance) than is apparent from a direct gate-level derivation.

$$T = 2 \left(n + \frac{n}{m} \right) + 1$$

So for the optimal block size:

$$\frac{\partial T}{\partial m} = 2 - \frac{2n}{m^2} = 0$$

$$m \approx \sqrt{n}$$

which gives an operational time of about $2\sqrt{n}\tau$. Thus, for example, a 16-bit adder divided into 4-bit blocks would have an operational time of 15τ—9τ in the block-0 adders and 6τ through the multiplexers—in contrast with the 33τ of a 16-bit carry-ripple adder.

For better performance, there are several variations on the basic carry-select adder described above. One variation is the use blocks of variable size. Another variation involves the application of techniques, such as parallel-prefix computation, used in the basic design of other types of high-performance adders. And a third is the "recursive" application of the basic carry-select technique—starting with very small blocks and increasing block sizes up to the adder size. We next briefly discuss these variations.

The optimal block size above is such as to ensure that the conditional-adder outputs and the multiplexer control signals arrive at the same time (or, practically, as nearly so as possible) at the last multiplexer and thus eliminate (or, practically, nearly eliminate) the "waiting time" at that multiplexer. If the same can be done with respect to all multiplexers, then the operational time will be as low as possible. On the basis of our timing assumptions, block 1 should be the same size as block 0 for the arrival times at the block-1 multiplexer to be the same. Thereafter, the each multiplexer adds 2τ in that chain; so each block should be 1 larger than the preceding block, which gives an adder with variable-size blocks. Thus, for example, a 16-bit adder with blocks of sizes 5, 4, 3, 2, and 2 will have an operational delay of 13τ, which is slightly better performance than one of four 4-bit blocks. In general, practical variations in block size will be determined by value of n and the actual delays through the various components, so consecutive block sizes may vary by values other than one.

In a "recursive" application, the basic technique is applied to increasingly smaller blocks. An n-bit adder is divided into two $n/2$-bit blocks, with a multiplexer; each of the two blocks is again divided, into $n/4$-bit blocks, with a multiplexer; and so on. With as much division as possible, the resulting adder is a *conditional-sum adder*, which consists of $\log_2 n$ levels: At the first level, conditional sum and carry bits at formed in one-bit groups; the least significant bit of the sum is completely determined at that level. At the second level, the bits from the first level are grouped into pairs, selections made, and at that level two bits of the sum are completely determined. And so on, with the number of completely determined sum bits doubling at each level. The conditional-sum adder therefore has a structure that is similar to that of the Ladner–Fisher adder but with a network of multiplexers instead of prefix cells.

The third variation in the design of carry-select adders is the application of the techniques used in the design of other high-performance adders. Such application may be "direct" or "indirect." The direct application consists of replacing the conditional carry-ripple adders with faster adders, e.g., parallel-prefix adders, thus speeding up the carry propagation within a block. The indirect application aims to speed up the carry propagation between blocks and consists of using fast carry networks to produce carries into blocks. Thus, for example, a parallel-prefix carry network may be used with conditional carry-ripple adders, and in such a design an even faster adder can be obtained by using adders that are faster than carry-ripple adders but not as costly as adders of the highest performance.[10] More details on such designs will be found in [20].

A final note is that although Fig. 1.9 shows two nominal conditional adders for each block, in practice they need not be distinct. With carry-ripple adders the logic for $x_i y_i$ and $x_i \oplus y_i$ can be shared, and similar sharing of other logic is possible with other types of adder, e.g., the parallel-prefix trees. Therefore, the logic required for a carry-select adder need not be twice that of one "ordinary" adder.

1.1.5   High Precision

The serial adder can be used easily for addition of any precision, but for high precisions the operational delay will be quite large. That is the case as well with the carry-ripple adder. Nevertheless some of the essential principles used in the design of these two adders can be used in the design of faster high-precision adders.

Let us suppose that we have an m-bit high-speed adder—e.g., a parallel-prefix adder—and that we wish to implement n-bit addition, where $n >> m$ and n is exactly divisible by m (an assumption made for simplicity). Then the m-bit adder can be used serially, as shown in Fig. 1.10. (A delay might be required between the c_{m-1} "output" and the c_{-1} "input.") For simplicity, we assume shift registers, but faster arrangements are possible, e.g., with ordinary registers and multiplexers. The adder operates in n/m cycles. In each cycle, m pairs of operand bits are shifted into the adder, an addition takes place, and the corresponding result bits are inserted into the sum register, whose contents are then shifted by m bit-positions to the right. The carry from one m-bit addition is the carry into the adder for the next m-bit addition.

The arrangement of Fig. 1.10 corresponds to a serial adder. If enough fast m-bit adders—e.g., parallel-prefix adders—are available, then an arrangement that corresponds to a carry-ripple adder can be obtained by stringing together the adders, as shown in Fig. 1.11. The adder is partitioned into $k \stackrel{\triangle}{=} n/m$ blocks of m bits each. $\mathbf{x}_i$, $\mathbf{y}_i$, and $\mathbf{s}_i$ denote the m bits of x, y, and s in block i.

---

[10]For example, *carry-skip adders*, which we have not covered.

Fig. 1.10 Parallel-serial adder

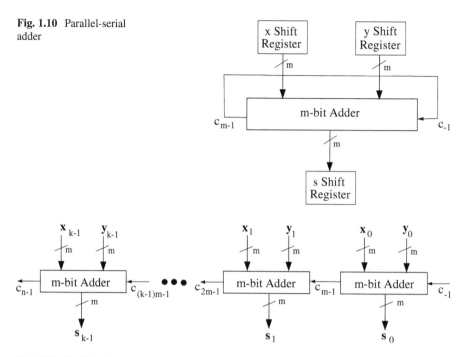

Fig. 1.11 Parallel-ripple adder

And one can imagine an arrangement that is between those of Figs. 1.10 and 1.11, i.e., a smaller number of ripple-connected adders used serially.

For very high precisions, the inherent limitations of serial and carry-ripple computation will have a significant effect on the operational delay in the adders of Figs. 1.10 and 1.11. Just as with the basic carry-ripple adder, here too the carry-lookahead principle can be used, at a higher level, to achieve high performance: "blocks" (groups of bit-positions) are grouped into another level (of "superblocks"), which in turn are grouped into "hyperblocks," and so forth, to the extent necessary. Good examples of how various techniques can be combined in the design of a fast, high-precision adder will be found in [20].

We may think of the arrangements of Figs. 1.10 and 1.11 as radix-2^m adders. Such a view has no practical implications, but it may be useful in understanding implementations for certain algorithms that are discussed in the subsequent chapters.

1.1.6   Signed Numbers and Subtraction

We have thus far assumed that the operands in addition are unsigned numbers. We now consider the effect of signed numbers in addition and also briefly discuss

Table 1.2 Signed-number representations

Binary Pattern	Decimal equivalent		
	Sign-magnitude	Ones' complement	Two's complement
000	0	0	0
001	1	1	1
010	2	2	2
011	3	3	3
100	−0	−3	−4
101	−1	−2	−3
110	−2	−1	−2
111	−3	−0	−1

subtraction. We start with a brief review of the most common systems for the representation of signed numbers.

The three major systems for the binary representation of signed numbers are[11]: *two's complement, ones' complement,* and *sign-and-magnitude*. With n-bit representations:

- In the sign-and-magnitude system the most significant bit represents the sign—0 for positive and 1 for negative—and the remaining bits represent the magnitude. The range of representable numbers is $[-(2^{n-1} - 1), \ 2^{n-1} - 1]$, with two representations for zero: $00 \cdots 0$ and $100 \cdots 0$.
- In the ones'-complement system too the range of representable positive numbers is $[-(2^{n-1}-1), \ 2^{n-1}-1]$. A negative number is represented by inverting each bit in the representation of the corresponding positive number; the most significant bit indicates the sign. There are two representations for zero: $00 \cdots 0$ and $11 \cdots 1$.
- In two's-complement system the representation of a negative number is obtained by adding 1 to the corresponding ones'-complement representation and ignoring any carry-out.[12] Here too the most significant bit indicates the sign. There is only one representation for zero, but the range is asymmetrical: $[-2^{n-1}, \ 2^{n-1} - 1]$.

Table 1.2 gives some examples of representations in the three notations.

Shifts are sometimes required in some arithmetic operations, such as multiplication and division. For the left shifting of representations of positive numbers in all three systems, 0s are inserted at the right-hand end, with only the magnitude affected in a sign-and-magnitude representations. That is the case too with representations of negative numbers in sign-and-magnitude and two's-complement representations; with ones'-complement representations of negative numbers, 1s are inserted since the bits for the corresponding positive number would be 0s. Thus, for example,

---

[11]Note the position of the apostrophe: *ones'* vs. *two's*. See [21] for an explanation.

[12]In manual arithmetic, the simplest method is this: scan the ones'-complement representation from the least significant bit to the most significant bit; copy every 0 and the first 1; thereafter invert every bit.

the five-bit representations of negative-five are 10101 (sign and magnitude), 11011 (two's complement), and 11010 (ones' complement), and the corresponding results of one-bit left shifts (the representations of negative ten) are 11010, 10110, and 10101. Numbers will be represented in finite storage (e.g., registers), so *overflow* can occur—if the most significant bit shifted out is 1 for unsigned representations or signed sign-and-magnitude representations and 0 for negatives numbers in ones'- and two's-complement representations.

With sign-and-magnitude representation a right shift is a straightforward shift of the magnitude. On the other hand, with one's-complement and two's-complement representations the shift must include sign extension. That is so because the sign bit in a ones'-complement or two's-complement representation is actually the truncation of an infinite string of 1s. Take, for example, the representation of negative five in two's complement. In four, five, six, and seven bits, the representations would be 1011, 11011, 111011, and 1111011; the sign is represented in one, two, three, and four bits.

Sign-and-magnitude representation is almost never used for integer (fixed-point) arithmetic in modern computers,[13] and as all cryptography arithmetic is on integers, we shall not consider the representation any further. Ones'-complement representation is never used of itself in ordinary integer arithmetic, but it is important in some modular arithmetic (Chap. 5); also, forming a ones'-complement representation is a step in forming a two's-complement representation. Two's complement is the standard system for signed integer arithmetic.

The computer representation of a negative number may also be interpreted as that of an unsigned number, and in what follows we shall make use of that fact. Let $-z$ (with z positive) be a negative number represented in n bits. The representation may also be interpreted as that of $2^n - z$ for two's complement and of $2^n - z - 1$ for ones' complement. Thus, for example, 101 in Table 1.2 is also the representation of five.

Ones'-Complement

There are three cases to consider in the addition of x and y.

Case 1: Both Operands Are Positive

The result of adding the two numbers is correct if $x + y \leq 2^{n-1} - 1$; otherwise, the result cannot be represented in n bits, and *overflow* is said to have occurred. Since $x + y > 2^{n-1} - 1$ if there is a carry into the sign position, checking for 1 in that position suffices to detect an overflow state: the sign bit of the result will be 1 instead of the correct 0 (the sign of the operands). Examples showing a no-overflow case and an overflow case are given in Table 1.3a.

---

[13]The representation is used in standard floating-point representations.

Table 1.3 Examples of ones'-complement addition

3	=	00011	7 =		00111
5	=	00101	11 =		01011
8	=	01000	−13 =		10010

(a)

−3 =		11100	−3 =		11100
−5 =		11010	−12 =		10011
	1 ←	10110		1 ←	01111
		1			1
−8 =		10111	−15 =		10000

(b)

−11 =		10100
−6 =		11001
	1 ←	01101
		1
14 =		01110

(c)

6 =		00110	11 =		01011
−11 =		10100	−6 =		11001
−5 =		11010		1 ←	00100
					1
			5 =		00101

(d)

Case 2: Both Operands Negative

Taken as representations of unsigned numbers, that of x represents $2^n - 1 - x'$, and that of y represents $2^n - 1 - y'$, where $x' = -x$ and $y' = -y$. The result of a correct addition is the negation of $x' + y'$, i.e., $2^n - 1 - (x' + y')$, which is obtainable only if $x' + y' \leq 2^{n-1} - 1$.

Let us suppose that $x' + y' \leq 2^{n-1} - 1$, and reformulate $x + y$ as $2^n + [2^n - 2 - (x' + y')]$. The sum in the square brackets is positive and requires at most $n - 1$ bits for representation; so the outer term in 2^n indicates a carry from the addition. Ignoring that carry is equivalent to subtracting 2^n, which leaves the sum in the square brackets. That sum is 1 less than the desired result, so the addition of a 1 will yield the correct result. Since the 1 is added only when there is a carry-out, we may view the carry and the added 1 as being the same; indeed, in implementation they will be the same, and the added 1 is therefore usually referred to as an *end-around carry*. An observation that is useful for overflow detection—the next sub-case, below—is that since the addition of just the most significant digits would leave a 0 in that position of the intermediate result, there must be a carry from bit-position $n-2$ that leaves the correct sign in the final result. This carry out of bit-position $n-2$ will always occur since $2^n - 1 - (x' + y') \geq 2^n - 1 - (2^{n-1} - 1) = 2^{n-1}$. The carry will be generated during the preliminary addition or during the addition of the end-around-carry; Table 1.3b shows corresponding examples.

Now suppose that instead of the preceding sub-case we have $x' + y' > 2^{n-1} - 1$. Then overflow will occur, and this can be detected by the absence, before and after the addition of the end-around carry, of a carry from bit-position $n-2$. The justification for this is as follows. Let $x' + y' = 2^{n-1} + u$, with $u \geq 0$. Then $x + y = 2^n + (2^{n-1} - 1 - u) - 1$. Since $(2^{n-1} - 1 - u) - 1$ is representable in $n-2$ bits, there is no carry from bit-position $n-2$, and the 2^n term represents the carry-out from adding the sign bits. Adding the end-around carry—i.e., effectively subtracting 2^n and adding 1—leaves $2^{n-1} - 1 - u$, which is still representable in $n-2$ bits. The sign bit of the result 0, which is evidently incorrect, differs from the sign of the operands. An example is shown in Table 1.3c.

Case 3: Operands of Unlike Sign

Without loss of generality, assume that x is positive and y is negative. If the representation of y is taken as that of an unsigned number, then $y = 2^n - 1 - y'$, where $y' = -y$. If $y' \geq x$, then the correct result is the negation of $y' - x$. The result of adding x and y in this case will be $2^n - 1 - (y' - x)$, which is the negation of $y' - x$. On the other hand, if $y' < x$, then the correct result is $x - y'$. Since $x - y' \geq 1$, we have $x + y = 2^n + (x - y') - 1 \geq 2^n$, and a carry will be produced from the sign position. If we ignore this carry—i.e., effectively subtract 2^n—and add 1, then the correct result of $x - y'$ is obtained. Overflow cannot occur when adding numbers of unlike sign. Examples are shown in Table 1.3d.

Algorithm

The actions in the discussion above describe an algorithm that consists of four main parts:

(i) Add x and y to produce an intermediate sum s'.

(ii) $x \geq 0$ and $y \geq 0$ (both sign bits are 0): If the sign bit of the result is 1, then overflow has occurred; otherwise, s' is the correct final result.

(iii) $x < 0$ and $y < 0$ (both sign bits are 1): Add the carry-out, end-around, to s'. If the sign bit of the result of that addition is 0, then overflow has occurred; otherwise, the result is correct.

(iv) If $x < 0$ and $y \geq 0$ or $x \geq 0$ and $y < 0$ (different sign bits): If there is no carry from the sign position of s', then s' is the correct result; otherwise, add the end-around carry to obtain the correct result. Overflow cannot occur in this case.

Fig. 1.12 Ones'-complement
adder

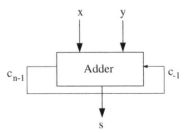

Figure 1.12 shows a generic architecture for a basic ones'-complement adder, with any one of the adders of Sects. 1.1.1–1.1.4 as the "unsigned adder"; the carry-out (end-around carry) is included directly as a carry-in to the adder. (A delay might be required between the c_{n-1} "output" and the c_{-1} "input.") We have omitted the logic for overflow-detection, the inclusion of which is left as an exercise for the reader; the necessary logic consists of two AND gates and an OR gate.

The underlying adder in Fig. 1.12 may be a parallel-prefix adder, in which case, as shown, the end-around-carry would be fed into the first level of the prefix network. But a much better arrangement is possible with a parallel-prefix adder. Suppose there is no carry into the adder and that the last level is removed in a prefix network such as one of those shown in Figs. 1.6, 1.7, and 1.8. The output of the modified prefix network will be the signals $[G_0^i, P_0^i]$. The carry-out c_{n-1} can then be added in by modifying the last two stages (carry and sum outputs) through the inclusion of another level of prefix operators at all positions, to implement the equation[14]

$$c_i = \left[G_0^i, P_0^i \right] \circ [c_{n-1}]$$

$$= G_0^i + P_0^i c_{n-1}$$

The arrangement is as shown in Fig. 1.13. Ones'-complement adders are important for modular addition (Chap. 5).

Two's Complement

There are three cases to consider in the addition of x and y.

Case 1: Both Operands Are Positive

This case is similar to the corresponding one for the ones'-complement representations. If there is no carry from bit-position $n - 2$ into the sign position, then the result is correct. Otherwise, the sign bit of the result differs from that of

---

[14]As indicated above, a prefix operator whose output is a final carry can be simplified. We use $\circ$ for the simplified version of $\bullet$.

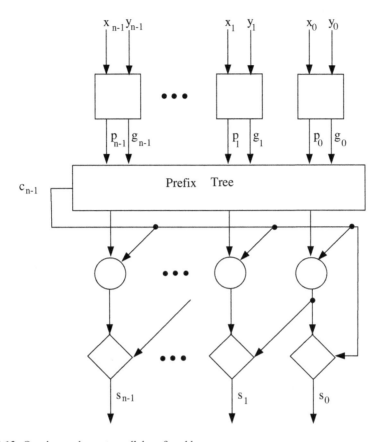

Fig. 1.13 Ones'-complement parallel-prefix adder

the operands—an indication of overflow. Table 1.4a shows examples of both no-overflow and overflow.

Case 2: Both Operands Negative

If the representations of x and y are taken as those of unsigned numbers, then $x = 2^n - x'$ and $y = 2^n - y'$, where $x' = -x$ and $y' = -y$. The result of a correct addition is $2^n - (x' + y')$, but the result of directly adding x and y is $2^n + [2^n - (x' + y')]$. Since the term in the square brackets is positive, the outer 2^n term represents a carry-out. If $x' + y' \leq 2^{n-1} - 1$, then ignoring the carry-out—i.e., effectively subtracting 2^n—leaves the correct final result. There is also a carry from bit-position $n - 2$ into the sign position, since $2^n - (x' + y') \geq 2^{n-1}$.

Table 1.4 Examples of two's-complement addition

13 =	00011		11 =	00111
5 =	00101		6 =	00110
8 =	01000		−15 =	01111
				10000

(a)

−3 =	11101		−11 =	10101
−5 =	11011		−6 =	11010
1 ←	11000		1 ←	01111
−8 =	11000		15 =	01111

(b)

11 =	01011			
−6 =	11010		−11 =	10101
1 ←	00101		6 =	00110
5 =	00101		−5 =	11011

(c)

On the other hand, if $x' + y' \geq 2^{n-1}$, then overflow occurs. There is, however, no carry from bit-position $n-2$ into the sign bit-position, since $x + y = 2^n + 2^{n-1} - u$ (with $u \geq 0$), and subtracting the 2^n term leaves $2^{n-1} - u$, which requires no more than $n-2$ bits for representation. The absence of a carry into the sign position leaves a sign bit different from that of the operands and indicates the overflow. Table 1.4b shows examples of both the no-overflow case and the overflow case.

Case 3: Operands of Unlike Sign

Without loss of generality, assume that x is positive and y is negative. If the representation of y is taken as that of an unsigned number, then $y = 2^n - y'$, where $y' = -y$, and $x + y = 2^n + x - y'$. So, if $x \geq y'$, then the correct result is $x - y'$; otherwise, it is the negation of $y' - x$. In the former case $2^n + x - y' \geq 2^n$, and the 2^n therefore represents a carry-out, ignoring which gives the correct result of $x - y'$. If, on the other hand, $x < y'$, then there is no carry-out, since $2^n + x - y' < 2^n$, and the result of adding the two operands is correct, since $x + y = 2^n - (y' - x)$, which is the negation of $y' - x$. Examples are shown in Table 1.4c.

Algorithm

The actions above describe an algorithm that consists of four main parts:

(i) Add x and y to produce an intermediate sum s'.

(ii) $x \geq 0$ and $y \geq 0$ (both sign bits are 0): If the sign bit of s' is 1, then overflow has occurred; otherwise, s' is the correct final result.

(iii) $x < 0$ and $y < 0$ (both sign bits are 1): If there is a carry-out, it is ignored. Then, if the sign bit of the result of that addition is 0, then overflow has occurred; otherwise, the result is correct.

(iv) $x < 0$ and $y \geq 0$ or $x \geq 0$ and $y < 0$ (different sign bits): Any carry-out is ignored. s' is the correct result. Overflow cannot occur in this case.

It is straightforward to modify, by including a few logic gates, any one of the adders of Sects. 1.1.1–1.1.4 so that overflow is detected if the operands are assumed to be in two's-complement representation.

Subtraction

It is possible to design subtractors by proceeding as we have done above for adders—i.e., starting with a *full subtractor*, in place of a full adder, and then developing more complex designs for faster implementations—but nowadays that is almost never done. There is little advantage in a "real" subtractor, as it is more cost-effective to employ a single unit for both addition and subtraction.

Subtraction through addition is effected by negating the subtrahend and adding to the minuend, i.e., $x - y = x + (-y)$. The negation is quite simple: with ones' complement notation, it is just bit inversion; and with two's-complement notation it is bit inversion and the addition of a 1. The 1 in the latter case is easily included by injecting it as the carry-in c_{-1}. Figure 1.14 shows the high-level design of a two's-complement adder-subtractor, with any of the adders of Sects. 1.1.1–1.1.4 as the Adder. The inclusion of the few gates that are required for overflow-detection is left as an exercise for the reader. The output of the unit is a sum (s) or difference (d), according to the control signal $\overline{\text{ADD}}$/SUB.

1.2 Multiplication

In ordinary paper-and-pencil multiplication one forms an array of multiples of the multiplicand and then adds them up. The multiples are products of the multiplicand and digits of the multiplier and are formed by "scanning" the multiplier from the least significant digit to the most significant or from most significant to least significant. Both possibilities are shown in the examples of Table 1.5. In ordinary arithmetic the former is standard, but in cryptography arithmetic the latter too is useful (Chap. 5).

Fig. 1.14 Two's-complement
adder-subtractor

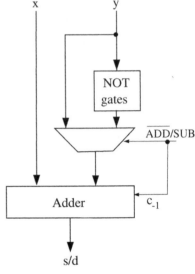

Table 1.5 Examples of
paper-and-pencil
multiplication

```
   2 1 3 4            2 1 3 4
   5 6 1 2            5 6 1 2
   -------            -------
   4 2 6 8          1 2 7 7 0
 2 1 3 4            1 2 8 0 4
1 2 8 0 4              2 1 3 4
1 2 7 7 0              4 2 6 8
-----------        -----------
1 4 0 7 6 0 0 8    1 4 0 7 6 0 0 8
```

Multiplication as shown in Table 1.5 can be reflected directly in the design of
a hardware multiplier that is appropriately known as a *parallel-array multiplier*
and whose basic elements correspond to the digits in the multiplicand-multiple
array. But the hardware requirements for such a multiplier will be quite large for
anything other than small and moderate precisions. A cheaper arrangement is to
form the multiples one at a time and add each, as it is formed, to a running *partial
product*[15] that is initially zero; the multiplier in this case is a *sequential multiplier*.
And sequential and parallel computations may be combined in a *sequential-parallel
multiplier*.

This section consists of a discussion on the design of the sequential multiplier,
parallel multiplier, and "hybrids."[16] We shall first discuss algorithms and multiplier
designs for basic multiplication and then consider variations for high performance—

---

[15] Some authors use "partial product" to refer to what we term "multiplicand multiple."

[16] One can also devise a *serial* multiplier [1], based on the serial adder of Sect. 1.1.1. Such a
multiplier will be extremely cheap but also extremely slow. Nevertheless, as with the serial adder,
it can be usefully employed in a massively parallel system.

variations that involve high-speed addition without the propagation of carries and variations that involve the scanning of several multiplier digits at a time. We shall assume that the two operands, x and y, are each represented in n bits, denoted $x_{n-1}x_{n-2}\cdots x_0$ and $y_{n-1}y_{n-2}\cdots x_0$; that is, $x = \sum_{i=0}^{n-1} x_i 2^i$ and $y = \sum_{i=0}^{n-1} y_i 2^i$. (The algorithms given are easily extended to an m-bit multiplicand and n-bit multiplier, with $m \neq n$.) We shall initially assume that the operational radix is two. Once the basics have been covered, we shall then consider larger radices–four, eight, and so forth. We shall also initially assume unsigned operands and later make some remarks to cover signed operands.

1.2.1 Sequential

For the binary computation of $z = xy$, the straightforward algorithm for sequential computation with a right-to-left scan of the multiplier bits may be expressed as

$$Z_0 = 0 \tag{1.19}$$

$$Z_{i+1} = Z_i + 2^i y_i x \qquad i = 0, 1, 2, \ldots, n-1 \tag{1.20}$$

$$z = Z_n \tag{1.21}$$

And for a left-to-right multiplier-scan the algorithm is

$$Z_0 = 0 \tag{1.22}$$

$$Z_{i+1} = 2Z_i + y_{n-i-1}x \qquad i = 0, 1, 2, \ldots, n-1 \tag{1.23}$$

$$z = Z_n \tag{1.24}$$

In both cases a multiplication by two reflects a left shift—of i bit-positions in the first case and one bit-position in the second case. Radix-r algorithms, for $r > 2$, are readily obtained by replacing 2 with r in Eqs. 1.20 and 1.23.

In implementation, the multiplication by 2^i in Eq. 1.20 does not strictly require the implied variable-length shifting, as this can instead be effected by multiplying the multiplicand by two (i.e., a one-bit shift) in each iteration:

$$Z_0 = 0 \tag{1.25}$$

$$X_0 = x \tag{1.26}$$

$$Z_{i+1} = Z_i + y_i X_i \qquad i = 0, 1, 2, \ldots, n-1 \tag{1.27}$$

$$X_{i+1} = 2X_i \tag{1.28}$$

$$z = Z_n \tag{1.29}$$

Table 1.6 Example of
computer binary
multiplication

5	=	0 0 1 0 1	Multiplicand
11	=	0 1 0 1 1	Multiplier
		0 0 0 0 0	Initial partial product
		0 0 1 0 1	Add 1st multiple
		0 0 1 0 1	
		0 0 0 1 0 1	Shift right
		0 0 1 0 1	Add 2nd multiple
		0 0 1 1 1 1	
		0 0 0 1 1 1 1	Shift right
		0 0 0 0 0	Add 3rd multiple
		0 0 0 1 1 1 1	
		0 0 0 0 1 1 1 1	Shift right
		0 0 1 0 1	Add 4th multiple
		0 0 1 1 0 1 1 1	
		0 0 0 1 1 0 1 1 1	Shift right
		0 0 0 0 0	Add 5th multiple
55	=	0 0 0 1 1 0 1 1 1	Final product

With both this variant and the original, the precision of the multiplicand multiples increases by one bit in each iteration, up to $2n$, and this precision must be allowed for in the additions. In *ordinary* multiplication,[17] however, the precisions in the additions can be limited to n, by shifting the partial products to the right instead of shifting the multiplicand to the left. In implementation, this shifting will be effected by employing a shift register with appropriate interconnections between multiplicand multiple and partial product. An example is shown in Table 1.6.

With the algorithm of Eqs. 1.22–1.24 all bits of a partial product must be included in each addition, and the additions must therefore be of $2n$ bits precision. For this reason, this algorithm is never used in ordinary arithmetic. In cryptography arithmetic—e.g., in the modular multiplication of Chap. 5—there are algorithms in which all bits of the partial products must be included in intermediate operations. In such cases both the algorithms of Eqs. 1.19–1.21 and Eqs. 1.22–1.24 are applicable, but the "optimized" version of the first algorithm is not. The discussions that follow here are of the "optimized" algorithm, but it is quite straightforward to modify them to obtain what is required for the other algorithm.

---

[17]What follows is not always possible with similar algorithms in the modular arithmetic of cryptography (Chap. 5).

Fig. 1.15 Sequential multiplier

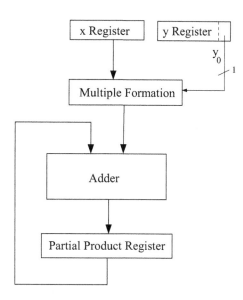

Figure 1.15 shows a basic architecture for the implementation of the first algorithm. It consists of an adder (of one of the types described in Sect. 1.1), a register that holds the multiplicand, a shift register that holds the multiplier, and a $2n$-bit shift register (initialized to zero) for the product. The Multiple Formation unit is just a set of AND gates whose output is either 0s or the bits of the multiplicand (x), according to the least significant bit of the multiplier (y). The multiplier operates in n cycles. In each cycle a multiplicand multiple is added to the top n bits of the partial product, the partial product is shifted one place to the right, and the multiplier too is shifted one place to the right. The bits shifted out of the adder position are final bits of the product.

The arrangement of Fig. 1.15 is useful if the only adder available is one of the types described in Sect. 1.1. Otherwise, there is a better alternative, based on the observation that, in sequence of additions, carries need not be propagated with each addition. Instead, the carries produced in one step may be "saved" and included, with appropriate displacement, as operands in the next addition. Such *carry-save addition* can be done until the last cycle, after which there is no "next addition," and the carries must then be propagated. So the adder of Fig. 1.15, which is known as a *carry-propagate adder* (CPA), may be replaced with a *carry-save adder* (CSA)[18] in the loop and a carry-propagate adder outside the loop. The CSA consists of just unconnected full adders and is therefore very fast. The new arrangement is shown in Fig. 1.16.

In each cycle of the multiplier of Fig. 1.16, a multiplicand multiple is added to the partial product to produce a new partial product in the form of *partial carry*

---

[18] A CSA is also known as a *3:2 compressor* because it "compresses" three inputs into two outputs.

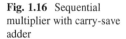

Fig. 1.16 Sequential
multiplier with carry-save
adder

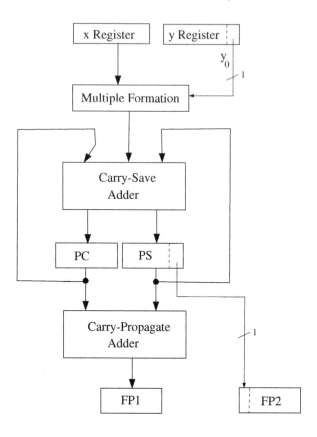

(PC) bits and *partial sum* (PS) bits, and the content of the multiplier and product
registers are shifted one place to the right. For the next addition, the PC bits are
shifted one place to the left, through appropriate interconnections, since the carries
should be added one bit-position up. At the end of the cycling, the remaining carries
are propagated, by combining PC and PS in the CPA, a process that we shall refer
to as *assimilation*.[19]

The PC bits are shifted one place relative to the PS bits, shown in the figure by the
slanted line; and the PS bit that is shifted out in each cycle is a "complete" bit of the
final product. The partial-product register now consists of two parts, each of *n* bits:
an "ordinary" register (FP1) that eventually holds the high half of the final product,
and a shift register (FP2) into which the bits of the lower half of the final product
are shifted as they are formed. An example computation is shown in Table 1.7.

---

[19]Note that a CPA is not absolutely necessary. The assimilation can be done by cycling *n* times
through the CSA, but that is a very slow method that, in practice, was abandoned in the late 1950s.

Table 1.7 Multiplication
with carry-save addition

7 =	0 0 1 1 1	Multiplicand (x)
15 =	0 1 1 1 1	Multiplier (y)

0 0 0 0 0	Initial partial sum
0 0 0 0 0	Initial partial carry
0 0 1 1 1	Add 1st multiple
0 0 1 1 1	1st partial sum
0 0 0 0 0	1st partial carry
0 0 0 1 1 1	Shift partial sum right
0 0 1 1 1	Add 2nd multiple
0 0 0 0 0	Add 1st partial carry
0 0 1 0 0 1	2nd partial sum
0 0 0 1 1	2nd partial carry
0 0 0 1 0 0 1	Shift partial sum right
0 0 1 1 1	Add 3rd multiple
0 0 0 1 1	Add 2nd partial carry
0 0 1 1 0 0 1	3rd partial sum
0 0 0 1 1	3rd partial carry
0 0 0 1 1 0 0 1	Shift partial sum right
0 0 1 1 1	Add 4th multiple
0 0 0 1 1	Add 3rd partial carry
0 0 1 1 1 0 0 1	4th partial sum
0 0 0 1 1	4th partial carry
0 0 0 1 1 1 0 0 1	Shift partial sum right
0 0 0 0 0	Add 5th multiple
0 0 0 1 1	Add 4th partial carry
0 0 0 0 0 1 0 0 1	5th partial sum
0 0 0 1 1	5th partial carry
0 0 0 0 0 0 1 0 0 1	Shift partial sum right
0 0 0 1 1	Propagate final carries

105 =	0 0 1 1 0 1 0 0 1	

1.2.2 High Radix

Multiplication as described above can be speeded up by scanning several bits of the multiplier in each cycle. This corresponds to the use of a larger operational radix: two bits at a time for radix four, three bits at a time for radix eight, and so forth. "High radix" will mean a radix larger than two.

The sequential algorithm of Eqs. 1.19–1.21 requires only a minor change for high-radix computation. In general, with radix-r operands the algorithm is

$$Z_0 = 0 \tag{1.30}$$

$$Z_{i+1} = Z_i + r^i y_i x \qquad i = 0, 1, 2, \ldots, m - 1 \tag{1.31}$$

$$z = Z_m \tag{1.32}$$

If the multiplier is represented in n bits, then $m = \lceil n/r \rceil$. In computer implementation, r will almost always be a power of two, and we shall assume that in what follows. We shall also assume that the algorithm is implemented in its "optimized" form, in which the shifting implied in Eq. 1.31 is that of the partial product to the right instead of the multiplicand multiple to the left.

Scanning several multiple bits of the multiplier in each cycle requires several changes to an implementation based on the architecture of Fig. 1.16 and to the corresponding operational procedure. A minor change is that the more bits have to be shifted per cycle in each of the product and multiplier registers: k bits per cycle for k-bit multiplier scanning. A second change—a major one—is the provision of more multiples of the multiplicand x, with one chosen in each cycle: $0, x, 2x$, and $3x$ for two-bit scanning; $0, x, 2x, \ldots, 7x$ for three-bit scanning; and so forth. And a third change is the inclusion of a k-bit CPA whose inputs are the unassimilated k; PC-PS bit pairs are shifted out in each cycle and whose outputs are the corresponding bits of the final product. A difficulty with the straightforward application of this method is that multiplicand multiples that are not powers of two are, relatively, not easy to compute, as each requires a full-length carry-propagate addition.

The idea of scanning of several multiplier bits in each cycle can be extended from a fixed number of bits per cycle to an arbitrary and variable number, with the potential to greatly improve performance. The most direct way to do this is to "skip" past a string of 0s or a string of 1s in the multiplier without performing any addition or at least performing far fewer additions than would otherwise be the case. The case of 0s requires no explanation, as the multiplicand multiple for each 0 is just zero; the explanation for the 1s case is as follows.

Consider the string of $\cdots 011 \cdots 10 \cdots$ in the multiplier operand, with the most significant 1 in position j and the least significant 1 in position i, counting from right to left and starting the count at zero. The string corresponds to $j - i + 1$ multiplicand multiples whose sum is

$$\left(2^j + 2^{j-1} + \cdots + 2^i\right) x = \left(2^{j+1} - 2^i\right) x \qquad (1.33)$$

So $j - i + 1$ additions may be replaced with one addition in position $j + 1$ and one addition (a subtraction) in position i; all the multiplier bits between those two positions are "skipped." As an example, multiplication by 0111111000 (decimal 248) would effectively be carried out as multiplication by $2^8 - 2^3$ ($j = 7$ and $i = 3$).

There are two major problems with both fixed-length multiplier scanning and variable-length multiplier scanning as described above. With the fixed-length scanning, only multiples that correspond to powers of two—$2x, 4x, 8x$, etc.—can be formed easily, by shifting; the other multiples require carry-propagate additions or some other less-than-straightforward procedure.[20] And the variable-length scanning requires logic—a shifter, a counter, etc.—whose cost, in both hardware and operational time, is generally not considered worthwhile with the precisions used in ordinary arithmetic. Both problems can be partially solved by combining the positive aspects of the two techniques involved. We next explain how. The solution is *multiplier recoding*, which we describe next. .

Suppose, for example, that the multiplier is being scanned two bits at a time and the pair of bits under consideration is 11. That would "normally" correspond to the addition of the multiple $3x$. Now suppose the technique of skipping past 1s is applied. If the 11 is in the middle of a string of 1s, then no action is required. If it is at the start of a string of 1s—i.e., the 1s in $\cdots 110 \cdots$—then we should subtract $2^0 x = x$. And if it is at the end of a string of 1s—i.e., the 1s in $\cdots 011 \cdots$—then we should add $2^1 x = 2x$. So the requirement for $3x$ may be replaced with that for 0, or $-x$, or $2x$, all of which multiples are easily formed.

Determining which of the preceding three cases—skip, subtract, and add—applies is easily done by examining the two bits on the "sides" of the bit pair being scanned; that is, y_{i+2} and y_{i-1}, if the bits under consideration are $y_{i+1} y_i$. It is, however, sufficient to always examine only one of y_{i+1} or y_{i-1}, because the low (high) end of a string in one step is the high (low) end of a string in another step. With only one "side bit" examined, it is necessary to include a 0 at a hypothetical position -1 of the multiplier (for y_{i-1}), or at a hypothetical position n (for y_{i+1}) for an n-bit operand, in order to start or end the process. In what follows we shall assume that it is y_{i-1} that is examined; the changes required for the alternative case are straightforward and are left to the reader. It may also be necessary to append additional bits at the most significant end of the multiplier in order to ensure that there are enough bits to be scanned in the last cycle: if the multiplier is of n bits, the scanning is k bits at a time, and n is not divisible by k, then l bits should be added so that $n + l$ is divisible by k. (The extra bits will be 0s for a positive number and 1s for a negative number.)

---

[20]One alternative is to use redundant representation for the "problematic" multiples.

Table 1.8 Actions in multiplier recoding

(a) radix 4: two bits per cycle

$y_{i+1}y_i$	y_{i-1}	Action	
00	0	Shift P two places	
00	1	Add x; shift P two places	
01	0	Add x; shift P two places	
01	1	Add $2x$; shift P two places	
10	0	Subtract $2x$; shift P two places	
10	1	Subtract x; shift P two places	
11	0	Subtract x; shift P two places	
11	1	Shift P two places	

(b) radix 8: three bits per cycle

$y_{i+2}y_{i+1}y_i$	y_{i-1}	Action	
000	0	Shift P three places	$[-, -, -]$
000	1	Add x; shift P three places	$[0, 0, 1]$
001	0	Add x; shift P three places	$[0, 2, -1]$
001	1	Add $2x$; shift P three places	$[0, 2, 0]$
010	0	Add $2x$; shift P three places	$[4, -2, 0]$
010	1	Add $3x$; shift P three places	$[4, -2, 1]$
011	0	Add $3x$; shift P three places	$[4, 0, -1]$
011	1	Add $4x$; shift P three places	$[4, 0, 0]$
100	0	Subtract $4x$; shift P three places	$[-4, 0, 0]$
100	1	Subtract $3x$; shift P three places	$[-4, 0, 1]$
101	0	Subtract $3x$; shift P three places	$[-4, 2, -1]$
101	1	Subtract $2x$; shift P three places	$[-4, 2, 0]$
110	0	Subtract $2x$; shift P three places	$[0, -2, 0]$
110	1	Subtract x; shift P three places	$[0, -2, 1]$
111	0	Subtract x; shift P three places	$[0, 0, -1]$
111	1	Shift P three places	$[-, -, -]$

As noted above, the effect of adding $2^i x y_i$ in Eq. 1.20 can be obtained by always shifting the partial product (one in each cycle) and adding $x y_i$; this is taken into account in what follows, with respect to the nominal additions and subtractions of 2^{j+1} and 2^i in Eq. 1.33. With two-bit scanning, $y_{i+1}y_i$ corresponds to the multiple $2^i(2^1 y_{i+1} + 2^0 y_i)$; the 2^i is accounted for in the shifting, and the weights associated with the bit pair are 2^1 and 2^0. Similarly, with three-bit recoding the weights associated with $y_{i+2}y_{i+1}y_i$ are 2^2, 2^1, and 2^0.

On the basis of the preceding remarks, the actions required when scanning two bits per cycle and three bits per cycle are as shown in Table 1.8. (P denotes the partial product.) The benefits of the modified scanning are particularly striking in the second case: the multiples $3x$, $5x$, and $7x$ are no longer required.

The actions in Table 1.8 are easily understood in terms of "start of string," "middle of string," and "end of string," as in the description above for 11 in some

Table 1.9 Example of radix-4 multiplication

$x = 5 = 00000101$		$y = 114 = 01110010$
$k = 3$		$r = 8$

0 0 0 0 0 0 0 0	Initial P
0 0 0 0 1 0 1 0	Add $2x$
0 0 0 0 1 0 1 0	
0 0 0 0 0 0 0 1 0 1 0	Shift P three places
1 1 1 0 1 1 1 0	Subtract $2x$ (Add $-2x$)
1 1 1 1 1 0 1 1 1 0 1 0	
1 1 1 1 1 1 1 0 1 1 1 0 1 0	Shift P three places
0 0 0 0 1 0 1 0	Add $2x$
570 = 0 0 0 0 1 0 0 0 1 1 1 0 1 0	Final product

string. Thus, for example, the third line in Table 1.8a is both the start of a string of 1s (subtract x) and the end of a string of 1s (add $2x$), which is equivalent to adding x; the fourth line is the end of a string of 1s (add $2x$); and the last line is the middle of a string of 1s (no arithmetic). The last column in Table 1.8b shows how the "action" arithmetic has been obtained: relative to position i, the weights associated with y_{i+2}, y_{i+1}, and y_i are 4, 2, and 1. For example, for the line with $y_{i+2} y_{i+1} y_i y_{i-1} = 1010$: $y_i = 1$ is the start of a string of 1, so x is subtracted; $y_{+1} = 0$ is the end of string of 1s, so $2x$ is added; and $y_{i+2} = 1$ is the start of a string of 1s, so $4x$ is subtracted. We thus get $-4x + 2x - x = -3x$.

An example computation is shown in Table 1.9. Subtraction is as the addition of the two's complement of the subtrahend. Recall (from Sect. 1.1.6) that right-shifting the representation of a negative number requires sign extension.

For implementation, having both shifting with arithmetic and shifting without arithmetic—the first and last lines on Table 1.8a, b—implies a variable operational time, which will be undesirable in many cases. That can be dealt with by having the "actions" include "Add Zero" for the multiplier string $00 \cdots 0$ and "Subtract Zero" for the multiplier string $11 \cdots 1$. This modification also simplifies the decoding, as the most significant bit scanned then indicates the required arithmetic operation: 0 for addition and 1 for subtraction.

We may view the actions in Table 1.8 as a specification for an on-the-fly *recoding* of the multiplier, from the binary digit set into a different digit set: $\{-2, -1, 0, 1, 2\}$ for Table 1.8a and $\{-4, -3 - 2, -1, 0, 1, 2, 3, 4\}$ for Table 1.8b. Thus, for example, in the computation of Table 1.9, the multiplier is, in essence, recoded into the radix-8 representation $2\bar{2}2$, which represents $2 \times 8^2 - 2 \times 8^1 + 2 \times 8^0 = 114$ in decimal. In effect, recoding with Table 1.8 changes Eq. 1.31 to

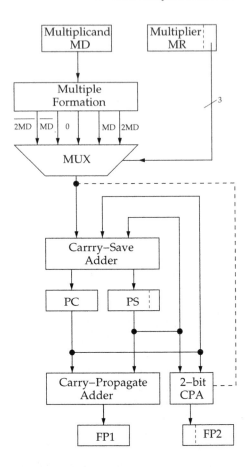

Fig. 1.17 Radix-4 sequential multiplier

$$Z_{i+1} = Z_i + r^i y_i'x \text{ where } y_i'x \in \{-2, -1, 0, 1, 2\} \text{ or } \{-4, -3-2, -1, 0, 1, 2, 3, 4\}$$

(Recall that in implementation, the multiplication by r^i is implicit; here it is effected by the shifting indicated in the tables.)

A digit set of the preceding type is usually written as $\{\overline{m}, \overline{m-1}, \ldots, \overline{2}, \overline{1}, 0, 1, 2, \ldots, m-1, m\}$, where $\overline{m}$ denotes $-m$, and so on. Because the number of digits in such a set exceeds the radix, a given number will have more than one representation. For example, with the digit set $\{\overline{1}, 0, 1\}$ two three-bit representations of the number three are $10\overline{1}$ and 011. Such a digit set is therefore known as a *redundant signed-digit set*.

Figure 1.17 shows an architecture for a sequential multiplier with two-bit recoding. The registers for the multiplier and lower half of the final product are shift registers that shift by two bit-positions in each cycle. The multiplicand multiples

$2x$ is formed by shifting through interconnections.[21] A two-bit carry-propagate adder (CPA) at the end of the carry-save adder (CSA) assimilates PC-PS bits that are shifted out from the CSA in each cycle. Subtraction (indicated by the most significant of the multiplier bits examined in a cycle) is performed by adding the negation of the subtrahend, which in two's-complement representation consists of the ones' complement and the addition of a 1 in the least significant bit-position. The Multiple Formation unit produces the required complement. The 1 is included in the small CPA, in the free slot created by a left shift of the carry bits relative to the sum bits. A carry output of the small CPA in one cycle is saved, say in a flip-flop, and becomes the carry input in the next cycle; in the last cycle the carry out of the small CPA becomes a carry in to the CPA used to complete the top half of the product. A detailed description of the timing will be found in [1].

1.2.3 Parallel and Sequential-Parallel

In an implementation of the architecture of Fig. 1.16, the PC-PS register delay will be a major factor in the operational time. High-radix recoding (as in Fig. 1.17) will reduce the effect of that delay, by reducing the number of times in which it is incurred, and thus give better performance. An alternative for better performance is to reduce the number of cycles required, by using several CSAs and adding several multiplicand multiples in each cycle. An example is shown in Fig. 1.18, for the addition of two multiples (M_{i+1} and M_i) at a time. A relative shift is required between the multiples, and this is done by wiring into the CSA inputs. The least significant bit out of each CSA is a "complete" bit of the final product (not shown).

The "logical extreme" from Fig. 1.18 is an arrangement in which the loop has been completely "unraveled," the PC-PS registers (and the inherent delay) are done away with, and there is one CSA for each multiplicand multiple. Such an arrangement reflects the multiplicand-multiple array in paper-and-pencil multiplication (Table 1.5). An example is shown in Fig. 1.19, for a case in which a total of five multiples are to be added. The required relative shifts between the multiples is through wired shifts into the CSA inputs. The least significant bit of M_0 is a "complete" bit of the final product, as is the least significant bit out of each CSA (Table 1.7). The carry out of the CPA is the most significant bit of the final product. (The reader might find it helpful to verify all this by drawing a detailed full-adder diagram of the array; alternatively, standard texts will provide the details.)

As addition is an associative operation, there are alternatives to the iterated-multiple-CSAs arrangement of Fig. 1.18: if several multiples are to be added in a cycle, then the multiplicand multiples may be taken in a variety of groups and the group additions carried out in parallel, with the potential for much higher

---

[21] With a higher radix multiples that are not powers of two may be "pre-computed" by addition—e.g., $3x$ as $2x + x$—or on-the-fly, in redundant representation [4].

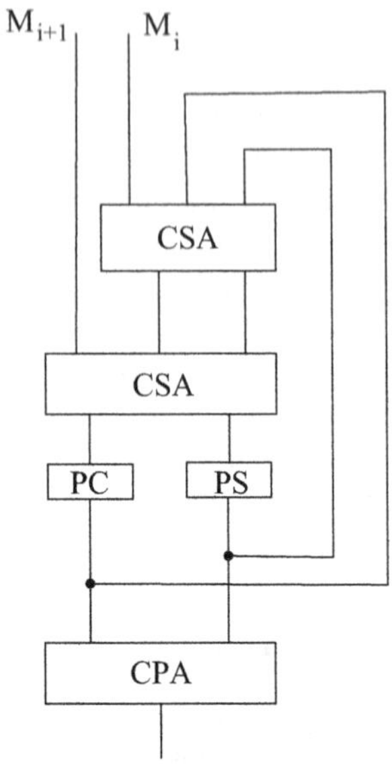

Fig. 1.18 Multiple-CSA
multiplier core

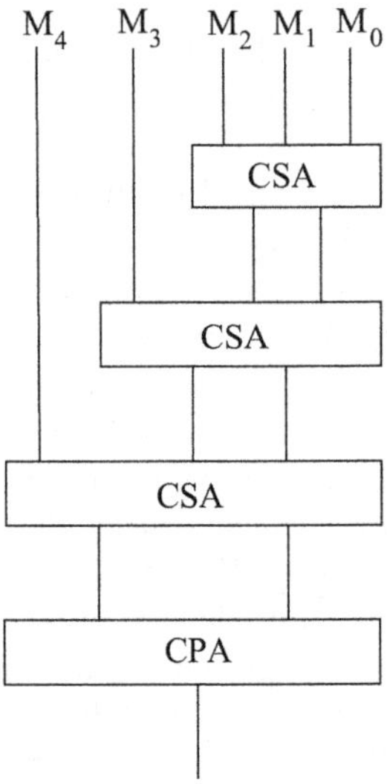

M_{i+1} $\quad$ M_i

CSA

CSA

PC $\quad$ PS

CPA

Fig. 1.19 Parallel-array
multiplier core

M_4 $\quad$ M_3 $\quad$ M_2 M_1 M_0

CSA

CSA

CSA

CPA

performance. Similarly, in "unraveling the loop," as is done in going from Figs. 1.18 and 1.19, there are numerous alternatives, of which the most straightforward is to take the multiplicand multiples and partial products in groups of threes (because a CSA is a three-input device) and add as many as possible concurrently. The multiplier thus obtained is a *Wallace-tree multiplier*. An example is shown in Fig. 1.20. As with Fig. 1.19, the relative shifts between multiplicand multiples is through wiring. More multiples are now added in the same number of CSA levels as in Fig. 1.19, so, for given operand precisions, the Wallace-tree multiplier has a small

Fig. 1.20 Wallace-tree
multiplier core

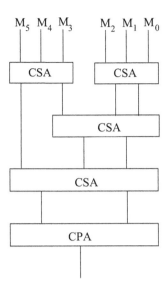

latency that the parallel-array multiplier. But the latter is better for high degrees of
pipelining and also has a more regular structure. Also, because the least significant
bits out of the CSA are not, as with Fig. 1.19, "neatly" bits of the final product, a
wider CPA is required. Thus there is a tradeoff between the reduction in the number
of CSA levels (and the delay thereof) and the extra delay in the CPA.

1.2.4 High Precision

We next briefly discuss the multiplication of high-precision numbers. We shall
assume that high-precision carry-propagate adders (CPAs) can be constructed as
described in Sects. 1.1.1–1.1.4. Carry-save adders (CSAs) are easily constructed to
any precision, by a simple replication of full adders.

Multiplier recoding can be used with arrangements such as those of Figs. 1.18,
1.19, and 1.20, which is what is done in almost all current high-performance
multipliers. The published literature will supply numerous detailed examples for
the interested reader.

The sequential multiplier of Fig. 1.16 will readily accommodate high-precision
multiplication, given the aforementioned assumptions, but it will be quite slow
in such a case. On the other hand, a highly parallel multiplier of the types shown
in Figs. 1.19 and 1.20 will be quite costly for very high precision. A "middle"
architecture is that of the type shown Fig. 1.18, in which a compromise between
cost and performance can be made for any precision, according to the desired
tradeoffs: increasing the number of CSAs decreases the number of cycles required
and therefore increases performance, but it also increases the cost. One can imagine

numerous straightforward variations on such an architecture—and the published literature will readily supply such variations—but we will not consider them here. Instead, we will broadly look at how high-precision multiplication can be carried out using multipliers of smaller precision than the target precision.

Suppose a multiplication is to be of n-bit operands and $n/2$-bit-by-$n/2$-bit multipliers are available. The two operands x and y may each be split into two equal parts, $\mathbf{x}_h$ and $\mathbf{x}_l$ and $\mathbf{y}_h$ and $\mathbf{y}_l$:

$$x = 2^{n/2}\mathbf{x}_h + \mathbf{x}_l \tag{1.34}$$

$$y = 2^{n/2}\mathbf{y}_h + \mathbf{y}_l \tag{1.35}$$

Then

$$xy = 2^n\mathbf{x}_h\mathbf{y}_h + 2^{n/2}(\mathbf{x}_h\mathbf{y}_l + \mathbf{x}_l\mathbf{y}_h) + \mathbf{x}_l\mathbf{y}_l \tag{1.36}$$

$$\stackrel{\triangle}{=} 2^n\mathbf{z}_h + 2^{n/2}\mathbf{z}_m + \mathbf{z}_l \tag{1.37}$$

where the powers of two represent relative shifts.

The multiplication xy is therefore carried out as four "half-precision" multiplications, some wired-shifting (multiplications by powers of two), and three additions. The additions may all be in CPAs or in CSAs with a single carry-propagate adder, the former being the case when only CPAs are available. With appropriate modifications, a multiplier built on such a basis can be used to carry out n-by-n-bit multiplications at one rate and $n/2$-by-$n/2$-bit multiplications at twice that rate. Examples will be found in [1].

The essential idea in Eqs. 1.34–1.37 can be applied "recursively," according to the size of multipliers available: the n-bit operands are each split into two parts, each of which is split into two parts, each of which ..., and so on. Taken to the extreme—i.e., down to one-bit operands—and with carry-save adders used, the end result is just the parallel-array multiplier. Also, in general, the splitting of an operand need not be into equal parts.

A closely related algorithm, known as the *Karatsuba-Ofman Algorithm*, computes xy from x and y decomposed as in Eqs. 1.34 and 1.35, but with fewer multiplications and more additions than in Eq. 1.36. The key here is the observation that the $\mathbf{z}_m$ may be computed as

$$(\mathbf{x}_h + \mathbf{x}_l)(\mathbf{y}_h + \mathbf{y}_l) - \mathbf{z}_h - \mathbf{z}_l \tag{1.38}$$

As in the preceding case, other splittings are possible, and the idea may be applied, "recursively," to smaller splittings.

Whether or not the Karatsuba-Ofman algorithm works out to be better that the algorithm of Eqs. 1.34–1.37 depends on the tradeoffs in cost and performance between multiplications and additions. And in this regard, it is important to note the additions in Eq. 1.36 may be carried out as three carry-save additions and a single carry-propagate addition, whereas using carry-save adders for the first two additions in Eq. 1.38 will lead to complexities.

In the both algorithms above, if each operand is split into m-bit pieces, then the computations may be viewed as involving radix-2^m digits. Such a view may be helpful in understanding the implementations of certain algorithms, but it has little practical implication.

1.2.5   Signed Numbers

Multiplication with signed operands can be handled in various ways, of which we describe only one: the use of multiplier recoding. Although we have above discussed multiplier recoding in the context of multiple-bit scanning of the multiplier, the idea arose in the context of one-bit scanning, with the sole objective being to facilitate multiplication with signed numbers.[22]

The main change that is required for signed numbers in multiplier recoding is the sign extension of partial products as they are shifted to the right. Sign extension is necessary, to ensure correctness, because the sign bit in a ones'-complement or two's-complement representation is actually the truncation of an infinite string of 1s (Sect. 1.1.6).

An example of multiplication with one-bit recoding is shown in Table 1.10. The reader can easily verify that the essential idea will work with operands of other signs and with multiple-bit scanning.

1.2.6   Squaring

Squaring may be regarded as just an instance of multiplication, with squares computed using the algorithms and implementations of the architectures described above. But if the operation is sufficiently frequent, then it might be worthwhile to devise specialized hardware. The basic idea is to take advantage of the fact that both operands in the nominal multiplication are the same, and so an implementation can be optimized to be less costly and faster than would otherwise be the case.

Take, for example, the computation of x^2 by a full multiplication and with x represented in five bits, $x_4x_3x_2x_1x_0$. The array of multiplicand multiples has the form shown in Figure 1.21a. This array can be optimized:

---

[22]One-bit recoding as originally devised is commonly known as *Booth's Algorithm*. The name is sometimes also applied to multiple-bit recoding.

Table 1.10 Signed-operand multiplication with Booth's algorithm

Multiplicand $(x) = 5 = 0101$, Multiplier $(y) = -10 = 10110$

0 0 0 0 0	Initial partial product
0 0 0 0 0	$y_0 y_{-1} = 00$; add zero
0 0 0 0 0	
0 0 0 0 0 0	Shift partial product right
1 1 0 1 1	$y_1 y_0 = 10$; subtract x
1 1 0 1 1 0	
1 1 1 0 1 1 0	Shift partial product right
0 0 0 0 0	$y_2 y_1 = 11$; subtract zero
1 1 1 0 1 1 0	
1 1 1 1 0 1 1 0	Shift partial product right
0 0 1 0 1	$y_3 y_2 = 01$; Add x
0 0 0 1 1 1 1 0	
0 0 0 0 1 1 1 1 0	Shift partial product right
1 1 0 1 1	$y_4 y_3 = 10$; subtract x

$-50 \quad = \quad 1\ 1\ 1\ 0\ 0\ 1\ 1\ 1\ 0$ Final product

- Every other term in the anti-diagonal is of the form $x_i x_i$, which is equivalent to x_i since x_i is 0 or 1. There is also a symmetry around the same diagonal, since $x_i x_j = x_j x_i$. Therefore, the two terms $x_i x_j$ and $x_j x_i$ may be replaced with their sum, $2x_i x_j$; and since multiplication by two is a 1-bit left-shift, that is just $x_i x_j$ moved into the next column to the left. Therefore, the matrix of Fig. 1.21a may be compressed into the equivalent one in Fig. 1.21b.
- Consider the terms x_i and $x_i x_j$ in the same column.

  - If $x_i = 0$, then $x_i + x_i x_j = 0$;
  - if $x_i = 1$ and $x_j = 0$, then $x_i + x_i x_j = x_i = x_i \overline{x_j}$;
  - if $x_i = 1$ and $x_j = 1$, then $x_i + x_i x_j = 2 = 2x_i x_j$.

So $x_i + x_i x_j = 2x_i x_j + x_i \overline{x_j}$, which corresponds to $x_i \overline{x_j}$ in the same column and $x_i x_j$ moved into the next column to the left. This gives the array of Fig. 1.21c from that of Fig. 1.21b.

Figure 1.22 shows an example of optimized squaring.

$$
\begin{array}{ccccc}
 & & x_4x_0 & x_3x_0 & x_2x_0 & x_1x_0 & x_0x_0 \\
 & & x_4x_1 & x_3x_1 & x_2x_1 & x_1x_1 & x_0x_1 \\
 & & x_4x_2 & x_3x_2 & x_2x_2 & x_1x_2 & x_0x_2 \\
 & & x_4x_3 & x_3x_3 & x_2x_3 & x_1x_3 & x_0x_3 \\
 & & x_4x_4 & x_3x_4 & x_2x_4 & x_1x_4 & x_0x_4
\end{array}
$$

(a)

$$
\begin{array}{ccccccccc}
x_3x_4 & x_2x_4 & x_1x_4 & x_0x_4 & x_0x_3 & x_0x_2 & x_0x_1 & 0 & x_0 \\
x_4 & & x_2x_3 & x_1x_3 & x_1x_2 & & x_1 & 0 & \\
 & & x_3 & & x_2 & & & &
\end{array}
$$

(b)

$$
\begin{array}{ccccccccc}
x_3x_4 & \overline{x_3}x_4 & x_2x_4 & x_1x_4 & x_0x_4 & x_0x_3 & x_0x_2 & \overline{x_0}x_1 & 0 & x_0 \\
 & & x_2x_3 & \overline{x_2}x_3 & x_1x_3 & \overline{x_1}x_2 & x_1x_0 & x_1 & & \\
 & & & & x_1x_2 & & & & &
\end{array}
$$

(c)

Fig. 1.21 Array compression in squaring

$$
x = 13 = 01101_2 \qquad x^2 = 169 = 0010101001_2
$$

$$
\begin{array}{cccccccccc}
x_3x_4 & \overline{x_3}x_4 & x_2x_4 & x_1x_4 & x_0x_4 & x_0x_3 & x_0x_2 & \overline{x_0}x_1 & 0 & x_0 \\
0 & 0 & 0 & 0 & 0 & 1 & 1 & 0 & 0 & 1 \\
 & & x_2x_3 & \overline{x_2}x_3 & x_1x_3 & \overline{x_1}x_2 & x_1x_0 & x_1 & & \\
 & & 1 & 0 & 0 & 1 & 0 & 0 & & \\
 & & & & x_1x_2 & & & & & \\
 & & & & 0 & & & & & \\
\hline
0 & 0 & 1 & 0 & 1 & 0 & 1 & 0 & 0 & 1
\end{array}
$$

Fig. 1.22 Optimized squaring

1.3 Division

Division is a fundamental operation in ordinary arithmetic, but, unlike addition and multiplication, ordinary integer division does not have much direct use in the algorithms of this book. Nevertheless, the *essence* of division will be found in several algorithms in the book, such as those for modular reduction (Chap. 4) and algorithms that require such reduction (Chaps. 5 and 6). Indeed, certain extensions of, and alternatives to, those algorithm will require division. We will also in other places make some comparative references to division.

Division may be considered as the inverse of multiplication, and this is partially reflected in the algorithms, in which certain actions in multiplication algorithms are replaced with their "inverses" in division algorithms. Such "inversion" starts with the operands: the multiplication algorithms above take an n-bit multiplicand and an n-bit multiplier and produce a $2n$-bit product; the (first) division algorithms that follow take a $2n$-bit dividend and n-bit divisor and yield an n-bit quotient and n-bit remainder.[23] We shall initially assume that the operands are unsigned and then later make some remarks on signed operands.

Let x and d be the dividend and divisor, q and r be the quotient and remainder from the division, x_h be the value represented by the high-order n digits of the dividend, and b be representation radix. In order to avoid overflow—i.e., a q that is too large to represent in n digits—we must have $q \leq b^n - 1$ and $d \neq 0$. Therefore, from

$$x = qd + r$$

we get

$$x \leq \left(b^n - 1\right) d + r$$
$$< b^n d \qquad \text{since } r < d$$

So the check for the possibility of overflow is $x_h < d$.

Basic integer multiplication consists of the addition of multiplicand multiples to a running partial product that is zero at the start and the sought product at the end. Direct division is the converse of that. It consists of subtractions of multiples of the divisor from a *partial remainder* that is initially the dividend and finally a remainder less than the divisor, with the quotient is formed according to multiples that are subtracted at the various steps. In ordinary multiplication, the product is formed from least significant digit to the most significant digit; in ordinary division, the quotient is formed from the most significant digit to the least significant digit.

An example of paper-and-pencil division is shown in Fig. 1.23a. Figure 1.23b is more explicit in showing what is usually omitted in the former—that the multiples of the divisor are weighted by powers of the radix: $9 * 124 * 10^2$, $8 * 124 * 10^1$, and $5 * 124 * 10^0$ in the examples. The trailing 0s are usually omitted because they are not included in the corresponding subtractions.

Let q_j denote the jth radix-b digit of the quotient; that is, $q = \sum_{j=0}^{n-1} b^j q_j = b(\cdots (bq_{n-1} + q_{n-2}) \cdots q_1) + q_0$. Then a direct algorithm for the division of Fig. 1.23b is

---

[23] Just as the multiplication algorithms can be easily modified for n-bit multiplicand, m-bit multiplier, and $n + m$-bit product, so too can the division algorithms be modified for $n + m$-bit dividend, m-bit divisor, n-bit quotient, and m-bit remainder.

Fig. 1.23 Ordinary
paper-and-pencil division

$$
\begin{array}{r}
985 \\
124 \enclose{longdiv}{122153} \\
1116 \\
\hline
1055 \\
992 \\
\hline
633 \\
620 \\
\hline
13
\end{array}
$$

(a)

$$
\begin{array}{r}
985 \\
124 \enclose{longdiv}{122153} \\
111600 \\
\hline
10550 \\
9920 \\
\hline
633 \\
620 \\
\hline
13
\end{array}
$$

(b)

$$R_0 = x$$

$$Q_0 = 0$$

$$R_{i+1} = R_i - b^j q_j d \qquad i = 0, 1, 2, \ldots, n - 1, \ j = n - 1 - i$$

$$Q_{i+1} = bQ_i + q_j$$

$$r = R_n$$

$$q = Q_n$$

Table 1.11a shows the correspondence of this algorithm to the example of Fig. 1.23b.

In a "naive" implementation of the preceding algorithm, forming $b^j q_j d$ would require variable-length shifting, which should be avoided if possible. In basic multiplication, variable-length shifting is avoided by holding the multiplicand multiple in a fixed position and shifting the partial products to the right. Here, variable-length shifting can be avoided by holding the divisor multiples in place and shifting the partial remainders to the left.

For the overflow check, the divisor is aligned with the top half of the dividend (i.e., the initial partial remainder). If the division is started at that point, then the divisor d is effectively replaced with $d^* = b^n d$. At that point a subtraction cannot be successful because $x < b^n d$, i.e., $x_h < d$; so the process starts with a left shift (multiplication by b) of the partial remainder, which is then similarly shifted at each step. At the end of the process, the last partial remainder is scaled by b^{-n}—an n-digit right shift—to account for the initial scaling of the divisor.

Putting together the preceding remarks, we have this algorithm:

$$R_0 = x \tag{1.39}$$

$$Q_0 = 0 \tag{1.40}$$

$$d^* = b^n d \tag{1.41}$$

$$R_{i+1} = bR_i - q_j d^* \qquad i = 0, 1, 2, \ldots, n - 1, \ j = n - 1 - i \tag{1.42}$$

Table 1.11 Example of mechanical division

(a)

$x = 122, 153, \ d = 124, \ d^* = 124, 000, \ b = 10$

i	R_i	q_j	$b^j q_j d$	Q_i
0	122,153	9	111,600	0
1	10,553	8	9920	9
2	633	5	620	98
3	13	–	–	985

(b)

i	R_i	q_j	$q_j d^*$	Q_i
	(bR_i)			
0	122,153	9	1,116,000	0
	(1,221,530)			
1	10,553	8	992,000	9
	(105,530)			
2	633	5	620,000	98
	(633,000)			
3	13,000	–	–	985

$r = 13000 * 10^{-3} = 13$

$$Q_{i+1} = bQ_i + q_j \tag{1.43}$$

$$r = R_n b^{-n} \tag{1.44}$$

$$q = Q_n \tag{1.45}$$

A slightly different but equivalent algorithm consists of shifting the scaled divisor one place down, after the overflow check, and then shifting reduced partial remainders. That is, Eqs. 1.41 and 1.42 are replaced with

$$d^* = b^{n-1} d$$

$$R_{i+1} = b \left(R_i - q_j d^* \right)$$

Table 1.11b shows the correspondence of the algorithm of Eqs. 1.39–1.45 to the example of Fig. 1.23b. Several points should be noted for an implementation of the algorithm. First, there need not be any real shifting in the initial scaling of the divisor; it can be simply hardwired to the appropriate position. Second, although the numbers in Table 1.11b are large, the least significant n digits of $q_j d^*$ are always 0s and so need not be explicitly represented. Third, as a consequence of the second point, the arithmetic to reduce the shifted partial remainder may be on only the most significant $n + 1$ digits. Fourth, the trailing 0s in shifting the partial remainders need not be explicitly represented, and there is thus no need for an explicit scaling of R_n. Lastly, there is no real addition in the computation of Q_{i+1}: q_i is "inserted" into the "empty slot" that is created by the left shift of Q_i. These points are reflected in the *restoring division* algorithm (below) and its implementation.

We have thus far not considered how a divisor multiple and the corresponding quotient digit are determined. This is generally the most challenging aspect of "subtractive" division,[24] and it is this that makes division inherently more complex than multiplication. In paper-and-pencil division, the determination is usually done through trial subtractions on the side and subtraction then carried out of the correct divisor multiple. That cannot be done easily in a computer. Fortunately, with binary representation there are only two possible multiples—zero and the divisor—so a straightforward solution to the problem is to proceed as follows. Subtract the divisor. If the result is negative, then the subtraction should not have taken place, and the partial remainder is *restored* (by adding back the divisor) and then shifted. The corresponding bit of the quotient is 1 or 0, according to whether or not a successful subtraction takes place. The algorithm is known as *restoring division*.

The binary restoring algorithm:

$$Q_0 = 0 \tag{1.46}$$

$$R_0 = x \tag{1.47}$$

$$d^* = 2^n d \qquad \text{(implicit)} \tag{1.48}$$

$$\widetilde{R}_{i+1} = 2R_i - d^* \qquad\qquad i = 0, 1, 2, \ldots, n-1, \ j = n-1-i \tag{1.49}$$

$$q_j = \begin{cases} 1 & \text{if } \widetilde{R}_{i+1} \geq 0 \\ 0 & \text{otherwise} \end{cases} \tag{1.50}$$

$$Q_{i+1} = 2Q_i + q_j \tag{1.51}$$

$$R_{i+1} = \begin{cases} \widetilde{R}_{i+1} & \text{if } \widetilde{R}_{i+1} \geq 0 \\ \widetilde{R}_{i+1} + d^* & \text{otherwise} \end{cases} \tag{1.52}$$

$$r = R_n \tag{1.53}$$

$$q = Q_n \tag{1.54}$$

A partial remainder satisfies the condition $R_i < d^*$.

It should be noted that some of the arithmetic suggested in the algorithm is not "real." The computation of $2R_i$ is just a one-bit left shift of R_i; the nominal multiplication $q_i d^*$ is just the selection of 0 or d^*; and the addition in the computation of Q_{i+1} is just the insertion of q_j in the "space" created by the left shift of Q_i.

An example application of the algorithm is given in Table 1.12. Negative numbers are in two's-complement representation, and subtraction is as the addition of the

---

[24]There are "multiplicative" algorithms, which the reader will find elsewhere [1–5].

Table 1.12 Example of restoring division

$x = 143 = 10001111_2,\ d = 14 = 1110_2$

i	R_i			q_j	Q_i	Action
0	00	1000	1111		0000	R_0, Q_0
	01	0001	1110		0000	Left shift R_0, Q_0
	11	0010				Subtract d^*
	00	0011		1		$\widetilde{R}_1 \geq 0, q_3 = 1$
1	00	0011	1110		0001	R_1, Q_1
	00	0111	1100		0010	Left shift R_1, Q_1
	11	0010				Subtract d^*
	11	1001		0		$\widetilde{R}_2 < 0, q_2 = 0$
	00	1110				Restore (Add d^*)
	00	0111	1100			
2	00	0111	1100		0010	R_2, Q_2
	00	1111	1000		0100	Left shift R_2, Q_2
	11	0010				Subtract d^*
	00	0001		1		$\widetilde{R}_3 \geq 0, q_1 = 1$
3	00	0001	1000		0101	R_3, Q_3
	00	0011	0000		1010	Left shift R_3, Q_3
	11	0010				Subtract d^*
	11	0101		0		$\widetilde{R}_4 < 0, q_0 = 0$
	00	1110				(Add d^*)
	00	0011	0000			
4	00	0011	0000		1010	$R_4 = r, Q_4 = q$
$r = 3 = 0011_2$			$q = 10 = 1010_2$			

two's complement of the subtrahend. So two extra bits are required in the arithmetic: one for sign, and one to accommodate the magnitude of $2R_i$.

The restoring-division algorithm will in some steps require two arithmetic operations in a single cycle. *Nonrestoring division* is a much better algorithm that requires only one arithmetic operation per cycle and is the basis of most current "subtractive" algorithms for division. In nonrestoring division, negative partial remainders are permitted, and the basic operation is subtraction or addition, according to whether a partial remainder is positive or negative. An intermediate partial remainder that is "incorrect" (i.e., negative) gets "corrected," by an addition, in the next cycle. The justification for this is as follows.

Suppose $\widetilde{R}_i$ in the restoring-division algorithm (Eqs. 1.46–1.54) is negative. Then the restoration would be the computation of $\widetilde{R}_i + d^*$, and the next cycle would be the computation of

$$R_{i+1} = 2\left(\widetilde{R}_i + d^*\right) - d^*$$
$$= 2\widetilde{R}_i + d^* \tag{1.55}$$

So, we may omit the restoration step and immediately proceed to compute R_{i+1} according to Eq. 1.55. And if $\widetilde{R}_i$ is not positive, then there is no restoration step, and

$$R_{i+1} = 2\widetilde{R}_i - d^* \tag{1.56}$$

Since negative partial remainders are now allowed, there is no need to compute a tentative partial remainder, and we may therefore replace $\widetilde{R}_i$ with R_i. The combination of Eqs. 1.55 and 1.56 then gives this rule: if the partial remainder is negative, then add; and if the partial remainder is positive, then subtract. As before, in the last cycle there is no "next cycle," but the partial remainder might be negative. In that case an explicit corrective addition is necessary, with a corresponding correction of the last quotient bit.

The binary nonrestoring algorithm:

$$Q_0 = 0 \tag{1.57}$$

$$R_0 = x \tag{1.58}$$

$$d^* = 2^n d \qquad \text{(implicit)} \tag{1.59}$$

$$R_1 = 2R_0 - d^* \tag{1.60}$$

$$q_j = \begin{cases} 1 & \text{if } R_i \geq 0 \qquad j = n-1, n-2, \ldots, 0 \\ 0 & \text{otherwise} \end{cases} \tag{1.61}$$

$$Q_{j+1} = 2Q_j + q_j \tag{1.62}$$

$$R_{i+1} = \begin{cases} 2R_i - d^* & \text{if } R_i \geq 0 \qquad i = 1, 2, \ldots, n-1 \\ 2R_i + d^* & \text{otherwise} \end{cases} \tag{1.63}$$

$$r = \begin{cases} R_n & \text{if } R_n \geq 0 \\ R_n + d^* & \text{otherwise} \end{cases}$$

$$q_0 = \begin{cases} 1 & \text{if } R_n \geq 0 \\ 0 & \text{otherwise} \end{cases}$$

$$q = Q_n \tag{1.64}$$

A partial remainder satisfies the condition $|R_i| < d^*$.

An example application of the algorithm is given in Table 1.13. Subtraction is as the addition of the two's complement of the subtrahend, so an extra bit is necessary for sign. And one more bit is necessary to accommodate the range of $2R_i$.

A basic architecture for nonrestoring division is shown in Fig. 1.24. The partial remainder and quotient registers are left-shift ones that shift once in each cycle. The Adder-Subtractor is as in Fig. 1.14. After each addition or subtractor, the most significant bit (i.e., sign bit) of the result is examined. If the bit is a 0 (for a positive

Table 1.13 Nonrestoring division

$x = 143 = 10001111_2$, $d = 14 = 1110_2$

i	R_i			q_j	Q_i	Action
0	00	1000	1111		0000	R_0, Q_0
	01	0001	1110		0000	Left shift R_0, Q_0
	11	0010				Subtract d^*
	00	0011		1		$R_1 \geq 0, q_3 = 1$
1	00	0011	1110		0001	R_1, Q_1
	00	0111	1100		0010	Left shift R_1, Q_1
	11	0010				Subtract d^*
	11	1001		0		$R_2 < 0, q_2 = 0$
2	11	1001	1100		0010	R_2, Q_2
	11	0011	1000		0100	Left shift R_2, Q_2
	00	1110				Add d^*
	00	0001		1		$R_3 \geq 0, q_1 = 1$
3	00	0001	1000		1010	R_3, Q_3
	00	0011	0000		1010	Left shift R_3, Q_3
	11	0010				Add d^*
	11	0001		0		$R_4 < 0, q_0 = 0$
4	11	0001			1010	R_4, Q_4
	00	1110				Correction (Add d^*)
	00	0011	0000		1010	$R_4 = r, Q_4 = q$

$r = 3 = 0011_2$, $q = 10 = 1010_2$

Fig. 1.24 Nonrestoring divider

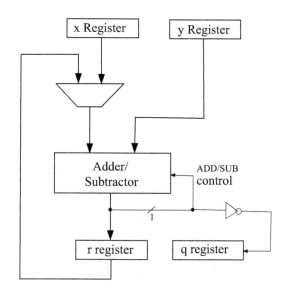

number), then a 1 is entered in the quotient, and the next operation is a subtraction; and if the bit is a 1 (for a negative number), then a 1 is entered in the quotient, and the next operation is an addition.[25]

To produce an n-bit quotient, an implementation based on Fig. 1.24 requires n cycles of the use of a carry-propagate adder. The process can be speeded up in three main ways. The first is to speed up each cycle, by using a carry-save adder in place of a carry-propagate adder. The second is to reduce the number of cycles in which arithmetic is required, by skipping over strings of 0s and 1s in the partial remainder. And the third is to reduce the number of cycles by using an implementation radix larger than two. Similar techniques are used in multiplication,[26] but here they cannot be applied in a straightforward way. All three techniques can be combined in a single algorithm: *SRT division*, of which we next describe the binary version.

For the following discussions we shall assume that the dividend x and the divisor d are such that $0 \le x < 1, 0 < d < 1$, and $x < d$. The quotient q too will then be a positive fraction:

$$q = 0.q_1 q_2 \cdots q_n$$

Algorithms for high-speed division are most often used to divide significands in floating-point arithmetic[27]; so there is no loss of generality in these assumptions. For integer arithmetic, it is convenient to first describe the algorithms for fractional operands, which is how they are almost-always formulated, and then make additional remarks to account for the differences.

We have seen that the number of cycles in a multiplication can be reduced by shifting past 0s and 1s in the multiplier operand without performing any arithmetic or by using a radix larger than two (Sect. 1.2.2). Similar techniques can be employed in division too, and we next explain how in division it is possible to past 0s in a positive partial remainder or 1s in a negative partial remainder without performing any arithmetic.

Since $x < 1$ and $d < 1$, their binary representations will be of the form $0.* * * \cdots *$. A shifted partial remainder can be large enough to be of the form $1.** \cdots *$, and it can be negative, which requires an additional bit for sign, so partial remainders will have the form $** . ** \cdots$. Now consider, for example, the partial remainder $00.000001 * * * \cdots *$ and the divisor 0.001011 in a nonrestoring division. A successful subtraction will not be possible until a shifted partial remainder of the form $00.01 * * \cdots$ has been obtained. And if the shifted partial remainder is of the form $00.001 * * * \cdots$, then a successful subtraction might or might not

---

[25] We have omitted some details; for example, the bit should be stored in, say, a flip-flop, as it is used to control the operation in the next cycle, not the current one.

[26] The second is skipping past 0s or 1s in the partial remainder (which in multiplication corresponds to skipping past 0s or 1s in the multiplier), and the third is taking several bits of the partial remainder at each step (which in multiplication corresponds to taking several bits of the multiplier at each step.)

[27] Arithmetic in "scientific" notation.

be possible. A similar situation occurs with a negative shifted partial remainder (in two's-complement representation), but with "addition" in place of "subtraction" and 1s in place of 0s. Where it is possible to shift past 0s or 1s, no arithmetic need take place, and the corresponding bits of the quotient can be entered immediately.

From the preceding, we have three cases of partial remainder: (1) the shifted partial remainder is positive and large enough to guarantee a successful subtraction, which is then carried out; (2) the shifted partial remainder is negative and of a magnitude large enough for a successful addition, which is then carried out; (3) a middle region in which a subtraction, or an addition, or no operation may be carried out.

To formulate an algorithm, a definite selection rule is required from the Robertson diagram—specifically, the determination of a value constant c such that $[-c, c]$ is the range in the middle region over which no arithmetic is necessary and with subtraction or addition according whether the partial remainder is above or below that range. Standard considerations in the past have included choosing a c that optimizes the probability of being able to shift over 0s and 1s; $c = d$ is not such a value, but it splits the range nicely, and it is also convenient for reasons that will become apparent in what follows. The rule for computing the next partial remainder is then

$$R_{i+1} = \begin{cases} R_i - d & \text{if } 2R_i \geq d \\ R_i & \text{if } -d \leq 2R_i < d \\ R_i + d & \text{if } 2R_i < -d \end{cases}$$

If we make this correspond to the recurrence of Eq. 1.42, i.e.,

$$R_{i+1} = 2R_i - q_i d$$

then we are "naturally" led to a redundant signed-digit (RSD) representation[28] for q_i:

$$q_i = \begin{cases} 1 & \text{if } 2R_i \geq d \\ 0 & \text{if } -D \leq 2R_i < d \\ \overline{1} & \text{if } 2R_i < -d \end{cases} \tag{1.65}$$

where $\overline{1}$ denotes -1. Thus in fast multiplication the multiplier is implicitly recoded into an RSD representation, and in fast division the quotient is explicitly recoded into an RSD representation.

The standard representation for the three cases above is a *Robertson diagram*, an example of which is shown in Fig. 1.25. ($U_i \equiv R_i$, $s_i \equiv q_i$, $D \equiv d$, etc.)

---

[28] See also Sect. 1.2.2.

Fig. 1.25 Roberston diagram
for radix-2 division

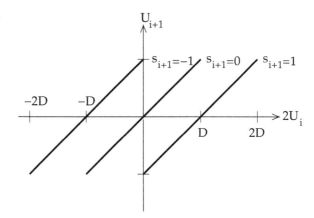

The use of signed digits explicitly shows the corrective actions during a subtraction.[29] Suppose, for example, that the correct quotient bits in some part of a division are 01, which correspond to an addition and a subtraction, but that there is instead a subtraction for the first, thus producing 1 for the first quotient-bit. The next operation would be a "corrective" addition, which would produce a quotient bit of $\bar{1}$. And $1\bar{1} = 01$, since, with the appropriate positional weighting $1\bar{1} = 2^1 - 2^0 = 01$. We will see that the redundancy obtained with the digit set $\{1, 0, \bar{1}\}$ is very useful.

The selection rule in Eq. 1.65 implies full comparisons against the divisor and its negation. It would obviously be better—and fortunately it is possible—to have comparisons against low-precision constants. To determine such constants, it is conventional to use a *P-D diagram*, which consists of plots of the shifted partial remainder ($P = 2R_i$) against the divisor; and it is standard to assume that the divisor has been "normalized" so that it is at least $1/2$, with the dividend and quotient scaled accordingly if necessary. The explanation for the latter is as follows.

A direct implementation of shifting over 0s and 1s, as described above, would require two shifters, since the most significant bit in each of the partial remainder and divisor can be at any position. But the divisor does not change for a given division. So, if it can be arranged that the divisor always has its most significant bit in a fixed position, then only one shifter will suffice. Requiring that the divisor always have the representation $0.1 * * \cdots *$ satisfies that condition, and the pattern represents a value of at least $1/2$. For the remainder of this discussion we assume such a *normalized* divisor.[30]

With the assumption that $d \geq 1/2$, the P-D diagram for binary division is as shown in Fig. 1.26. ($s_i \equiv q_i$, $D \equiv d$, etc.) To formulate a rule for choosing between $q_i = 0$ and $q_i = 1$ and between $q_i = 0$ and $q_i = \bar{1}$, one or more separating lines are

---

[29]The basic nonrestoring algorithm can be formulated with the nonredundant digit set $\{\bar{1}, 1\}$, although there is little practical benefit in doing so.

[30]Note that the basic nonrestoring algorithm also "normalizes" the divisor d to $d^* = 2^n d$.

Fig. 1.26 P-D diagram for
binary SRT division

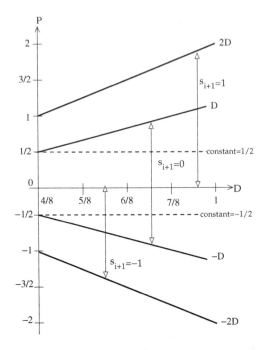

required in each region of overlap in the figure. The constant line $P = 1/2$ gives
such a separation, and, by symmetry, so does $P = -1/2$ on the other side. So we
may take the comparison constants to be $-1/2$ and $1/2$, and the q_i-selection rule
is then

$$q_i = \begin{cases} 1 & \text{if } 2R_i \geq 1/2 \\ 0 & \text{if } -1/2 \leq 2R_i < 1/2 \\ \bar{1} & \text{if } 2R_i < -1/2 \end{cases} \qquad (1.66)$$

The corresponding Roberston diagram is shown in Fig. 1.27. ($U_i \equiv R_i$, $s_i \equiv q_i$,
$D \equiv d$, etc.) The comparison is independent of d and requires only three bits of
$2R_i$: one sign bit, one integer bit, and one fraction bit.

Putting together all of the preceding remarks, we end up with the *binary SRT
division algorithm* (for an *m*-bit quotient):

$$Q_0 = 0 \qquad (1.67)$$

$$R_0 = x \qquad\qquad x < d, \; 1/2 \leq d \qquad (1.68)$$

$$q_i = \begin{cases} 1 & \text{if } 2R_i \geq 1/2 \\ 0 & \text{if } -1/2 \leq 2R_i < 1/2 \\ \bar{1} & \text{if } 2R_i < -1/2 \end{cases} \qquad i = 0, 1, 2, \ldots, m-1 \quad (1.69)$$

$$R_{i+1} = 2R_i - q_i d \qquad (1.70)$$

Fig. 1.27 Modified radix-2
Robertson diagram

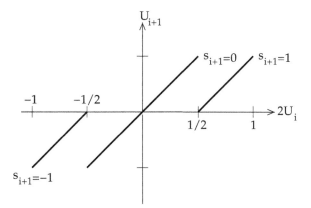

$$Q_{i+1} = 2Q_i + q_i \tag{1.71}$$

$$r = \begin{cases} R_m & \text{if } R_m \geq 0 \\ R_m + d & \text{otherwise} \end{cases}$$

$$q_0 = \begin{cases} 1 & \text{if } R_m \geq 0 \\ 0 & \text{otherwise} \end{cases} \tag{1.72}$$

$$q = Q_m \tag{1.73}$$

An example of SRT division is given in Table 1.14. The quotient is now produced in RSD form and at the end of the iterations must be converted into conventional form.[31] Negative numbers are in two's-complement representation, and subtraction is as the addition of the negation of the subtrahend. In each of the first three cycles, $0 < 2R_i \leq 1/2$, so 0s are entered in the quotient and shifting without any arithmetic takes place. In the fourth cycle, $2R_i > 1/2$, so a 1 is entered in the quotient and a shift and subtraction take place. In each of the next three cycles, $-1/2 \leq 2R_i < 0$, so 0s are entered in the quotient, and shifting without arithmetic again takes place. In the eighth cycle $2R_i < -1/2$ and a $\bar{1}$ is entered in the quotient, followed by an addition and a shift. And so on. The number of iterations is determined by the desired precision in the quotient.

If it is desirable to have a predictable operational time—and this would be the norm—then "no arithmetic" would be replaced with "Add Zero" or "Subtract Zero," according to the sign of the partial remainder. As in multiplier recoding, such a change would mean no speed-advantage (over the basic algorithm), but only if arithmetic is in non-redundant form. We next discuss fast arithmetic.

---

[31]A simple way to do this is to subtract the negative digits from the positive ones. For example $1\bar{1}001\bar{1} = 100010 - 010001 = 010001$. The conversion can also be done on-the-fly as the quotient digits are produced [3, 16], so the implied delay need not occur.

Table 1.14 SRT division

$x = 0.000110 = 3/32,\ d = 0.1101 = 13/16,\ x/d \approx 0.1154$				
i	R_i	q_i	Q_i	Action
0	00.000110	0	0·0000000000	Shift
1	00.001100	0	·0000000000	Shift
2	00.011000	0	0·0000000000	Shift
3	00.110000 11.001100 1.1111000	1	0·0000000001	Subtract
4	11.111000	0	0·0000000010	Shift
5	11.110000	0	0·0000000100	Shift
6	11.100000	0	0·0000001000	Shift
7	11.000000 00.110100 11.110100	$\overline{1}$	0·000001000$\overline{1}$	Add
8	11.101000	0	0·000010001$\overline{0}$	Shift
9	11.010000 00.110100	$\overline{1}$	0·00010001$\overline{1}$01$\overline{1}$	Add
10	00.001000		0·001000$\overline{1}$01$\overline{1}$0	

$Q_{10} = 2^{-3} - 2^{-7} - 2^{-9} \approx 0.1152$

We have thus far assumed the use of a carry-propagate adder (CPA) for the reductions of the magnitude of the partial remainder. As with multiplication, using a carry-save adder (CSA) in the iterations would give a substantial improvement in performance, but there is a significant difficulty here: with a CSA, the partial remainder consists of a partial carry (PC) and partial sum (PS), and its sign is known definitely only after assimilation (which requires a CPA). This is where the redundancy in the RSD notation is especially beneficial: a given quotient will have more than one representation; so an "error" in the choice of one quotient bit—an "error" that corresponds to an "incorrect" arithmetic operation—can be corrected by appropriately choosing the bits in subsequent cycles. The practical implication is that only an approximation, consisting of a few leading bits of the assimilated partial remainder, is needed in order to select the next quotient bit. The approximation is obtained by assimilating a few bits of the CSA's outputs.[32] The final remainder will now be in PC-PS form and at the end of the iterating must be converted into conventional form. We next give the details of the approximation.

Let $\widetilde{2R}_i$ denote the approximation that is obtained by truncating the PC-PS representation of $2R_i$ and then assimilating PC and PS. If the truncation is to p fraction bits, then the error in the value represented by PC is bounded by 2^{-p}, and the error in the value represented by PS is similarly bounded—a total of 2^{-p+1}. Figure 1.26 shows that at $d = 1/2$, $P = d$, so there is no margin for error with

---

[32]This can be done in a small CPA or by using a lookup table.

the selection rule of Eq. 1.47. Replacing $1/2$ and $-1/2$ with $1/4$ and $-1/4$ would provide such a margin and will work with $p \geq 3$. An alternative, which gives a smaller value of p, is to make use of the asymmetry in the effect of truncating twos-complement PC-PS representations: truncating a twos-complement representation "reduces" the number represented. (If the number is positive, then the magnitude is decreased; otherwise, the magnitude is increased.) This has the following effect.

In Fig. 1.26, the separation between $q_i = 0$ and $q_i = 1$ has to be made between the lines $P = 0$ and $P = d$. If $2R_i$ is positive, then $2\widetilde{R}_i \geq 0$, which thus gives an error tolerance of at most $1/2$ (at $d = 1/2$). Therefore, if we take $p = 2$, then we may take $P = 0$ as the separating line. On the negative side, the truncation can only increase the magnitude; so its effect is that where before the separating line had to be found in $[0, -d]$, now the range is $[0, -d - 2^{-p}]$. $P = -1/2$ will still do, with $p = 2$, which gives the selection rule

$$q_i = \begin{cases} 1 & \text{if } 2\widetilde{R}_i \geq 0 \\ 0 & \text{if } -1/2 \leq 2\widetilde{R}_i < 0 \\ \bar{1} & \text{if } 2\widetilde{R}_i < -1/2 \end{cases}$$

With the values $-1/2$ and $1/2$, only one fraction bit of $\widetilde{R}_i$ is needed to make the selection. And with one fraction bit, the only representable number in the range $[-1/2, 0)$ is $-1/2$. Therefore, the selection rule may be modified to

$$q_i = \begin{cases} 1 & \text{if } 2\widetilde{R}_i \geq 0 \\ 0 & \text{if } 2\widetilde{R}_i = -1/2 \\ \bar{1} & \text{if } 2\widetilde{R}_i < -1/2 \end{cases}$$

We can also arrive at a slightly different rule on the basis that the truncation results in a reduction by $1/2$: from

$$q_i = \begin{cases} 1 & \text{if } 2R_i \geq 0 \\ 0 & \text{if } -1/2 \leq 2R_i < 0 \\ \bar{1} & \text{if } 2R_i < -1/2 \end{cases}$$

subtracting $1/2$ from the ends of the intervals gives

$$q_i = \begin{cases} 1 & \text{if } 2\widetilde{R}_i \geq 0 \\ 0 & \text{if } 1 \leq 2\widetilde{R}_i < 0 \\ \bar{1} & \text{if } 2\widetilde{R}_i < -1 \end{cases}$$

In either case, $|2R_i| < 2$, and we have $-5/2 \leq \widetilde{R}_i \leq 3/2$, which requires one sign bit and two integer bits for representation.

In summary, with carry-save representation $\widetilde{R}_i$ will be formed by assimilating four bits of the representation of $2R_i$—one sign bit, two integer bits, and one

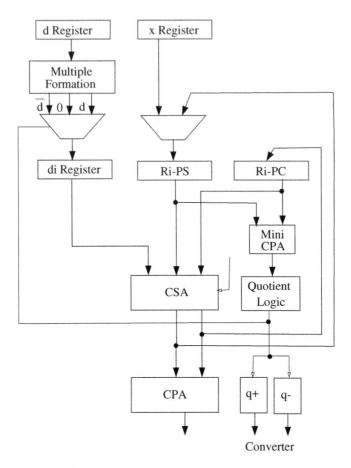

Fig. 1.28 Radix-2 SRT divider

fraction bit. Figure 1.28 shows an architecture for a radix-2 SRT divider.[33] The additions and subtractions during the iterations are done in a carry-save adder (CSA), and the partial remainder is generated in partial-carry/partial form (in the registers Ri-PC and Ri-PS). Subtraction is done by adding ones' complement and a 1 (i.e., two's-complement arithmetic). The multiplication by two of the partial remainder is done through a wired left-shift. The Mini CPA is a small carry-propagate adder (CPS) that assimilates the top bits of the partial remainder, for the determination of the next quotient digit. The quotient digits may be stored separately, according to sign, and the conversion to conventional form then done

---

[33]This is a simple diagram that omits some details that are related to control and timing. For example, in a given cycle, except the first, the quotient digit is used to select the divisor multiple (and arithmetic operation) in the *subsequent* cycle.

by subtracting the "negative" sequence from the "positive" one, or the conversion may be done on-the-fly [3, 16].

The SRT algorithm can be extended to radices larger than two [1–5]. In practice, the complexities are such that direct implementation for radices larger than four are not worthwhile. Larger-radix implementations are rare and are usually obtained by overlapping small-radix units—e.g., a radix-16 unit from two radix-4 units.

We now briefly consider integer division with the SRT algorithm; more details will be found in [3]. Let x be the n-bit dividend, d be the n-bit divisor, and q and r be the quotient and remainder from the division. The operands for the algorithm must be fractions, and the divisor must be normalized (i.e., n with 1 as the most significant bit). Considering the latter first, the operand is normalized, through a k-bit shift, to obtain

$$d^* = 2^k d$$

The quotient is then

$$q = \left\lfloor \frac{x}{d} \right\rfloor$$

$$= \left\lfloor \frac{x}{d^*} 2^k \right\rfloor$$

This quotient will be of up to $k + 1$ bits, so $k + 1$ iterations are required in the algorithm.

The required fractional operands, x_f and d_f, are obtained by scaling x and d^*:

$$x_f = 2^{-n} x$$

$$d_f = 2^{-n} d^*$$

That is, in essence, a binary point is placed immediately to the left of the representations of x and d^*.

At the end of the iterations q and r are nominally in fractional form and an implied "reverse" scaling is required to obtain integer results. The representation of the quotient does not have an implied binary point; so the bits may be taken as properly positioned, and no scaling is required. That is not the case for the remainder, which must therefore be explicitly scaled, by right-shifting and ignoring the binary point. The remainder can also be negative, in which case it should be corrected, with any necessary change in the last bit of the quotient.

For our purposes here, a simple example that does not involve carry-save arithmetic will suffice; this is given in Table 1.15. Subtraction is as the addition of the two's complement of the subtrahend. Note that the initial partial remainder is $x_f / 2$, in order to satisfy the bounds on partial remainders.

Table 1.15 SRT integer division

$x = 53 = 00110101_2$, $d = 5 = 00000101_2$, $d^* = 10100000$, $k = 5$				
$x_f = 0.00110101$, $d_f = 0.10100000$				
i	R_i	q_i	Q_i	Action
	00.00010101		000000	R_0, Q_0
0	00.00110101	0	000000	Left shift R_0, Q_0
	<u>00.00000000</u>			Subtract zero; add q_0
	00.00110101		000000	R_1, Q_1
1	00.01101010	0	000000	Left shift R_1, Q_1
	<u>00.00000000</u>			Subtract zero; add q_1
	00.01101010		000000	R_2, Q_2
2	00.11010100	1	000000	Left shift R_1, Q_1
	<u>11.01100000</u>			Subtract d_f; add q_2
	00.00110100		000001	R_3, Q_3
3	00.01101000	0	000010	Left shift R_3, Q_3
	<u>00.00000000</u>			Subtract zero; add q_3
	00.01101000		000010	R_4, Q_4
4	00.11010000	1	000100	Left shift R_4, Q_4
	<u>11.01100000</u>			Subtract d_f; add q_4
	00.00110000		000101	R_5, Q_5
5	00.01100000	0	001010	Left shift R_5, Q_5
	<u>00.00000000</u>			Subtract zero; add q_5
	00.01100000		001010	
$q = 001010_2 = 10, r = 2^5 * 0.0110 = 00000011_2 = 3$				

Signed Operands

For the changes required to accommodate signed operands, it suffices to consider only the nonrestoring algorithm; modifications similar to what we will describe are straightforward for the restoring algorithm (which is unlikely to be used in practice) and the SRT algorithm (which is just a special case of the nonrestoring algorithm).

The essence of the division algorithms above is that the magnitude of the partial remainder R_i is repeatedly reduced, through a sequence of additions or subtractions, to a value that is less than the divisor, with bits of the quotient are determined according to the result at each step.

The four cases according to the signs of the operands are as follows, with two's-complement arithmetic in the additions and subtractions.

Case (i): Positive Dividend and Positive Divisor

This is the case assumed in the discussions above. The magnitude of a positive partial remainder is reduced by subtracting the divisor, and that of a negative operand is reduced by adding the divisor. The quotient bit is 1 for a positive result and 0 for a negative result. At the end of the cycling a remainder that is negative is corrected by adding the divisor. An example is shown in Table 1.16.

Table 1.16 Positive dividend and positive divisor

$x = 33 = 00100001_2$, $d = 5 = 0101_2$				q_j	i	Action
i	R_i					
	00	0010	0001		0000	R_0, Q_0
0	00	0100	001*		0000	Left shift R_0, Q_0
	11	1011				Subtract d^*
	11	1111	001*	0	0000	R_1, Q_1
1	11	1110	01**		0000	Left shift R_1, Q_1
	00	0101				Add d^*
	00	0011	01**	1	0001	R_2, Q_2
2	00	0110	1***		0010	Left shift R_2, Q_2
	11	1011				Subtract d^*
	00	0001	1***	1	0011	R_3, Q_3
3	00	0011	****		0110	Left shift R_3, Q_3
	11	1011				Subtract d^*
	11	1110	****	0	0110	R_4, Q_4
	11	1110	****		0110	R_4, Q_4
	00	0101				Correction: add d^*
	00	0011			0110	r, q
$r = 0011_2 = 3$, $q = 0110_2 = 6$						

Case (ii): Negative Dividend and Positive Divisor

The reduction of partial remainders is as in Case (i), but the quotient bit is 0 for a positive result and 1 for a negative result. This generates the quotient in ones'-complement representation, $\overline{q}$, to which a 1 is added to obtain the two's-complement representation. At the end of the cycling a remainder that is positive is corrected by adding the divisor. An example is shown in Table 1.17.

Case (iii): Negative Dividend and Negative Divisor

The magnitude of a positive partial remainder is reduced by adding the divisor, and that of a negative operand is reduced by subtracting the divisor. The quotient bit is 1 for a positive result and 0 for a negative result. At the end of the cycling a remainder that is positive is corrected to a negative one. An example is shown in Table 1.18.

Case (iv): Positive Dividend and Negative Divisor

The magnitude of a positive partial remainder is reduced in the same manner as in Case (iii); the quotient bit is 0 for a positive result and 1 for a negative result. As in Case (ii), the quotient is produced in ones'-complement representation, $\overline{q}$, to which a 1 is added to get the two's complement. At the end of the cycling a remainder that is negative is corrected to a positive one. An example is shown in Table 1.19.

Table 1.17 Negative dividend and positive divisor

i	R_i			q_j	i	Action
	11	1101	1111		0000	R_0, Q_0
0	11	1011	111*		0000	Left shift R_0, Q_0
	00	0101				Add d^*
	00	0000	001*	1	0001	R_1, Q_1
1	00	0001	11**		0010	Left shift R_1, Q_1
	11	1011				Subtract d^*
	11	1100	11**	0	0010	R_2, Q_2
2	11	1001	1***		0100	Left shift R_2, Q_2
	00	0101				Add d^*
	11	1110	1***	0	0100	R_3, Q_3
3	11	1101	****		1000	Left shift R_3, Q_3
	00	0101				Add d^*
	00	0010	****	1	1001	R_4, Q_4
	00	0010	****		1001	R_4, Q_4
	00	0101				Correction: add d^*
	11	1101			1001	$r, \overline{q}$

$r = 1101_2 = -3$, $q = \overline{q} + 1 = 1001_2 + 1_2 = 1010_2 = -6$

Table 1.18 Negative dividend and negative divisor

i	R_i			q_j	i	Action
	11	1101	1111		0000	R_0, Q_0
0	11	1011	111*		0000	Left shift R_0, Q_0
	00	0101				Subtract d^*
	00	0000	001*	0	0000	R_1, Q_1
1	00	0001	11**		0010	Left shift R_1, Q_1
	11	1011				Add d^*
	11	1100	11**	1	0001	R_2, Q_2
2	11	1001	1***		0010	Left shift R_2, Q_2
	00	0101				Subtract d^*
	11	1110	1***	1	0011	R_3, Q_3
3	11	1101	****		0110	Left shift R_3, Q_3
	00	0101				Subtract d^*
	00	0010	****	0	0110	R_4, Q_4
	00	0010	****		0110	R_4, Q_4
	00	1011				Correction: add d^*
	11	1101			1110	r, q

$r = 1101_2 = -3$, $q = 0110_2 = 6$

Table 1.19 Positive dividend and negative divisor

$x = 33 = 00100001_2$, $d = -5 = 1011_2$						
i	R_i			q_j	i	Action
	00	0010	0001		0000	R_0, Q_0
0	00	0100	001*		0000	Left shift R_0, Q_0
	11	1011				Add d^*
	11	1111	001*	1	0001	R_1, Q_1
1	11	1110	01**		0010	Left shift R_1, Q_1
	00	0101				Subtract d^*
	00	0011	01**	0	0010	R_2, Q_2
2	00	0110	1***		0100	Left shift R_2, Q_2
	11	1011				Add d^*
	00	0001	1***	0	0100	R_3, Q_3
3	00	0011	****		1000	Left shift R_3, Q_3
	11	1011				Add d^*
	11	1110	****	1	1001	R_4, Q_4
	11	1110	****		1001	R_4, Q_4
	00	0101				Correction: subtract d^*
	00	0011			1001	$r, \widetilde{q}$

$r = 0011_2 = 3$, $q = \widetilde{q} + 1 = 1001_2 + 1_2 = 1010_2 = -6$

References

1. A. R. Omondi, 1994. *Computer Arithmetic Systems*. Prentice Hall, Hemel Hempstead, UK.
2. B. Parhami. 2000. *Computer Arithmetic*. Oxford University Press, Oxford, UK.
3. M. Ercegovac and T. Lang. 2004. *Digital Arithmetic*. Morgan Kaufmann, San Francisco, CA, USA.
4. M. J. Flynn and S. F. Oberman. 2001. *Advanced Computer Arithmetic Design*. Wiley-Interscience, New York, USA.
5. A. R. Omondi. 2015. *Computer-Hardware Evaluation of Mathematical Functions*. Imperial College Press, London, UK.
6. R. W. Doran, 1988. Variants on an improved carry-lookahead adder. *IEEE Transactions on Computers*, 37(9):1110–1113.
7. S. V. Lakshmivaran and S. K. Dhall, 1994. *Parallel Computation Using the Prefix Problem*. Oxford University Press, Oxford, UK.
8. R. P. Brent and H. T. Kung, 1982. A regular layout for parallel adders. *IEEE Transactions on Computers*, C-31(3):260–264.
9. R. E. Ladner and M. J. Fischer. 1980. Parallel prefix computation. *Journal of the ACM*, 27:831–838.
10. P. M. Kogge and H. S. Stone. 1973. A parallel algorithm for the efficient computation of a general class of recurrence relations. *IEEE Transactions on Computers*, 22:786–793.
11. S. Knowles. 1999. A family of adders. *Proceedings, 14th Symposium on Computer Arithmetic*, pp. 30–34.
12. T. Han and D. A. Carlson. 1987. Fast area-efficient VLSI adders. *Proceedings, 8th International Symposium on Computer Arithmetic*, pp. 49–56.
13. A. Beaumont-Smith and C.-C. Lim, 2001. "Parallel prefix adder design". *Proceedings, 15th International Symposium on Computer Arithmetic*, pp. 218–225.

14. A. Tyagi. 1993. A reduced area scheme for carry-select adders. *IEEE Transactions on Computers*, 42(10):1163–1170.
15. G. A. Ruiz and M. Granda. 2004. An area-efficient static CMOS carry-select adder based on a compact carry look-ahead unit. *Microelectronics Journal*, 35:939–944.
16. M. D. Ercegovac and T. Lang. 1987. On-the-fly conversion of redundant into conventional representations. *IEEE Transactions on Computers*, C-36(7):895–897.
17. N. H. E. Weste and K. Eshraghian. 1994. *Principles of CMOS VLSI Design*. Addison-Wesley. Reading, Massachusetts, USA.
18. J. M. Rabaey. 1996. *Digital Integrated Circuits*. Prentice Hall, Saddle River, New Jersey, USA.
19. N. Burgess. 2013. Fast ripple-carry adders in standard-cell CMOS VLSI. *Proceedings, 20th International Symposium on Computer Arithmetic*, pp. 103–111.
20. X. Cui, W. Liu, S. Wang, E. E. Swartzlander, and E. Lombardi. 2018. Design of high-speed wide-word hybrid parallel-prefix/carry-select and skip adders. *Journal of Signal Processing Systems*, 90(3):409–419.
21. D. E. Knuth. 1998. *The Art of Computer Programming, Vol. 2*. Addison-Wesley, Reading, Massachusetts, USA.

Part II

Chapter 2
Mathematical Fundamentals I: Number Theory

Abstract There are many cryptosystems that are based on *modular arithmetic* (also known in some contexts as *residue arithmetic*); examples of such systems are given in the next chapter. This chapter covers some of the fundamentals of modular arithmetic and will be a brief review or introduction, according to the reader's background. The first section of the chapter gives some basic definitions and mathematical properties. The second section is on the basic arithmetic operations, squares, and square roots. The third section is on the *Chinese Remainder Theorem*, a particularly important result in the area. And the last section is on *residue number systems*, unconventional representations that can facilitate fast, carry-free arithmetic.

More comprehensive discussions of modular arithmetic and residue number systems will be found in [1, 2].

We shall make use of several mathematical results, of which proofs and other details will be found in most standard texts on elementary number theory [3, 4]. For the reader's convenience, some of the proofs are included in the Appendix.

2.1 Congruences

Modular arithmetic is based on the *congruence* relation. We shall start with a definition of this relation and then proceed to certain relationships that follow from the definition.

Definition For a given nonzero positive integer m, two integers x and y are *congruent modulo m* if $x - y = km$ for some integer k. This is usually denoted by writing

$$x \equiv y \pmod{m} \tag{2.1}$$

Evidently, if $x \equiv y \pmod{m}$, then $x - y$ is divisible by m.

© Springer Nature Switzerland AG 2020

A. R. Omondi, *Cryptography Arithmetic*, Advances in Information Security 77,
https://doi.org/10.1007/978-3-030-34142-8_2

As examples:

$$10 \equiv 7 \pmod{3}$$

$$10 \equiv 1 \pmod{3}$$

$$10 \equiv 4 \pmod{3}$$

$$10 \equiv -2 \pmod{3}$$

The number m in Eq. 2.1 is known as the *modulus*. We shall assume that its values exclude 1, which produces only trivial congruences. The number y is said to be a *residue* of x *with respect to* (or *modulo*) m.

The standard definition of integer division leads naturally to modular congruence. Let q and r be the quotient and remainder from the integer division of x by m. Then

$$x = qm + r \qquad 0 \le r < m \tag{2.2}$$

and by definition

$$x \equiv r \pmod{m}$$

We shall sometimes express this as

$$r = x \bmod m$$

and refer to process of computing r as the *reduction* of x *modulo* m. An alternative expression for the reduction is

$$r = |x|_m$$

Note that $x \equiv y \pmod{m}$ if and only if $x \bmod m = y \bmod m$.

r in Eq. 2.2 is the smallest positive residue of x with respect to m. Since $0 \le r \le m - 1$, the set $\{0, 1, 2, \ldots, m - 1\}$ is accordingly called the set of *least positive residues modulo* m. For a given m, every integer is congruent to exactly one member of this set, which accordingly is known as a *complete set of residues*. We may therefore consider properties of arbitrary residues (modulo m) in terms of those properties on this set. So, unless otherwise indicated, or it is evident from the context, we shall assume that for a given modulus these are the only residues.

It is straightforward to verify that for a fixed integer m and arbitrary integers x, y, z, w:

- $x \equiv y \pmod{m}$ if and only if x and y leave the same remainder on division by m
- if $x \equiv y \pmod{m}$, then $y \equiv x \pmod{m}$
- if $x \equiv y \pmod{m}$ and $y \equiv z \pmod{m}$, then $x \equiv z \pmod{m}$

- if $x \equiv y \pmod{m}$ and $z \equiv w \pmod{m}$, then $x + z \equiv y + w \pmod{m}$ and $xz \equiv yw \pmod{m}$
- if $x \equiv y \pmod{m}$, then $x + z \equiv y + z \pmod{m}$ and $xz \equiv yz \pmod{m}$
- if $x \equiv y \pmod{m}$, then $x^k \equiv y^k \pmod{m}$ for any positive integer k

2.2 Modular-Arithmetic Operations

We next consider briefly the basic arithmetic operations of addition, subtraction, multiplication, and division and then also discuss squares and square roots. The first three are very straightforward; the others are less so.

2.2.1 Addition, Subtraction, and Multiplication

Modular addition requires little explanation, but modular subtraction requires some "special" handling. We have indicated above that we will mostly be working with positive residues. In order to obtain such results for subtraction, and also facilitate hardware implementation, an appropriate definition of subtraction is necessary. As in ordinary computer arithmetic, we may here define subtraction as the addition of the "negation" of the subtrahend. In residue arithmetic that "negation" is the *additive inverse*.

Definition The *additive inverse* of a residue x with respect to (or modulo) a modulus m, denoted $|-x|_m$, is defined by the equation

$$(x + |-x|_m) \equiv 0 \pmod{m}$$

When the modulus is evident from the context, we will write $-x$ for $|-x|_m$. (The expression $-x \bmod m$ is also sometimes used for $|-x|_m$.)

For each residue and modulus, the additive inverse always exists and is unique. Since $(m - x) \bmod m = -x$, the inverse may be obtained as

$$|-x|_m = m - x \tag{2.3}$$

where the "$-$" on left denotes inversion, and the "$-$" on the right denotes ordinary subtraction. Thus, for example, if the modulus is 5, then the additive inverses of 1, 2, 3, and 4 are 4, 3, 2, and 1, respectively.

We may now define the subtraction of two residues x and y, modulo m, as

$$(x - y) \bmod m = (x + |-y|_m) \bmod m$$

As an example, with $m = 13$, we have

$$(8 - 12) \bmod 13 = (8 + |-12|_{13}) \bmod 13$$
$$= (8 + 1) \bmod 13$$
$$= 9$$

We could also have arrived at the same result by first computing $8 - 12 = -4$ with "ordinary" subtraction and then using Eq. 2.3 to obtain a positive reside: $|-4|_{13} = 13 - 4 = 9$.

On the basis of what has been given so far, certain properties can be verified easily for addition, subtraction, and multiplication. Let x, y, and k be arbitrary integers, and m a given modulus. Then

$$(x + y) \bmod m = (x \bmod m + y \bmod m) \bmod m$$
$$(x - y) \bmod m = (x \bmod m - y \bmod m) \bmod m$$
$$(x * y) \bmod m = (x \bmod m * y \bmod m) \bmod m$$
$$(x \pm k * m) \bmod m = x \bmod m$$

2.2.2 Division

It is sometimes useful in ordinary computer arithmetic to consider division as the multiplication of the dividend by the reciprocal of the divisor. A similar approach is useful here too, with the notion of reciprocal replaced with an appropriate analogue: the *multiplicative inverse*.

Definition If x is a nonzero integer and m is a modulus, then the **multiplicative inverse** of x with respect to (or modulo) a modulus m, denoted $|x^{-1}|_m$, is defined by

$$x * |x^{-1}|_m \equiv 1 \pmod{m}$$

As an example, if $m = 11$ and $x = 7$, then $|x^{-1}|_{11} = 8$.

When the modulus is evident from the context, we will write x^{-1} for $|x^{-1}|_m$. (The expression $x^{-1} \bmod m$ is also sometimes used for $|x^{-1}|_m$.)

Whereas every residue has an additive inverse (with respect to any modulus), it is not the case that every residue has a multiplicative inverse.[1] The multiplicative inverse of x with respect to m exists if and only if x and m are relatively prime, i.e., if

---

[1] See the corollary to Theorem 6.3, proof in the Appendix.

Table 2.1 Multiplicative inverses

$m = 7$		$m = 8$	
x	x^{-1}	x	x^{-1}
1	1	1	1
2	4	2	–
3	5	3	3
4	2	4	–
5	3	5	5
6	6	6	–
		7	7

their greatest common divisor (gcd) is 1. If m is prime, then evidently $\gcd(x, m) = 1$ for every $1 < x < m$; so every residue has an inverse. Table 2.1 gives some examples.

With the multiplicative inverse as the analogue of the reciprocal, we may now define the division of x and y, with respect to a modulus, m, as

$$\frac{x}{y} = \begin{cases} \left(x * y^{-1}\right) \bmod m & \text{if } y^{-1} \text{ exists} \\ \text{undefined} & \text{otherwise} \end{cases}$$

There is no general expression that will readily yield the multiplicative inverse, but for a prime modulus, Fermat's Little Theorem, which is given next, can be useful. Otherwise, the standard computational method for obtaining multiplicative inverses is the *Extended Euclidean Algorithm* (Section 6.2).

Theorem 2.1 (Fermat's Little Theorem) *If p is a prime and a is an integer such that $\gcd(a, p) = 1$, then*

$$a^{p-1} \equiv 1 \pmod{p} \tag{2.4}$$

Corollary 2.1 *If p is prime, then $a^p \equiv a \pmod{p}$.*

Fermat's Little Theorem yields a simple expression for inversion: multiplying both sides of Eq. 2.4 by a^{-1}, we get

$$a^{-1} \equiv a^{p-2} \pmod{p} \tag{2.5}$$

2.3 Generators and Primitive Roots

Given a modulus m, repeated addition (i.e., multiplication) or repeated multiplication (i.e., exponentiation) modulo m on some single element in a subset of $\{0, 1, 2, \ldots, m - 1\}$ will produce elements of the same set. If the resulting subset is the same as the initial one, then we say that the element is a *generator* and that it *generates* the subset.

We shall use the notation

$$k * x = \overbrace{x + x + \cdots + x}^{k \text{ times}} \tag{2.6}$$

$$x^k = \overbrace{x * x * \cdots * x}^{k \text{ times}} \tag{2.7}$$

and sometimes write kx for $k * x$.

Example 2.1 5 is generator for $\{0, 1, 2, 3, 4, 5\}$ under addition $(+)$ modulo 6:

$$1 * 5 \mod 6 = 5 \qquad 4 * 5 \mod 6 = 2$$
$$2 * 5 \mod 6 = 4 \qquad 5 * 5 \mod 6 = 1$$
$$3 * 5 \mod 6 = 3 \qquad 6 * 5 \mod 6 = 0$$

and so on, in cyclic repetition. 1 is another generator for the same set.

On the other hand, 4 is not a generator in this case:

$$1 * 4 \mod 6 = 4 \qquad 4 * 4 \mod 6 = 4$$
$$2 * 4 \mod 6 = 2 \qquad 5 * 4 \mod 6 = 2$$
$$3 * 4 \mod 6 = 0 \qquad 6 * 4 \mod 6 = 0$$

and so on, in cyclic repetition.

□

Example 2.2 2 is a generator for $\{1, 2, 4\}$ under multiplication $(*)$ modulo 7:

$$2^1 \mod 7 = 2 \qquad 2^4 \mod 7 = 2$$
$$2^2 \mod 7 = 4 \qquad 2^5 \mod 7 = 4$$
$$2^3 \mod 7 = 1 \qquad 3^6 \mod 7 = 1$$

and so on, in cyclic repetition.

□

Example 2.3 3 is a generator for $\{1, 2, 3, 4, 5, 6\}$ under multiplication $(*)$ modulo 7:

$$3^1 \mod 7 = 3 \qquad 3^4 \mod 7 = 4$$
$$3^2 \mod 7 = 2 \qquad 3^5 \mod 7 = 5$$
$$3^3 \mod 7 = 6 \qquad 3^6 \mod 7 = 1$$

and so on, in cyclic repetition.

On the other hand, Example 2.2 shows that 2 is not a generator in this case.

$\square$

Example 2.3, in which the "multiplicative" generator produces all the nonzero (least positive) residues, is an instance of a general case that is of special interest.

Definition Let a and m be positive integers such that $\gcd(a, m) = 1$. Then a is of *order k modulo m* if k is the smallest positive integer such that $a^k \bmod m = 1$.

Thus Examples 2.2 and 2.3 show that 2 is of order 3 modulo 7, and 3 is of order 6 modulo 7.

We shall make use of the following result.

Theorem 2.2 *Let a, n and m be such that $n > 0$ and $\gcd(a, m) = 1$, and let k be the order of a modulo m. Then*

$$a^n \equiv 1 \ (\bmod \ m) \quad \text{if and only if } n \text{ is divisible by } k$$

Definition For $m \geq 1$, *Euler's Totient function*, denoted $\phi(m)$, gives the number of integers that are not greater than m and are relatively prime to m.

For example, $\phi(5) = 4$, $\phi(6) = 2$, $\phi(9) = 6$, and $\phi(16) = 8$. Note that if m is prime, then $\phi(m) = m - 1$

Definition Let a and m be positive integers such that $m > 1$ and $\gcd(a, m) = 1$. Then a is said to be a *primitive root* of m if a is of order $\phi(m)$ modulo m.

Another way to state the definition is that a is a primitive root of p if the set $\{a \bmod p, a^2 \bmod p, a^3 \bmod p, \ldots, a^{p-1} \bmod p\}$ is exactly $\{1, 2, 3, \ldots, p - 1\}$. Thus Example 2.3 shows that 3 is a primitive root of 7.

Primitive roots exist for every prime, and it can be shown that in general m has a primitive root if m is 2, or 4, or p^k, or $2p^k$, for an odd prime p and integer $k \geq 1$. An integer can have more than one primitive root; for example, 3 and 5 are primitives root of 7. If they exist, then the number of primitive roots of m is $\phi(\phi(m))$. It is of special interest that under multiplication modulo an odd prime p, the set $\{1, 2, 3, \ldots, p - 1\}$ is generated by a primitive root of p, and in this case there are exactly $\phi(p - 1)$ primitive roots.

We can now state an important problem that is the basis of many cryptosystems, some of which are described in Chap. 3.

Discrete Logarithm Problem

If g is a generator under multiplication modulo p and $0 < y < p$, then the equation

$$g^x \equiv y \ (\bmod \ p)$$

has a solution in x. In such a case x is known as the *discrete logarithm* of y (*with respect to the base* g). With an appropriate choice of parameters, determining whether one number is the discrete logarithm of another number with respect to a given modulus is an extremely difficult problem that is known as the *Discrete-Logarithm Problem*.

2.4 Quadratic Residues and Square Roots

The following is a brief discussion on modular squares and square roots, which are important in some cryptosystems.

In ordinary integer arithmetic, a number has proper square roots only if it is a *perfect square*. The analogue here of that is the *quadratic residue*.

Definition[2] Let a and m be integers such that $\gcd(a, m) = 1$. If the equation

$$x^2 \equiv a \pmod{m} \tag{2.8}$$

has a solution in x, then a is said to be a ***quadratic residue***[3] of (or modulo) m, and the solution x is a ***square root*** of a. Otherwise a is ***quadratic nonresidue***(see footnote 3) of (or modulo) m.

Given that the set $\{0, 1, 2, \ldots, m - 1\}$ is a complete set of residues modulo m, in looking for quadratic residues modulo m, it suffices to consider just the elements of this set for the values of a in Eq. 2.8.

Example 2.4 The quadratic residues of 15 are 1, 4, 6, 9, and 10; and the quadratic nonresidues are 2, 3, 5, 7, 8, 9, 11, 12, 13, and 14.

a	1	2	3	4	5	6	7	8	9	10	11	12	13	14
a^2 mod 15	1	4	9	1	10	6	4	4	6	10	1	9	4	1

The quadratic residues of 13 are 1, 3, 4, 9, 10, and 12; and the quadratic nonresidues are 2, 5, 6, 7, 8, and 11.

a	1	2	3	4	5	6	7	8	9	10	11	12
a^2 mod 13	1	4	9	3	12	10	10	12	3	9	4	1

---

[2]Some texts give this definition with respect to only a prime modulus. Our purposes require this more general definition.

[3]One could just as well use the terms "square" and "nonsquare," but standard terminology is what it is.

A check on two of the values of a:

- $x^2 \equiv 3 \pmod{13}$ has the solutions $x = 4$ and $x = 9$
- $x^2 \equiv 5 \pmod{13}$ has no solutions

$\square$

The second case in Example 2.4, in which the modulus is a prime p, is especially interesting for our purposes. Half of the residues are quadratic residues, and half are quadratic nonresidues.[4] This "half-half" split is the case in general[5]: for prime modulus p, there are exactly $(p - 1)/2$ quadratic residues and $(p - 1)/2$ are quadratic nonresidues in $\{1, 2, 3, \ldots, p - 1\}$. So it is not particularly difficult to find a quadratic residue: one may as well just make a random selection and then check its status.

The following theorem and corollary give some useful information on the determination of quadratic residuosity.

Theorem 2.3 (Euler's Criterion) *Let p be an odd prime and a be an integer such that $\gcd(a, p) = 1$. Then a is a quadratic residue of p if and only if*

$$a^{(p-1)/2} \equiv 1 \pmod{p}$$

Whence

Corollary 2.2 *Let p be a prime and a be an integer such that $\gcd(a, p) = 1$. Then a is a quadratic nonresidue of p if and only if*

$$a^{(p-1)/2} \equiv -1 \pmod{p}$$

Example 2.5 Take $p = 11$, so $(p - 1)/2 = 5$.

a	1	2	3	4	5	6	7	8	9	10
$a^5 \bmod 11$	1	10	1	1	1	10	10	10	1	10

Since $10 \equiv -1 \pmod{11}$, the quadratic residues of 11 are 1, 3, 4, 5, and 9.

$\square$

For certain forms of a and odd prime p, a solution to $x^2 \equiv a \pmod{p}$, i.e., a square root, can be found readily, as shown in Table 2.2. We verify one of the forms—that of $p = 4k + 3$—and leave the verification of the others to the reader.

---

[4]Note the symmetry in both tables of the example, which symmetry indicates that it suffices to consider just half of the residues.

[5]A proof of the split is given in the Appendix (Corollary A.4).

Table 2.2 Solutions of $x^2 \equiv y \pmod{p}$

Form of p	Form of a	Solution (x)
$4k + 3$	–	$a^{k+1} \bmod p$
$8k + 5$	$a^{2k+1} \equiv 1 \pmod{p}$	$a^{k+1} \bmod p$
$8k + 5$	$a^{2k+1} \equiv -1 \pmod{p}$	$\frac{1}{2}(4a)^{k+1}(p + 1) \bmod p$

k a positive integer

$$x^2 = \left(a^{k+1}\right)^2 = \left(a^{(p+1)/4}\right)^2$$

$$= a^{(p+1)/2}$$

$$= a^{2k+2}$$

$$= a^{2k+1}a$$

$$= a^{(p-1)/2}a$$

$$\equiv a \pmod{p} \qquad \text{by Theorem 2.3}$$

In ordinary integer arithmetic, if x_1 is a solution to $x^2 = a$, then so is its negation, $x_2 = -x_1$. A similar situation holds here with quadratic residues: if x_1 is a solution to $x^2 \equiv a \pmod{m}$, then so is its additive inverse, $x_2 = m - x_1$.

Example 2.6 Take $p = 11$. From Example 2.5, the quadratic residues are 1, 3, 4, 5, and 9. With the first p-form of Table 2.2, we have $k = 2$, and the relevant equations and their solutions, x_1 and x_2, are

$$
\begin{array}{llll}
x^2 \equiv 1 \pmod{11}: & x_1 = 1^3 \bmod 11 = 1 & x_2 = 10 \\
x^2 \equiv 3 \pmod{11}: & x_1 = 3^3 \bmod 11 = 5 & x_2 = 6 \\
x^2 \equiv 4 \pmod{11}: & x_1 = 4^3 \bmod 11 = 9 & x_2 = 2 \\
x^2 \equiv 5 \pmod{11}: & x_1 = 5^3 \bmod 11 = 4 & x_2 = 7 \\
x^2 \equiv 9 \pmod{11}: & x_1 = 9^3 \bmod 11 = 3 & x_2 = 8
\end{array}
$$

□

Some cryptosystems are based on the fact that in certain cases it is not easy to determine whether or not a given number is a quadratic residue. A specific instance that we shall return to is in the following result for the product of two primes.

Theorem 2.4 *For an integer a and distinct primes p and q, a is a quadratic residue of pq if and only if a is a quadratic residue of p and a is a quadratic residue of q.*

(The generalization of this result is to any prime factorization.)

A cryptosystem can be based on the fact that, with appropriate choices, it is quite difficult to determine efficiently—i.e., within a reasonable time—if an integer a is a quadratic residue modulo a composite integer m. This is the *Quadratic Residuosity Problem*, whose formal statement is given below.

Two additional facts, given in the next result, will be useful later on.

Theorem 2.5 *With respect to a given modulus:*

(i) *the product of two quadratic residues or two quadratic nonresidues is a quadratic residue, and*

(ii) *the product of a quadratic residue and a quadratic nonresidue is a quadratic nonresidue.*

We next introduce some standard notation and terminology that are useful in the description of cryptosystems of the type mentioned above.

Whether or not an integer a is a quadratic residue with respect to a prime modulus p may be expressed in terms of the *Legendre symbol*.

Definition Let a be an integer and p be an odd prime such that $\gcd(a, p)$. The ***Legendre symbol***, $\left(\frac{a}{p}\right)$, is defined as

$$\left(\frac{a}{p}\right) = \begin{cases} 1 & \text{if } a \text{ is a quadratic residue of } p \\ -1 & \text{if } a \text{ is a quadratic nonresidue of } p \end{cases} \tag{2.9}$$

Euler's Criterion (Theorem 2.2) and its corollary may now be expressed as

$$a^{(p-1)/2} \equiv \left(\frac{a}{p}\right) \pmod{p}$$

We shall make use of the two properties of the Legendre symbol that are expressed in the following result.

Theorem 2.6 *For integers a and b and prime modulus p:*

(i)

$$\left(\frac{ab}{p}\right) = \left(\frac{a}{p}\right)\left(\frac{b}{p}\right)$$

(ii)

$$\left(\frac{a}{p}\right) = \left(\frac{b}{p}\right) \qquad \text{if } a \equiv b \pmod{p}$$

The *Jacobi symbol* is a generalization of the Legendre symbol to a modulus m that is not necessarily a prime.

Definition Let $m \geq 3$ be an odd integer such that for some primes $p_1, p_2, \ldots, p_k$ and integer exponents $e_1, e_2, \ldots, e_k$

$$m = p_1^{e_1} p_2^{e_2} \cdots p_k^{e_k}$$

Then for an integer a, the **Jacobi symbol**, $\left(\frac{a}{m}\right)$, is defined, in terms of Legendre symbols $\left(\frac{a}{p_i}\right)$, as

$$\left(\frac{a}{m}\right) = \left(\frac{a}{p_1}\right)^{e_1} \left(\frac{a}{p_2}\right)^{e_2} \cdots \left(\frac{a}{p_k}\right)^{e_k}$$

The (..) on the left is the Jacobi symbol, and a (..) on the right is a Legendre symbol; the difference will always be clear from the context.

A particular case we shall be interested in is when the modulus m is the product of two primes p and q. In that case

$$\left(\frac{a}{m}\right) = \left(\frac{a}{p}\right) \left(\frac{a}{q}\right)$$

We can now state the *Quadratic Residuosity Problem*, in a manner relevant to how it will be used later.

Quadratic Residuosity Problem

Let m be a product of two distinct odd primes and x be an integer such that $1 \leq x \leq m - 1$ and $\left(\frac{x}{m}\right) = 1$. The problem is to determine whether or not x is a quadratic residue of m.

In general, solving this problem is very difficult—in that it cannot be done in a reasonable time—as it depends on being able to factor m, which in turn is also generally a difficult problem. The essence of the difficulty lies in determining which of the two situations in the first part of Theorem 2.4 is the case; and it is of significance that if the quadracity with respect to one of the two primes is known, then the problem is easily solvable.

2.5   The Chinese Remainder Theorem

The Chinese Remainder Theorem (CRT) is a very important result that has several applications in modular arithmetic.

Chinese Remainder Theorem *Let $m_1, m_2, \ldots, m_n$ and $a_1, a_2, \ldots, a_n$ be some integers such that $0 \leq a_i < m_i$ and $\gcd(m_i, m_j) = 1$, $i = 1, 2, \ldots, n$ and*

$j = 1, 2, \ldots, n$. *Then the set of equations*

$$x \equiv a_1 \pmod{m_1}$$

$$x \equiv a_2 \pmod{m_2}$$

$$\cdots$$

$$x \equiv a_n \pmod{m_n}$$

has the unique solution

$$x \equiv a_1 M_1 \left| M_1^{-1} \right|_{m_1} + a_2 M_2 \left| M_1^{-1} \right|_{m_1} + \cdots + x_n M_n \left| M_n^{-1} \right|_{m_n} \pmod{M} \qquad (2.10)$$

where

$$M = \prod_{i=1}^{n} m_i$$

$$M_i = \frac{M}{m_i} \qquad i = 1, 2, \ldots, n$$

$$\left| M_i^{-1} \right|_{m_i} = \textit{the multiplicative inverse of } M_i \textit{ modulo } m_i$$

With the previously-stated assumption of least positive residues, we may take $x \bmod M$ as the unique solution.

Example 2.6 For the equations

$$x \equiv 1 \pmod{3}$$

$$x \equiv 2 \pmod{5}$$

$$x \equiv 3 \pmod{7}$$

we have $m = 105$ and

$$M_1 = 35 \qquad M_2 = 21 \qquad M_3 = 35$$

$$\left| M_1^{-1} \right|_{m_1} = 2 \qquad \left| M_2^{-1} \right|_{m_2} = 1 \qquad \left| M_3^{-1} \right|_{m_3} = 1$$

Therefore

$$x = (1 * 35 * 2 + 2 * 21 * 1 + 3 * 15 * 1) \bmod 105$$

$$= 52$$

The confirmation: $52 \bmod 3 = 1$, $52 \bmod 5 = 2$, and $52 \bmod 7 = 3$.

□

Instances of the CRT in which the modulus is the product of two distinct and odd primes are of special interest in cryptography. We next describe one such case that we will return to in Chap. 3.

Suppose our task is to to solve the equation

$$x^2 \equiv a \pmod{m} \tag{2.11}$$

where $m = pq$, with p and q distinct and odd primes.

If x_1 as a solution to

$$x^2 \equiv a \pmod{p}$$

and x_2 is a solution to

$$x^2 \equiv a \pmod{q}$$

then we can use the CRT to deduce a general form for solutions of Eq. 2.11:

$$x = \left(x_1 \left| q^{-1} q \right|_p + x_2 \left| p^{-1} p \right|_q \right) \bmod m \tag{2.12}$$

Given that there are two possible values for x_1 and two possible values for x_2, this gives us four possible values for x.

Example 2.7 Suppose $p = 7$ and $q = 11$, and $a = 23$. That is, our task is to solve

$$x^2 \equiv 23 \pmod{77}$$

From Table 2.2, for the first modulus we have

$$23^{(7+1)/4} \bmod 7 = (23 \bmod 7)^2 \bmod 7 = 2^2 \bmod 7 = 4$$

and the other square root is $7 - 4 = 3$.

And for the second modulus we have

$$23^{(11+1)/4} \bmod 11 = (23 \bmod 11)^3 \bmod 11 = 1^3 \bmod 11 = 1$$

and the other square root is $11 - 1 = 10$.

For the inverses: $|7^{-1}|_{11} = 8$ and $|11^{-1}|_7 = 2$.

Applying Eq. 2.12 to the pairs 4 and 1, 4 and 10, 3 and 1, and 3 and 10, we get the four square roots of 23 with respect to the modulus 77:

$$(4 * 11 * 2 + 1 * 7 * 8) \bmod 77 = 67$$

$$(4 * 11 * 2 + 10 * 7 * 8) \bmod 77 = 32$$

$$(3 * 11 * 2 + 1 * 7 * 8) \bmod 77 = 45$$

$$(3 * 11 * 2 + 10 * 7 * 8) \bmod 77 = 10$$

$\square$

2.6   Residue Number Systems

Conventional number systems are *positional* and *weighted*: the position of each digit is significant and is associated with a weight that is a power of the radix, r, that is employed—$r^0, r^1, r^2, \ldots$, proceeding from left to right in a representation. The lack of positional independence in the digits leads to the carry-propagation problem of Sect. 1.1, which is the limiting factor in the performance of ordinary adders (and, therefore, of multipliers and other related arithmetic units). In *residue number systems* (RNS) the digits in the representation are independent, and in the addition of two numbers there is no notion of carries from one digit position to the next. So addition and multiplication can be quite fast.

The possibility of fast carry-free arithmetic has led to numerous proposals for the use of RNS in many types of applications, including cryptosystems. There are, however, certain difficulties inherent in RNS, and, as a consequence, many such proposals are of questionable practical worth. Nevertheless, there might be new developments in the future; and for that reason, as well as for "completeness," we shall in later chapters mention a few of the proposed uses of RNS in cryptosystems. The following provides some necessary background; the interested reader will find more details in [1, 2].

Suppose we have a set $\{m_1, m_2, \ldots, m_n\}$ of n positive and pairwise relatively prime moduli. Let M be the product of the moduli. Then every number $0 \le x < M$ has a unique representation as the set of residues $\{x \bmod m_i : 1 \le i \le n\}$, and we may take these residues as the "digits" in the representation. We shall denote this by writing $x \cong \langle x_1, x_2, \ldots, x_n \rangle$, where $x_i = x \bmod m_i$. As an example, with $x = 52, m_1 = 3, m_2 = 5$, and $m_3 = 7$, we have $x \cong \langle 1, 2, 3 \rangle$.

Negative numbers can be represented using extensions of the conventional notations of sign-and-magnitude, ones' complement, and two's complement. For our purposes it suffices to consider only positive numbers.

Addition (and, therefore, subtraction) and multiplication are easy with RNS representation. In both cases the operation is on individual digit pairs, relative to the modulus for their position. There is no notion of a carry that is propagated from one digit position to the next digit position.

If the moduli are $m_1, m_2, \ldots, m_n$, $x \cong \langle x_1, x_2, \ldots, x_n \rangle$ and $y \cong \langle y_1, y_2, \ldots y_n \rangle$, then addition is defined by

$$x + y \cong \langle x_1, x_2, \ldots, x_n \rangle + \langle y_1, y_2, \ldots y_n \rangle$$

$$= \langle (x_1 + y_1) \bmod m_1 , (x_2 + y_2) \bmod m_2 , \ldots , (x_n + y_n) \bmod m_n \rangle$$

and multiplication by

$$x * y \cong \langle x_1, x_2, \ldots, x_N \rangle * \langle y_1, y_2, \ldots y_N \rangle$$

$$= \langle (x_1 * y_1) \bmod m_1, \ (x_2 * y_2) \bmod m_2, \ \ldots, (x_n * y_n) \bmod m_n \rangle$$

Example 2.8 With the moduli-set $\{2, 3, 5, 7\}$, the representation of seventeen is $\langle 1, 2, 2, 3 \rangle$, that of nineteen is $\langle 1, 1, 4, 5 \rangle$, and adding the two residue numbers yields $\langle 0, 0, 1, 1 \rangle$, which is the representation for thirty-six in that system:

$$\langle 1, 2, 2, 3 \rangle + \langle 1, 1, 4, 5 \rangle = \langle (1+1) \bmod 2, (2+1) \bmod 3, (2+4) \bmod 5, (3+5)$$

$$\bmod 7 \rangle$$

$$= \langle 0, 0, 1, 1 \rangle$$

And the product of the same operands is three hundred and twenty-three, whose representation is $\langle 1, 2, 3, 1 \rangle$:

$$\langle 1, 2, 2, 3 \rangle * \langle 1, 1, 4, 5 \rangle = \langle (1 * 1) \bmod 2, (2 * 1) \bmod 3, (2 * 4) \bmod 5,$$

$$(3 * 5) \bmod 7 \rangle$$

$$= \langle 1, 2, 3, 1 \rangle \qquad \qquad \square$$

$$\square$$

The carry-free benefits of RNS do not come for free: the lack of magnitude information in the digits means that comparisons and related operations will be difficult. For example, with the moduli-set $\{2, 3, 5, 7\}$, the number represented by $\langle 0, 0, 1, 1 \rangle$ is almost twice that represented by $\langle 1, 1, 4, 5 \rangle$, but that is far from apparent and is not easily confirmed.

Division in RNS is considerably more difficult that addition and multiplication. Basic division consists, essentially, of a sequence of additions and subtractions, magnitude comparisons, and selections of the quotient digits. But, as indicated above, comparison in RNS is a difficult operation, because RNS is not positional or weighted. One way in which division can be readily implemented is to convert the operands to a conventional notation, use a conventional division procedure, and then convert the result back into residue notation. That of course would do away with most of the benefits of RNS, unless division is very infrequent. Moreover, conversions, especially from RNS representation, are generally costly operations.

For all operations, the implementation of RNS arithmetic requires the initial conversion of operands from conventional form and finally the conversion of results into conventional form. The former is known as *forward conversion* and the latter as *reverse conversion*.

Forward conversion is just modular reduction with respect to each moduli used, and it can be done with reasonable efficiency, especially for certain forms of moduli (Chap. 4). Reverse conversion on the other hand is much more difficult, regardless of the moduli used, and it is this that has largely limited the practical uses of RNS. Therefore, as with division, RNS arithmetic will be worthwhile only if the numbers of additions and multiplications are very large relative to the number of conversions so that the cost of the latter is amortized.

There are two standard methods for reverse conversion, and both are costly to implement. The first is a direct application of the Chinese Remainder Theorem. Thus, for example, Example 2.6 is effectively a conversion from the representation $\langle 1, 2, 3 \rangle$ in the system with the moduli 3, 5, and 7: the x that is computed is the conventional form of the number represented by the RNS digits. The other method for reverse conversion is *mixed-radix conversion*, in which the essential idea is to assign weights to the digits of an RNS representation and then directly obtain a conventional form. The reader will find the relevant details in the published literature [1, 2].

References

1. A. R. Omondi and B. Premkumar. 2007. *Residue Number Systems: Theory and Implementation.* Imperial College Press, London, UK.
2. P. V. A. Mohan. 2016. *Residue Number Systems: Theory and Applications.* Birkhauser, Basel, Switzerland.
3. D. M. Burton. 2010. *Elementary Number Theory.* McGraw-Hill Education, New York, USA.
4. G.H. Hardy and E.M. Wright. 2008. *An Introduction to the Theory of Numbers.* Oxford University Press, Oxford, UK.

Chapter 3
Modular-Arithmetic Cryptosystems

Abstract This chapter consists of a few examples of cryptosystems that are based on modular arithmetic; the reader will find many more examples in the published literature. The descriptions of the various cryptosystems are not intended to be complete and are given only to provide a context for the arithmetic. The focus is on the essence of the algorithms, and the reader who requires them can readily find the details elsewhere.

Modular exponentiation is a key operation in many modular-arithmetic cryptosystems. Exponentiation is essentially a sequence of multiplications; multiplication is essentially a sequence of additions; and, relative to ordinary computer arithmetic, modular arithmetic also requires reductions. Accordingly, most of the cryptography arithmetic covered in the first part of the book is of modular reduction, addition, multiplication, and exponentiation. Subtraction and division are as the addition of additive inverses and multiplication by multiplicative inverses.

Examples of three types of cryptosystems are given. The first type is *message encryption*, in which problem is the standard one of sending a message that is encrypted in such a way that decryption can be carried out only by the intended recipient. The second type is *key agreement*, in which the basic problem is that two entities seek to agree on a "secret" to be used as a key, or to generate a key, for secure communications, but the two communicate through a channel that might be insecure, allowing a third party to eavesdrop; so it is necessary to agree on the secret in such a way that any information readily acquired by the third party would be insufficient to determine the secret. And the third type of cryptosystem is that of *digital signatures*, in which the problem is that of "signing" communications—i.e., appending some information to a message—in such a way that the "signature" can be verified as being from the true sender, cannot be forged, cannot be altered, cannot be associated with a message other than the one to which it is attached, and cannot be repudiated by the sender.

The security of many modular-arithmetic cryptosystems is generally based on the difficulty of solving the *Discrete Logarithm Problem* (Sect. 2.3) and the *Factoring Problem*, which is that of factoring a composite integer into smaller

© Springer Nature Switzerland AG 2020 89
A. R. Omondi, *Cryptography Arithmetic*, Advances in Information Security 77,
https://doi.org/10.1007/978-3-030-34142-8_3

(prime) integers.[1] For very large numbers, there are no known algorithms to solve the Factoring Problem efficiently; so for *practical purposes* it may be considered unsolvable [1, 4]. The Quadratic Residuosity Problem (Sect. 2.4), which is the basis of some cryptosystems, can also be reduced to the Factoring Problem. For further details on all these, the reader should consult standard texts on cryptography [2].

The most significant difference between ordinary modular arithmetic and practical modular arithmetic in cryptography is that the latter generally involves very large numbers—represented in hundreds or thousands of bits. Nevertheless, the numerical examples given here are all "unrealistic" small-numbers ones, so that the reader can manually check the correspondence to the general descriptions they exemplify. Also, unless otherwise indicated, all numbers indicated in the various expressions and statements will be assumed to be positive and nonzero.

Note Recall from Chap. 2 that we use $|x^{-1}|_m$ to denote the multiplicative inverse of x with respect to the modulus m but simply write x^{-1} when the modulus is evident from the context.

3.1   Message Encryption

In each of the systems described, a message to be encrypted and transmitted is represented as a binary string that is interpreted as an unsigned number. There will be an upper bound—imposed by some basic parameters of the algorithm at hand—on such a number. If the original message does not satisfy such conditions, then it is modified appropriately—for example, by splitting a long message into several smaller ones and then separately encrypting each small piece.

3.1.1   RSA Algorithm

The best known cryptosystem that exemplifies the use of modular arithmetic is the RSA algorithm [2, 3]. The algorithm is an example of a *public-key cryptosystem*: for encryption, a sender uses a key that is made publicly available by a receiver; and for decryption, the receiver uses a key that is private (secret). The essence of the system is as follows.

Let p and q be two large and distinct random prime numbers. The modulus n to be used is

$$n = pq$$

---

[1] We shall be interested primarily in factoring the product of two large prime numbers.

From p and q, the number ϕ is obtained:

$$\phi = (p - 1)(q - 1)$$

All three numbers p, q, and ϕ are kept secret.

A number e is then obtained—say, by applying the Extended Euclidean GCD algorithm (Chap. 6)—such that

$$\gcd(e, \phi) = 1 \quad \text{with} \quad 1 < e < \phi$$

That is, e and ϕ are relatively prime.

The pair (n, e) constitutes the receiver's public key. The corresponding private key used by the receiver is the pair (n, d), where

$$d = \left| e^{-1} \right|_{\phi} \tag{3.1}$$

Let M be the message to be sent, with $M < n$. Then the encryption consists of computing

$$C = M^e \bmod n \tag{3.2}$$

and the decryption consists of computing

$$X = C^d \bmod n$$

If all is well, then $X = M$.

Example 3.1 With $p = 3$ and $q = 11$, we have $n = 33$ and $\phi = 20$. If we select $e = 7$, then $d = 3$. The encryption of $M = 17$ consists of the computation

$$C = M^e \bmod pq$$
$$= 17^7 \bmod 33$$
$$= 8$$

and the decryption consists of the computation of

$$C^d \bmod pq = 8^3 \bmod 33$$
$$= 17 \qquad \qquad \square$$

To establish the correctness of the algorithm, we will make use of the following two theorems.

Theorem 3.1 $(a \bmod km) \bmod k = a \bmod k.$

Theorem 3.2 *If* $\gcd(k, m) = 1$, *then* $a \equiv b \pmod{km}$ *if and only if* $a \equiv b \pmod{k}$ *and* $a \equiv b \pmod{m}$.

From the Theorem 3.1, it follows that if we can establish that

$$C^d \bmod p = M \bmod p \tag{3.3}$$

$$C^d \bmod q = M \bmod q \tag{3.4}$$

then we may conclude that

$$C^d \bmod n = M \bmod n \qquad n = pq \tag{3.5}$$

And the Chinese Remainder Theorem (Sect. 2.5) guarantees that the value obtained in the decryption is unique.

In addition to the two theorems, we will also make use of Fermat's Little Theorem (Theorem 2.1 in Sect. 2.2).

Fermat's Little Theorem *If* p *is a prime and* a *is an integer such that* $\gcd(a, p) = 1$, *then*

$$a^{p-1} \equiv 1 \pmod{p} \tag{3.6}$$

The correctness proof is then as follows.
From Eq. 3.1,

$$ed \equiv 1 \pmod{\phi}$$

and, therefore,

$$ed = 1 + k\phi \qquad \text{for some integer } k$$

From Eq. 3.1

$$C^d \bmod p = \left(M^e \bmod pq\right)^d \bmod p$$

$$= \left(M^{ed} \bmod pq\right) \bmod p$$

$$= M^{ed} \bmod p \qquad \text{by Theorem 3.1}$$

$$= M^{1+k\phi} \bmod p$$

$$= M^{1+k(p-1)(q-1)} \bmod p$$

$$= \left[M\left(M^{p-1} \bmod p\right)^{k(q-1)}\right] \bmod p$$

If $\gcd(M, p) = 1$, then by Fermat's Little Theorem

$$\left[M\left(M^{p-1} \bmod p\right)^{k(q-1)}\right] \bmod p = \left[M * 1^{k(q-1)}\right] \bmod p$$

$$= M \bmod p$$

And if $\gcd(M, p) \neq 1$, then M is a multiple of p, so

$$M^{ed} \bmod p = 0$$

$$= M \bmod p$$

Therefore, in either case of the value of $\gcd(M, p)$, we have Eq. 3.3:

$$C^d \bmod p = M^{ed} \bmod p$$

$$= M \bmod p$$

Similarly, replacing p and q in the preceding line of reasoning, we have Eq. 3.4:

$$C^d \bmod p = M^{ed} \bmod q$$

$$= M \bmod q$$

And thus, by Theorem 3.2, we have Eq. 3.5.

The basis of security in the algorithm is as follows. In order to decrypt a message, a receiver without prior knowledge of d (but perhaps with knowledge of n and e) would have to first determine the value of ϕ. But that requires knowledge of p and q, which can be acquired only by factoring n. And, as indicated earlier, if the values of p and q are chosen appropriately, then that task is a very difficult one.

3.1.2 Rabin Algorithm

The security in this algorithm too is essentially based on the difficulty of factoring a number that is the product of large primes—the Factoring Problem stated in the introduction [4]. The keys are produced from two large and distinct random prime numbers, p and q. The private key for the decryption is the pair (p, q), and the public key for the encryption is the product $n = pq$. The message is an integer M such that $M < n$. With appropriate choices for p and q, the system is secure because it is not easy to obtain either p or q from n.

The encryption consists of computing

$$C = M^2 \bmod n \tag{3.7}$$

and the decryption consists of computing the modulo-m square roots of C and then determining which of those is the message. Modular square roots are discussed in Sect. 2.4. Here, two pairs of square roots are computed: one pair with respect to the modulus p, and the other pair with respect to the modulus q. The Chinese Remainder Theorem is then used, as described at the end of Sect. 2.5 (Eq. 2.12), to obtain four square roots with respect to the modulus m. The security of the system is based on the fact that it is necessary to know p and q, which requires factoring, except for the possessor of the private key.

As an example, if we have $p = 7$, $q = 11$, and $M = 32$, then $n = 77$, $M^2 \bmod n = 23$. Decryption produces the square roots 67, 32, 45, and 10. (See Example 2.7 in Sect. 2.5.) Some means is required to identify one of the four roots as the correct message. This can be done in various ways. For example, the receiver might make a choice according to context. Another way would be for the sender to add some redundancy (known to the receiver) to original message—e.g., replicate part of the message—and for the receiver to then check for the existence of the redundant information. Suppose, for example, that we wish to transmit three-bit messages and that the redundancy involves the replication of those bits. If the basic message consists of the bits 101 (decimal 5), then replication gives 101101 (decimal 45). Taking $M = 45$, we again have $M^2 \bmod n = 23$ in encryption and the same four roots in decryption, but only one of these has the required redundancy.

3.1.3   El-Gamal Algorithm

Security in this algorithm is based on the difficulty of computing discrete logarithms, as described in Sect. 2.2 [5]. The keys are generated as follows. A large random prime p is selected. If g is a primitive root of p, then a random number k is obtained such that $1 \leq k < p - 1$, and the value

$$x = g^k \bmod p$$

is computed. The public key is (p, g, x), and the private key is k.

To encrypt a message M such that $M < p$, a sender selects a random number j such that $1 \leq j < p - 1$ and then computes

$$y = g^j \bmod p$$

$$z = x^j M \bmod p$$

The encrypted message is sent as the pair (y, z).

For the decryption, the receiver uses the private key, k, to compute $y^{p-1-k} z \bmod p$. This yields the original M:

$$y^{p-1-k}z \bmod p = \left(y^{p-1} \bmod p\right)\left(y^{-k}z \bmod p\right) \bmod p$$

$$= y^{-k}z \bmod p \qquad \text{by Fermat's Little Theorem}$$

$$= \left[\left(\left(g^j\right)^{-k} \bmod p\right)\left(x^j M \bmod p\right)\right] \bmod p$$

$$= \left[\left(g^j\right)^{-k} x^j M\right] \bmod p$$

$$= \left[\left(g^k\right)^{-j}\left(g^k\right)^j M\right] \bmod p$$

$$= M \bmod p$$

$$= M \qquad \text{since } M < p$$

(A term of the form x^{-i} is interpreted as $|(x^i)^{-1}|_p$.)

The security of the system thus arises from the fact that decryption requires k, but this cannot be determined easily from x.

Example 3.2 Suppose $p = 17$, $g = 6$, and $k = 5$. Then

$$x = 6^5 \bmod 17 = 7$$

If, say, $M = 13$ and $j = 3$, then the encryption is the computation of

$$y = 6^3 \bmod 17 = 12$$
$$z = 7^3 * 13 \bmod 17 = 5$$

and the decryption is the computation of

$$y^{p-1-k}x \bmod p = 12^{11} * 5 \bmod 17$$
$$= 13$$

$\square$

3.1.4 Massey-Omura Algorithm

The algorithm is described in [9]. Both sender and receiver compute private keys on the basis of a large prime number, p. The sender randomly selects an integer e such that $0 < e < p - 1$ and $\gcd(e, p - 1) = 1$ and then computes $d = |e^{-1}|_{p-1}$; the pair (d, e) is the private (i.e., secret) key. The receiver similarly computes its private key (d_*, e_*).

To send a message M, with $M < p$:

(i) The sender computes $C_1 = M^e \bmod p$ and forwards that to the receiver.
(ii) The receiver computes $C_2 = C_1^{e_*} \bmod p$ and forwards that to the sender.
(iii) The sender computes $C_3 = C_2^d \bmod p$ and forwards that to the receiver.

The receiver can then recover the message by computing $C_3^{d_*} \bmod p$: Since $d = |e^{-1}|_{p-1}$, we have $ed \equiv 1 \pmod{p-1}$; i.e., $d = 1 + k(p-1)$ for some positive integer k. Therefore

$$M^{ed} \bmod p = M^{1+k(p-1)} \bmod p$$

$$= M^1 \left(M^{p-1}\right)^k \bmod p$$

$$= M \left(M^{p-1} \bmod p\right)^k \bmod p$$

$$= M \bmod p \qquad \text{by Fermat's Little Theorem}$$

$$= M \qquad \text{since } M < p$$

And, similarly

$$M^{e_* d_*} \bmod p = M$$

So

$$C_3^{d_*} \bmod p = M^{ede_* d_*} \bmod p$$

$$= \left(M^{ed}\right)^{e_* d_*} \bmod p$$

$$= \left(M^{ed} \bmod p\right)^{e_* d_*} \bmod p$$

$$= M^{e_* d_*} \bmod p$$

$$= M \qquad \text{since } M < p$$

Example 3.3 Suppose $p = 37$, $e = 25$ ($d = 31$), $e_* = 11$ ($d_* = 23$), and $M = 17$. Then

$$C_1 = 17^7 \bmod 37$$

$$= 15$$

$$C_2 = 15^{11} \bmod 37$$

$$= 19$$

$$C_3 = 19^{31} \bmod 37$$

$$= 32$$

and

$$C_3^{d_*} = 32^{23} \bmod 37$$

$$= 17$$

□

The system is secure because the only information that a third part can intercept consists of C_1, C_2, and C_3, but from only those it is practically impossible to obtain the private keys (The Discrete Logarithm Problem).

3.1.5 Goldwasser-Micali Algorithm

Security in this algorithm is based on the difficulty of the Quadratic Residuosity Problem (Sect. 2.4); i.e., in determining whether or not a given number is a quadratic residue with respect to a modulus that is a product of primes. The algorithm also includes a "randomization" aspect whose effect is that every encryption of a given message yields a different result. The details are as follows [6].

Let p and q be two large and distinct random primes, and let n be their product. A number x is chosen such that the Legendre symbols are

$$\left(\frac{x}{p}\right) = \left(\frac{x}{q}\right) = -1$$

Finding such an x is easy, given the profusion of quadratic nonresidues with respect to primes. For example, it can be done by random selections and checks until a suitable value is found.

Observe that the Jacobi symbol relating x and n is

$$\left(\frac{x}{n}\right) = \left(\frac{x}{p}\right)\left(\frac{x}{q}\right) = 1$$

but x is not a quadratic residue of n (Theorem 2.4 in Sect. 2.4). The significance of this point will become apparent below.

The receiver's public key used for encryption by a sender is the pair (x, n), and the private key used by the receiver for decryption of the sent message is the pair (p, q).

The encryption is on a bit-by-bit basis. Suppose the binary representation of the message M is $M_k M_{k-1} \cdots M_1$, where M_i is 0 or 1 ($i = 1, 2, \ldots, k$). For each bit M_i, a random integer y_i is obtained such that $1 \le y_i \le n - 1$ and $\gcd(y_i, n) = 1$. The corresponding bit, C_i, of the encrypted message is then computed as

$$C_i = \begin{cases} y_i^2 \bmod n & \text{if } M_i = 0 \\ y_i^2 x \bmod n & \text{if } M_i = 1 \end{cases}$$

and the entire encrypted message is $(C_k, C_{k-1}, \ldots, C_1)$.

The decryption consists of determining whether each C_i is a quadratic residue or not. The value[2]

$$u_i = \left(\frac{C_i}{p}\right)$$

is computed. If $u_i = 1$—i.e., C_i is a quadratic residue of p—then M_i is set to 0; otherwise M_i is set to 1. At the end, $M_k M_{k-1} \cdots M_1$ is the original message.

Example 3.4 Suppose $p = 7$ and $q = 11$; so $n = 77$. For x take 6, whose nonquadratic residuosity with respect to p and q is easily confirmed (through Corollary 2.2 in Sect. 2.4):

$$6^{(7-1)}/2 \bmod 7 = 6$$

$$\equiv -1 \ (\bmod \ 7)$$

$$6^{(11-1)}/2 \bmod 11 = 10$$

$$\equiv -1 \ (\bmod \ 11)$$

The public key is $(6, 77)$, and the private key is $(7, 11)$.

Now, suppose we wish to encrypt the message $M_3 M_2 M_1 = 101$. If we take $y_3 = 2$, $y_2 = 3$, and $y_3 = 5$, then the result of the encryption is (C_3, C_2, C_1):

$$C_3 = 2^2 * 6 \bmod 77 = 24$$

$$C_2 = 3^2 \bmod 77 = 9$$

$$C_1 = 5^2 * 6 \bmod 77 = 73$$

And the decryption will show that 24 and 73 are not quadratic residues of 7 but 9 is, which therefore yields 101.

□

The basis of security in the algorithm: If $M_i = 0$, then C_i is a quadratic residue modulo n; and if $M_i = 1$, then C_i is a quadratic nonresidue modulo n. The Jacobi symbol in both cases is 1, so without additional information it is not possible to distinguish between the two cases. The distinction can be made only if p or q is known and a check made with respect to either of them. Suppose p is chosen for the check. If $M_i = 0$, then

$$\left(\frac{C_i}{p}\right) = 1$$

---

[2]We could equally well use $\left(\frac{C_i}{q}\right)$.

and if $M_i = 1$, then

$$\left(\frac{C_i}{p}\right) = -1$$

But only the intended receiver knows p and q; otherwise, factoring an appropriately chosen n is a difficult task. In sum, the security of the algorithm is based on the fact that it is difficult to determine whether or not a randomly selected number is a quadratic residue with respect to a composite modulus if the factors of the modulus are not known.

3.2 Key Agreement

Two partners, A and B, wish to agree on a shared, secret key and to do so by communicating certain information in such a way that a third party that intercepts that communication cannot determine what the key is. The following is a brief description of the Diffie–Hellman key-agreement algorithm [7].

A selects a large prime p, selects g a primitive root of p, and makes p and g available to the B. A then selects a random number $j < p$, and B selects a random number $k < p$. Both j and k are kept secret.

In the next steps, A computes

$$x = g^j \bmod p$$

and sends that to B; and B computes

$$y = g^k \bmod p$$

and sends that to A.

A then computes $y^j \bmod p$, and B computes $x^k \bmod p$. The two computations yield the same result, which is the shared key:

$$y^j \bmod p = \left(g^k\right)^j \bmod p$$

$$= \left(g^j\right)^k \bmod p$$

$$= x^k \bmod p$$

A third party might be able to obtain p, g, x, and y, but it is practically impossible to compute the shared key from those. One would also have to know j and k, and that requires solving the Discrete Logarithm Problem.

Example 3.5 Suppose $p = 17$ and $g = 6$. If $j = 3$ and $k = 2$, then

$$x = 6^3 \bmod 17 = 12$$
$$y = 6^2 \bmod 17 = 2$$

and

$$y^j \bmod p = 2^3 \bmod 17 = 8$$
$$x^k \bmod p = 12^2 \bmod 17 = 8$$

$\square$

3.3   Digital Signatures

The basic requirements for digital signatures are as stated in the chapter's introduction; of those, we shall consider only verification, which is arguably the most
important. When an encrypted signature is appended to a message, the message
itself need not be encrypted; this is reflected in several algorithms.

The essential aspects of some of the algorithms described above can easily be
adapted for use in digital signatures. For example, with the RSA algorithm of
Sect. 3.1, the C in Eq. 3.2 may be taken as the signature and the pair (C, M) sent
to the receiver; signature verification then consists of the decryption of Eq. 3.2 and
a check that the decryption-result X is indeed the M that goes with the signature
C. The Rabin algorithm of Sect. 3.1 too can be used in a similar manner. We next
describe two digital-signature algorithms in more detail.

3.3.1   El-Gamal Algorithm

The sender starts with a large prime p and g a primitive root of p, selects a random
number k such that $0 < k < p-1$ for use as a private (i.e., secret) key, and computes

$$z = g^k \bmod p$$

The triplet (p, g, z) is made available as the public key.

To sign a message M, the sender selects a random number j such that $0 < j < p - 1$ and $\gcd(j, p - 1) = 1$ and computes

$$x = g^j \bmod p$$

$$y = \left| j^{-1} \right|_{p-1} (M - kx) \bmod (p - 1) \tag{3.8}$$

If $y = 0$, then the process is repeated; otherwise (x, y) is the signature that accompanies M to the receiver.

To validate the signature, the receiver uses the sender's public key, (p, g, z), to compute

$$u = (z^x x^y) \bmod p$$

$$v = g^M \bmod p$$

The signature is valid if $u = v$.

Example 3.6 Suppose $p = 17$, $g = 6$, $k = 5$, and $j = 11$ (i.e., $|j^{-1}|_p = 3$). Then

$$z = 6^5 \bmod 17 = 7$$

If $M = 15$, then the signature is the pair

$$x = 6^{11} \bmod 17$$
$$= 5$$

$$y = 3 * (15 - 5 * 5) \bmod 16$$
$$= 3 * (-10) \bmod 16$$
$$= 3 * 6 \bmod 16$$
$$= 2$$

The validation of the signature consists of the computations of

$$u = \left(7^5 * 5^2\right) \bmod 17$$
$$= 3$$

$$v = 6^{15} \bmod 17$$
$$= 3$$

□

To verify the correctness of the validation step, we will make use of the following theorem.

Theorem 3.3 *If* $0 < x < p$, *with* p *prime, and* $a \equiv b \pmod{p - 1}$, *then* $x^a \equiv x^b \pmod p$.

Now, from Eq. 3.7

$$M \equiv jy + kx \pmod{p - 1}$$

So, by Theorem 3.3

$$g^M \equiv g^{jy+kx} \pmod{p}$$

and therefore

$$v = g^M \bmod p$$

$$= g^{jy+kx} \bmod p$$

$$= \left(g^j\right)^y \left(g^k\right)^x \bmod p$$

$$= x^y z^x \bmod p$$

$$= u$$

More details on the algorithm will be found in [5].

3.3.2 NIST Algorithm

The algorithm used in the NIST *Digital Signature Standard* has some similarities with the El-Gammal algorithm and indeed may be regarded as a variant of that [8].
    The sender selects

- two large primes, p and q, with q a factor of $p - 1$
- a generator $g < p$ such that $g^q \bmod p = 1$
- a random number $k < q - 1$

and then computes

$$z = g^k \bmod p$$

The quadruplet (p, q, g, z) is made available as the public key; the sender's private key is k.
    To send a signed message M, the sender selects a random number s such that $0 < s < q - 1$ computes

$$x = \left(g^s \bmod p\right) \bmod q$$

$$y = \left|s^{-1}\right|_q (h(M) + kx) \bmod q \tag{3.9}$$

$$(with\ h\ \text{an approved hash function})$$

and sends (x, y, M).

To verify the signature (x, y), the receiver computes

$$t = \left| y^{-1} \right|_q$$

$$u = h(M)t \bmod q$$

$$v = xt \bmod q$$

$$w = \left(g^u z^v \bmod p \right) \bmod q$$

The signature is valid if $w = x$; i.e., if $g^s \bmod p = g^u z^v \bmod p$.

The confirmation of the correctness of the validation step makes use of the following result.

Theorem 3.4 *Let p and q be primes with q a factor of $p - 1$ and g be a generator such that $g^q \bmod p = 1$. If $a \equiv b \pmod q$, then $g^a \bmod p = g^b \bmod p$.*

From Eq. 3.9:

$$s \equiv h(M) \left| y^{-1} \right|_q + kx \left| y^{-1} \right|_q \pmod q$$

$$\equiv h(M)t + kxt \pmod q$$

and therefore, by Theorem 3.4:

$$g^s \bmod p = g^{h(M)t+kxt} \bmod p$$

$$= g^{h(M)t} g^{kxt} \bmod p$$

$$= g^{h(M)t} \left(g^k \right)^{xt} \bmod p$$

$$= g^{h(M)t} z^{xt} \bmod p$$

$$= g^u z^v \bmod p$$

References

1. A. Lenstra. 2000. Integer factoring. *Designs, Codes and Cryptography*, 19:101–128.
2. A. J. Menezes, P. C. van Oorschot, and S. A. Vanstone. 2001. *Handbook of Applied Cryptography*. CRC Press, Boca Raton, USA.
3. R. Rivest, R, A. Shamir, and L. Adleman. 1978. A method for obtaining digital signatures and public-key cryptosystems. *Communications of the ACM*, 21(2):120–126.
4. M. O. Rabin. 1979. Digital signatures and public-key functions as intractable as factorization. Technical Report No. MIT/LCS/TR-212, Laboratory of Computer Science Technical Report, MIT.

5. T. El-Gamal. 1985. A public-key cryptosystem and a signature scheme based on discrete logarithms. *IEEE Transactions on Information Theory*, 31(4):469–472.
6. S. Goldwasser and S. Micali. 1984. Probabilistic encryption. *Journal of Computer and System Sciences*, 28(2):270–299.
7. W. Diffie and M. Hellman. 1976. New directions in cryptography. *IEEE Transactions on Information Theory*, 22(6):644–654.
8. National Institutes of Standards and Technology. 2013. Digital Signature Standard, FIPS PUB 186-4. Gaithersburg, Maryland, USA.
9. J. L. Massey and J.K. Omura. 1983. A new multiplicative algorithm over finite fields and its applicability in public key cryptography. In: *Proceedings, EUROCRYPT 83*.

Chapter 4
Modular Reduction

Abstract Modular reduction is the computation of x mod m. Such computation is implicit in all modular-arithmetic operations, because it is necessary to ensure that results are within range, even though some intermediate value will most likely might not be. The first two sections of the chapter cover two well-known methods for modular reduction that are commonly used in cryptography arithmetic: *Montgomery reduction* and *Barrett reduction* . The third section—a short one–is on a method that might be suitable in certain case when an "isolated" reduction—i.e., not a sequence of reductions—is required. And the last section—another short one—is on reduction with respect to certain specific moduli, such as those that are significant because of their inclusion in some cryptography standards; the moduli all have forms that facilitate efficient reduction.

Both Montgomery reduction and Barrett reduction require costly, but "one-off," pre-processing, and are generally worthwhile only if the results of some pre-processing can be used in several reductions or several arithmetic computations that require reductions. Montgomery is particularly well suited to modular exponentiation, which consists of repeated multiplications; and Barrett reduction is well suited to multiple reductions with the same modulus, even if those are not all in a single, more-complex operation. In other cases, direct integer division may be the most reasonable method for reduction. Additional discussions are in Chaps. 5 and 6.

A straightforward way to carry out modular reduction is to proceed directly from the definition of x mod m. That is, divide x by m and take the remainder as the result: x mod $m = x - qm$, where q is the quotient from the division. That, however, is not necessarily the most efficient method, given that the quotient from the division is not really required and, more significantly, division will be an especially costly where reduction is required repeatedly—for example, in operations such as multiplication and exponentiation. Nevertheless, the *essence* of division is, in one form or another, at the core of almost all modular-reduction algorithms.

If the arithmetic operation is addition and the operands are within range, then reduction, if necessary, is simple: it is just the subtraction of m. That is not so with multiplication and exponentiation, each of which may be realized as a sequence of basic operations—additions and multiplications, respectively—and for which

© Springer Nature Switzerland AG 2020

A. R. Omondi, *Cryptography Arithmetic*, Advances in Information Security 77,

https://doi.org/10.1007/978-3-030-34142-8_4

we may have a reduction after each basic operation or a single reduction on the result of a sequence of basic operations. A significant point to note is that some of the methods used in ordinary modular arithmetic do not work well with the high precisions of cryptography arithmetic.

In ordinary floating-point arithmetic there are several high-speed algorithms in which division is implemented as the multiplication of the dividend by the reciprocal of the divisor. In integer and modular arithmetic there is no real concept of reciprocal, but the basic idea can nevertheless be applied usefully in two ways. The first, which is the basis of Montgomery reduction, is to consider modular multiplicative inverses as some sort of approximations to reciprocals. The second, which is the basis of Barrett reduction, is to use scaled reciprocals: instead of $1/m$, use an integer obtained from u/m, for some large integer u, and then scale the result. In both cases approximate remainders are produced that are easily corrected to obtain the correct results. The simplicity of the corrections is fundamental in the algorithms.

For the reader's "visual ease," all the examples given in what follows are of small numbers and mostly in decimal. But it is to be understood that in practical computer implementation the radix will be a power of two and the numbers will be much larger.

4.1   Barrett Reduction

The main idea in Barrett reduction is to replace the division of x by m with the multiplication of x by $1/m$ [1, 3]. Given that integers are involved here, it is necessary to scale $1/m$ in order to obtain an integer value and to also similarly scale x. So, briefly setting aside the modular-arithmetic aspects, the essence of Barrett reduction is the formulation of an expression for (an approximation of) x/m.

If $1/m$ is scaled by some integer u, then

$$\frac{x}{m} = \frac{x}{u} * \frac{u}{m}$$

If u is large enough, then an integer approximation is easily obtained for u/m. But if u is very large, then x/u will be very small, which implies that the u in that term should be also scaled—by some integer z. We thus have

$$\frac{x}{m} = \left(\frac{1}{z} * \frac{x}{u/z} \right) \frac{u}{m} \tag{4.1}$$

$$= \frac{[x/(u/z)](u/m)}{z}$$

As an example, consider the integer division of $x = 193$ by $m = 13$, which gives the quotient $q = 14$ and remainder $r = 11$. If we take $u = 200$, then

$$\frac{193}{13} = \frac{193}{200} * \frac{200}{13}$$

$200/13 = 15.385$, and from that we get an integer, 15, by truncating the fractional part; i.e., we compute[1] $\lfloor 200/13 \rfloor$. For the $193/200$ part, suppose we take $z = 5$. Then

$$\frac{193}{200} = \frac{1}{5} * \frac{193}{40}$$

$193/40 = 4.825$, from which, by truncation, we get the integer 4, which is $\lfloor 193/40 \rfloor$. So, from

$$\frac{193}{13} = \frac{(193/40)(200/13)}{5}$$

an approximation to the quotient of the integer division of 193 by 13 is

$$\tilde{q} = \frac{(4)(15)}{5} = 12$$

To get the exact remainder that would be obtained from the corresponding exact division, we start with the approximation

$$\tilde{r} = x - \tilde{q}m = 193 - 13 * 12 = 37$$

This value is larger than m, so a correction is required:

$$r = \tilde{r} - km$$

$$= 37 - 2 * 13 = 11 \qquad (k = 2)$$

What is significant about the algorithm is that k is always just 1 or 2; so any correction required is always a simple one.

The other key aspect of Barrett reduction is related to its implementation: if the scaling factors u and z above are chosen appropriately, then in practice the nominal divisions by u and z need not be effected as such. For example, if the factors are powers of the implementation radix, then the divisions are reduced to simple shifts. The division of u by m (Eq. 4.1) remains a real division, but it is a one-time

---

[1] $\lfloor .... \rfloor$ is the *floor function*, which gives the largest integer not less than or equal to ... Note that $x - 1 < \lfloor x \rfloor \le x$ for any x, so $0 \le x - \lfloor x \rfloor < 1$. Also, $\lfloor x + y \rfloor = \lfloor x \rfloor + y$ for any integer y. We will, shortly below, make use of these properties.

division of constants. If several reductions are to be carried out with the same m—for example, during the repeated additions in a modular multiplication—then the cost of that one-time division will be amortized over several reductions.

4.1.1   Basic Algorithm

The procedure consists of three main steps for the computation of $y = x \mod m$: the computation of an approximate quotient, the computation of an approximate remainder, and, perhaps, a correction of the approximate remainder (to get an exact value):

(i)   Compute an approximation $\tilde{q}$ to $\lfloor x/m \rfloor$.
(ii) Compute an approximation $\tilde{y}$ to y as $\tilde{y} = x - \tilde{q}m$.
(iii) If $\tilde{y} < m$, then $y = \tilde{y}$; otherwise, perform the corrective subtraction $y = \tilde{y} - km$.

We will show that $k = 1$ or $k = 2$; i.e., at most two subtractions, each or m, are ever required for the correction. And it can be shown that $\tilde{q} = q$ in 90% of the cases, and $k = 1$ in 1% of the cases.

In principle, any approximate-division algorithm may be employed in the first step of the algorithm. What is special about the Barrett reduction algorithm is that an approximate quotient is computed in such a way that the corresponding approximate remainder is either correct or can be corrected easily.

Suppose x is represented in $2n$ bits—i.e., $x < 2^{2n}$—and m is represented in n bits, where $n = \lfloor \log_2 m \rfloor + 1$. Then $2^{n-1} \le m < 2^n$ and $x < m^2$. Let q be the exact quotient from the integer division of x by m. If in Eq. 4.1 we take $u = 2^{2n}$ and $z = 2^{n+1}$, then

$$q = \left\lfloor \frac{x}{m} \right\rfloor$$

$$= \left\lfloor \frac{1}{2^{n+1}} * \frac{x}{2^{n-1}} * \frac{2^{2n}}{m} \right\rfloor$$

$$= \left\lfloor \frac{(x/2^{n-1})(2^{2n}/m)}{2^{n+1}} \right\rfloor$$

The approximation to q is then taken to be

$$\tilde{q} = \left\lfloor \frac{\lfloor x/2^{n-1} \rfloor * \lfloor 2^{2n}/m \rfloor}{2^{n+1}} \right\rfloor \tag{4.2}$$

of which we note that $\lfloor 2^{2n}/m \rfloor$ is a constant that may be "pre-computed," and in binary implementation the nominal divisions by powers of two are just right shifts of $n-1$ and $n+1$ bit-positions or some other arrangement that discards $n-1$ and $n+1$ bits.

The complete algorithm is as follows.

$$\tilde{q} = \left\lfloor \frac{\lfloor x/2^{n-1} \rfloor * \lfloor 2^{2n}/m \rfloor}{2^{n+1}} \right\rfloor \tag{4.3}$$

$$\tilde{y} = x - \tilde{q}m \tag{4.4}$$

$$y = \begin{cases} \tilde{y} & \text{if } \tilde{y} < m \\ \tilde{y} - m & \text{if } m \leq \tilde{y} < 2m \\ \tilde{y} - 2m & \text{otherwise} \end{cases} \tag{4.5}$$

Example 4.1 Let $x = 193, m = 1011_2 = 11$, and $n = 2$. The computation of $y = x \bmod m$:

$$\left\lfloor \frac{x}{2^{n-1}} \right\rfloor = \left\lfloor \frac{193}{2^3} \right\rfloor = 11000_2 = 24$$

$$\left\lfloor \frac{2^{2n}}{m} \right\rfloor = \left\lfloor \frac{2^8}{11} \right\rfloor = 10111_2 = 23$$

$$24 * 23 = 551 = 1000101000_2$$

$$\tilde{q} = \left\lfloor \frac{552}{2^5} \right\rfloor = 17 = 10001_2$$

$$\tilde{y} = 193 - 17 * 11 = 6$$

$$y = 6$$

(The bits in bold font are discarded in the nominal division.)

$\square$

Example 4.2 Let $x = 201 = 11001001, m = 11$, and $n = 4$. To compute $x \bmod m$, we have

$$\left\lfloor \frac{x}{2^{n-1}} \right\rfloor = \left\lfloor \frac{201}{2^3} \right\rfloor = 011001_2 = 25$$

$$\left\lfloor \frac{2^{2n}}{m} \right\rfloor = \left\lfloor \frac{2^8}{11} \right\rfloor = 23$$

$$25 * 23 = 575 = 1000\mathbf{101100}_2$$

$$\tilde{q} = \left\lfloor \frac{575}{2^5} \right\rfloor = 17 = 10001_2$$

$$\tilde{y} = 201 - 17 * 11 = 14$$

$$y = \tilde{y} - m = 14 - 11 = 3 \qquad \text{(correction)}$$

(The bits in bold font are discarded in the nominal division.)   □

The operational radix in the examples above is two, but any other value would do. Nevertheless, it should be noted that in a practical computer implementation any larger radix will almost always be a power of two—four, eight, sixteen, etc.—so the choice above is not particularly restrictive. The generalization of Eq. 4.3 to radix r is

$$\tilde{q} = \left\lfloor \frac{\lfloor x/r^{n-1} \rfloor * \lfloor r^{2n}/m \rfloor}{r^{n+1}} \right\rfloor \qquad (4.6)$$

with $x < r^{2n}$ and $r^{n-1} \leq m < r^n$.

As an example, the following decimal example corresponds to Example 4.1.

Example 4.3 Let $x = 193$, $m = 11$, $r = 10$, and $n = 2$. The computation of $x \bmod m$ is

$$\left\lfloor \frac{x}{r^{n-1}} \right\rfloor = \left\lfloor \frac{193}{10} \right\rfloor = 19$$

$$\left\lfloor \frac{r^{2n}}{m} \right\rfloor = \left\lfloor \frac{10^4}{11} \right\rfloor = 909$$

$$909 * 19 = 17271$$

$$\tilde{q} = \left\lfloor \frac{17271}{10^3} \right\rfloor = 17$$

$$\tilde{y} = 193 - 17 * 11 = 6$$

$$y = 6$$

□

We conclude by showing that no more than two subtractions, each of m, are ever required for the corrections, whence the claim of "easy" corrections in the third step of the algorithm.

From the definition of $\lfloor \cdots \rfloor$ (see the last footnote), it is evident in the approximation of Eq. 4.2 that $\tilde{q} \leq q$. Now, let α and β be the differences between the exact values of the main terms in the equation and the approximations of those values:

$$\alpha = \frac{x}{2^{n-1}} - \left\lfloor \frac{x}{2^{n-1}} \right\rfloor$$

$$\beta = \frac{2^{2n}}{m} - \left\lfloor \frac{2^{2n}}{m} \right\rfloor$$

Then

$$q = \frac{\left(\lfloor x/2^{n-1} \rfloor + \alpha \right) \left(\lfloor 2^{2n}/m \rfloor + \beta \right)}{2^{n+1}}$$

$$= \frac{\lfloor x/2^{n-1} \rfloor \lfloor 2^{2n}/m \rfloor + \beta \lfloor x/2^{n-1} \rfloor + \alpha \lfloor 2^{2n}/m \rfloor + \alpha\beta}{2^{n+1}}$$

$$\leq \frac{\lfloor x/2^{n-1} \rfloor \lfloor 2^{2n}/m \rfloor}{2^{n+1}} + \frac{\lfloor x/2^{n-1} \rfloor + \lfloor 2^{2n}/m \rfloor + 1}{2^{n+1}}$$

by definition of $\lfloor \cdots \rfloor$, $0 \leq \alpha, \beta < 1$

$$\leq \left[\frac{\lfloor x/2^{n-1} \rfloor \lfloor 2^{2n}/m \rfloor}{2^{n+1}} + \frac{(2^{n+1} - 1) + 2^{n+1} + 1}{2^{n+1}} \right]$$

since $x < 2^{2n}$, and $m \geq 2^{n-1}$

$$= \left[\frac{\lfloor x/2^{n-1} \rfloor \lfloor 2^{2n}/m \rfloor}{2^{n+1}} + 2 \right]$$

$$= \left[\frac{\lfloor x/2^{n-1} \rfloor \lfloor 2^{2n}/m \rfloor}{2^{n+1}} \right] + 2$$

$$= \tilde{q} + 2$$

So, $q - 2 \leq \tilde{q} \leq q$ and $0 \leq x - qm < m$. Since $y = qm$ and $\tilde{y} = \tilde{q}m$, we have

$$0 \leq y - \tilde{y} \leq 2m$$

and at most two subtractions, each of m, from $\tilde{y}$ will be required to bring the intermediate result into the correct range.

4.1.2 Extension of Basic Algorithm

Barrett reduction as presented above can be generalized through the inclusion of an additional parameter, in the following way [10].

Express $q = \lfloor x/m \rfloor$ as

$$\left\lfloor \frac{x}{2^{n+k}} \cdot \frac{1}{2^{j-k}} \cdot \frac{2^{n+j}}{m} \right\rfloor \qquad j, k, n \text{ integers}$$

and then use the approximation

$$\tilde{q} = \left\lfloor \frac{\lfloor x/2^{n+k} \rfloor \cdot \lfloor 2^{n+j}/m \rfloor}{2^{j-k}} \right\rfloor \qquad\qquad (4.7)$$

Example 4.4 Let $x = 193$, $m = 11$, $n = 4$, $j = 5$ and $k = -2$. Then

$$\tilde{q} = \left\lfloor \frac{\lfloor 193/2^2 \rfloor * \lfloor 2^9/11 \rfloor}{2^7} \right\rfloor$$

$$= \left\lfloor \frac{48 * 46}{2^7} = 17 \right\rfloor$$

$$\tilde{y} = x - \tilde{q}m = 193 - 17 * 11 = 6$$

$$y = 6$$

$\square$

If j and k in Eq. 4.7 are chosen appropriately, then at most one subtraction will be required to correct the intermediate value, $\tilde{y}$; that is, step (iii) in the basic algorithm now becomes

If $\tilde{y} < m$, then $y = \tilde{y}$; otherwise, $y = y - m$.

This is shown by the following reasoning.

$$\tilde{q} > \frac{\lfloor x/2^{n+k} \rfloor \cdot \lfloor 2^{n+j}/m \rfloor}{2^{j-k}} - 1$$

$$\geq \frac{1}{2^{j-k}} \left(\frac{x}{2^{n+k}} - 1 \right) \left(\frac{2^{n+j}}{m} - 1 \right) - 1$$

$$= \frac{1}{2^{j-k}} \left(\frac{x}{m2^{-j+k}} - \frac{x}{2^{n+k}} - \frac{2^{n+j}}{m} + 1 \right) - 1$$

$$= \frac{x}{m} - \frac{x}{2^{n+j}} - \frac{2^{n+k}}{m} + \frac{1}{2^{j-k}} - 1$$

$$\geq \left\lfloor \frac{x}{m} \right\rfloor - \frac{x}{2^{n+j}} - \frac{2^{n+k}}{m} + \frac{1}{2^{j-k}} - 1$$

$$= q - \frac{x}{2^{n+j}} - \frac{2^{n+k}}{m} + \frac{1}{2^{j-k}} - 1$$

If x is represented in $n+p$ bits, and m is presented in n bits (where $n = \lfloor \log_2 m + 1 \rfloor$) and $p \leq n$, then $x < 2^{n+p}$ and $m \geq 2^{n-1}$. Therefore,

$$q - \tilde{q} \leq 1 + \frac{x}{2^{n+j}} + \frac{2^{n+k}}{m} + \frac{1}{2^{j-k}}$$

$$\leq 1 + 2^{p-j} + 2^{k+1} - \frac{1}{2^{j-k}}$$

And since $q - \tilde{q}$ is integral, we may approximate it with

$$\left\lfloor 1 + 2^{p-j} + 2^{k+1} - \frac{1}{2^{j-k}} \right\rfloor$$

For $j \geq p + 1$ and $k \leq -2$, we have $q - \tilde{q} \leq 1$. So at most one subtraction will be required to correct $\tilde{y}$, the intermediate value of y.

Also, by choosing j and k appropriately, $\tilde{q}$ may be evaluated as

$$\tilde{q} = \begin{cases} \lfloor x/2^n \rfloor & \text{if } m \in \{2^n - l : 0 \leq l \leq \lfloor 2^n/(2^k + 1) \rfloor\} \\ \lfloor x/2^{n-1} \rfloor & \text{if } m \in \{2^n + l : 0 \leq l \leq \lfloor 2^{n-1}/(2^{k+1} - 1) \rfloor\} \end{cases}$$

So for these sets of moduli, the computation can be speeded up by eliminating a multiplication [3].

Fig. 4.1 Sequential
Barrett-reduction unit

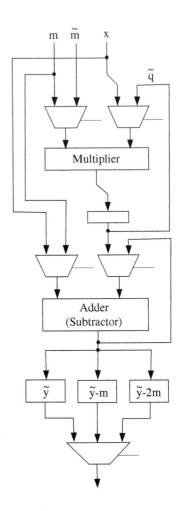

4.1.3   Implementation

Let us initially suppose that the implementation is to consist of "pre-built" individual
components—adders, multipliers, and so forth. The sketch of a basic, minimal
hardware architecture for Barrett reduction is shown in Fig. 4.1. A single multiplier
is used for the multiplications, and a single subtractor (an adder for the two's
complement of the subtrahend) is used for the subtractions. The value $\widetilde{m} = \lfloor 2^{2n}/m \rfloor$
is assumed to be available as a "pre-computed" constant.[2] The operational sequence
is as follows.

---

[2]It may be computed using a divider (Sect. 1.3) or in a simpler way (depending on the value of m).

Fig. 4.2 Sequential-parallel
Barrett-reduction unit

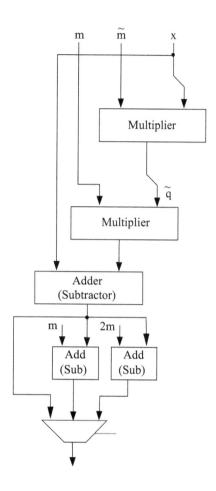

x is shifted to the right by $n - 1$ bit-positions to obtain $\lfloor x/2^{n-1} \rfloor$. (The nominal shift is just a wired shift that discards the low-order $n - 1$ bits.) The result of that shifting and the value of $\widetilde{m}$ then go through the first cycle of the multiplier, with the product then shifted to the right by $n + 1$ bit-positions (i.e., $n + 1$ bits discarded in a wired shift), to yield $\widetilde{q}$. The second cycle of the multiplier consists of the multiplication of $\widetilde{q}$ and m. Following the multiplications, the subtractor operates in three cycles and computes $\widetilde{y} = x - \widetilde{q}m$, $\widetilde{y} - m$, and $\widetilde{y} - 2m$, one of which is then chosen as the result of the reduction.

Still assuming "pre-built" individual components, a faster implementation than one based on Fig. 4.1 can be obtained by using two multipliers and three subtractors, as shown in Fig. 4.2. The operational procedure for this is straightforward and corresponds to a single-cycle version of that of Fig. 4.1. An implementation on the basis of Fig. 4.2 will evidently be more costly than one on the basis of Fig. 4.1, but it will be faster and also better for pipelining.

Let us now suppose that a unit for Barrett reduction is to be designed "from scratch." Consider the two multipliers in Fig. 4.2. An ordinary multiplier of reasonable speed will consist of one or more carry-save adders (CSAs) and a carry-propagate adder (CPA) to assimilate the partial-carry (PC) and partial-sum (PS) outputs of the CSA—Figs. 1.16, 1.17, 1.18, 1.19, and 1.20—with the delay through a CSA being much smaller than that through a CPA. Therefore, in "direct translation" the architecture of Fig. 4.2 essentially has four levels of CPAs—a substantial operational time. The CPA delay can be reduced to one level, through three observations:

- The first-level CPA can be eliminated if the corresponding input to the second multiplier is in partial-carry/partial-sum (PC-PS) form.
- The second-level CPA can be eliminated if its assimilation function is included in the following CPA (nominally the Adder/Subtractor).
- Assimilation functions can be combined with the additions (subtractions) to compute $\tilde{y}$, $\tilde{y} - m$, and $\tilde{y} - 2m$.

The resulting arrangement is shown in Fig. 4.3. To keep the diagram simple, we have left out certain minor details: if a first-level CSA-Subtractor, the subtrahend is subtracted adding its ones' complement and a 1; similarly, $-m$ is the ones' complement of m and a 1. The diagram is largely straightforward, except for the second multiplier, for which the multiplier operand is m and the multiplicand is the PC-PS output of the first multiplier. In a normal multiplier the running partial product is in PC-PS form, and that together with the next multiplicand multiple make up the three inputs into a CSA (a 3:2 compressor). On the other hand, with the arrangement of Fig. 4.3 the multiplicand multiple in one multiplier will also be in PC-PS form, and reducing the four inputs to two nominally requires two CSAs that form a 4:2 compressor. Thus in a sequential multiplier (corresponding to Fig. 1.17) the single CSA in the loop gets replaced with two, and in a multiple-CSA multiplier (corresponding to Figs. 1.18, 1.19, and 1.20) the top-level CSA gets replaced with two CSAs.

It should be noted—and this is especially significant for a sequential multiplier—that two successive 3:2 compressors can be replaced with a 4:2 compressor whose operational delay is substantially less than twice that of a 3:2 compressor [8, 9]. This applies to both the compressors used in the multipliers and those used in the subtractions.

The use of a "CSA-only" multiplier provides an additional opportunity to improve performance. Multiplier recoding helps solve the problem of "difficult" multiples of the multiplicand (x), but, depending on the radix employed, that might not completely eliminate them; for example in ordinary radix-8 multiplication, $3x$ is still required, which implies an initial carry-propagate addition ($x + 2x$). In the new multiplier, the multiplicand multiple will be in PC-PS form; so a carry-save addition, which is much faster, will suffice.

Fig. 4.3 High-performance
Barrett-reduction unit

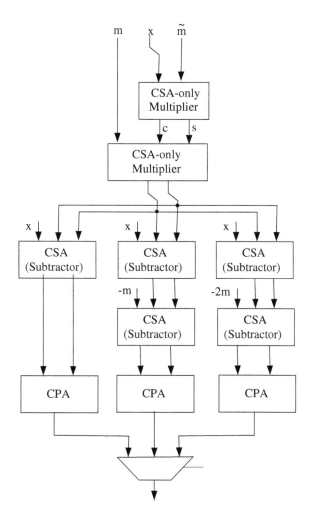

The core computations in Barrett reduction are

- an $(n + 1)$-bit $\times$ n -bit multiplication to compute $\tilde{q}$,
- an n-bit $\times$ n-bit multiplication to compute $\tilde{q}m$,
- a $(2n + 1)$ bit subtraction to compute $\tilde{y}$,
- two subtractions, of precisions and $n + 1$ bits each, for corrections of $\tilde{y}$, and
- the "precomputation" of the constant $\lfloor 2^{2n}/m \rfloor$

The last of these will be equivalent to the division that would be required for a single reduction. So, Barrett reduction will be costly if only single reduction or a very small number[3] of reductions is to be carried out. If, however, many reductions

---

[3]Relative to the number of primary computations.

are to be carried out with the same modulus, then the one-time cost of computing the constant may be worthwhile, because the other operations can all be carried out much faster than division.

4.2   Montgomery Reduction

From the x, m, and R, the Montgomery reduction algorithm computes the value $xR^{-1} \bmod m$, where R^{-1} is the multiplicative inverse of R with respect to m [2]. The form of the result is "unusual," but we will see that it is very useful—in, say, modular exponentiation (Chap. 6).

The algorithm can be used to compute $x \bmod m$ in two reductions, with an intervening multiplication by $R^2 \bmod m$:

$$u = xR^{-1} \bmod m \qquad \text{first reduction} \tag{4.8}$$

$$z = u\left(R^2 \bmod m\right) \tag{4.9}$$

$$y = \left(zR^{-1}\right) \bmod m \qquad \text{second reduction} \tag{4.10}$$

This gives $y = x \bmod m$:

$$zR^{-1} \bmod m = u\left(R^2 \bmod m\right) R^{-1} \bmod m$$

$$= \left(xR^{-1} \bmod m\right)\left(R^2 \bmod m\right) R^{-1} \bmod m$$

$$= xR^{-1}R^2R^{-1} \bmod m$$

$$= x \bmod m$$

The values of x, m, and R must satisfy three conditions:

- $\gcd(m, R) = 1$
- $m < R$
- $0 \le x < mR$

The first condition is required to ensure the existence of multiplicative inverses; the second is necessary if the algorithm is to be useful in multiplication (Sects. 5.2.2 and 5.2.3); and the third is, as explained below, necessary to ensure the correctness of the algorithm. For computer implementation, R will almost always be a power of two, in which case m must be odd, to satisfy the first condition.

It might appear that the algorithm may (generally) be used to compute $x \bmod m$ by first computing $u = xR$ and then applying the algorithm to u and R:

$$uR^{-1} \bmod m = (xR)R^{-1} \bmod m$$

$$= x \bmod m$$

But the range constraints on the argument of the reduction algorithm would not always be satisfied.

4.2.1 Algorithm

Let

- R^{-1} denote the multiplicative inverse of R with respect to m,
- m^{-1} denote the multiplicative inverse of m with respect to R, and
- $\widetilde{m}$ denote the additive inverse of m^{-1} with respect to R.

The algorithm is

$$\widetilde{m} = -m^{-1} \tag{4.11}$$

$$\widetilde{q} = x\widetilde{m} \bmod R \tag{4.12}$$

$$= [(x \bmod R)(\widetilde{m} \bmod R)] \bmod R \tag{4.13}$$

$$\widetilde{y} = \frac{x + \widetilde{q}m}{R} \tag{4.14}$$

$$y = \begin{cases} \widetilde{y} & \text{if } \widetilde{y} < m \\ \widetilde{y} - m & \text{otherwise} \end{cases} \tag{4.15}$$

y is the result, $xR^{-1} \bmod m$.

An important point to note is that although the computation of $\widetilde{y}$ nominally requires divisions by R, in practice R can be chosen so that these are trivial operations. This is discussed further below.

Example 4.5 Let $x = 169$, $m = 17$, and $R = 20$. Then:

$$R^{-1} = 6$$

$$m^{-1} = 13$$

$$\widetilde{m} = 7$$

$$\widetilde{q} = (169 \bmod 20) * 7 \bmod 20 = 3$$

$$\widetilde{y} = (169 + 3 * 17)/20 = 11$$

$$y = \widetilde{y}$$

$xR^{-1} \bmod m = 169 * 6 \bmod 17 = 11$.

□

Example 4.6 Let $x = 194$, $m = 17$, and $R = 20$. Then:

$$R^{-1} = 6$$
$$m^{-1} = 13$$
$$\widetilde{m} = 7$$
$$\widetilde{q} = (194 \bmod 20) * 7 \bmod 20 = 18$$
$$\widetilde{y} = (194 + 18 * 17)/20 = 25$$
$$y = 25 - 17 = 8 \qquad \text{(correction)}$$

$x R^{-1} \bmod m = 194 * 6 \bmod 17 = 8.$

$\square$

Suppose that in Example 4.6 we wished to compute $x \bmod m$—i.e., 194 mod 17—instead of $x R^{-1} \bmod m$. Then we would multiply y by $R^2 \bmod m$ and carry out another reduction (Eqs. 4.8–4.10):

$$y \left(R^2 \bmod m \right) = 8 * (400 \bmod 17)$$
$$= 72$$
$$\widetilde{q} = 72 * 7 \bmod 20$$
$$= 4$$
$$y = (72 + 4 * 17)/20$$
$$= 7$$

194 mod 17 = 7.

The value $R^2 \bmod m$ has to be computed separately, by other means, and that computation will be a relatively costly one.[4] Therefore, such use of Montgomery reduction is useful only if there are several reductions, and the one-off cost of computing the value gets amortized over those reductions, or if there is already a Montgomery-reduction hardware and using it is the best available option.

We now turn to the correctness of the algorithm. For that, it suffices to show that:

(a) $x + \widetilde{q}m$ is exactly divisible by R, since an integer result is required in computing $(x + \widetilde{q}m)/R$;
(b) $(x + \widetilde{q}m)/R$ is an approximation of $x R^{-1} \bmod m$; and
(c) correcting the intermediate result, $\widetilde{y}$, requires no more than a single subtraction.

---

[4]The computation may be a division or a simpler reduction, depending on the value of m.

For (a), we have

$$\tilde{q}m \bmod R = (x\tilde{m} \bmod R)m \bmod R$$
$$= -xm^{-1}m \bmod R$$
$$= -x \bmod R$$

whence

$$\tilde{q}m \equiv -x \pmod{R}$$
$$x + \tilde{q}m \equiv 0 \pmod{R}$$

So, $x + \tilde{q}m$ is exactly divisible by R.

For (b), let

$$t = \frac{x + \tilde{q}m}{R}$$

Then:

$$tR = x + \tilde{q}m$$
$$tRR^{-1} = xR^{-1} + \tilde{q}mR^{-1}$$
$$t \equiv xR^{-1} \pmod{m}$$

That is;

$$\frac{x + \tilde{q}m}{R} = xR^{-1} + km \qquad\qquad \text{for some integer } k \qquad\qquad (4.16)$$

which shows that $(x + \tilde{q}m)/R$ is an approximation of $xR^{-1} \bmod m$ and a good one if k is small.

Lastly, for (c), we show that k in Eq. 4.16 is either 0 or 1. $\tilde{q} = (x\tilde{m}) \bmod R$ implies that $\tilde{q} < R$ and $\tilde{q}m < mR$. And one of the conditions on x, m and R is that $x < mR$. Therefore,

$$\frac{x + \tilde{q}m}{R} < \frac{mR + mR}{R} = 2m$$

From this, by Eq. 4.16, either

$$\frac{x + \tilde{q}m}{R} = \left(xR^{-1}\right) \bmod m$$

or

$$\frac{x + \widetilde{q}m}{R} = \left(x R^{-1}\right) \bmod m + m$$

So, either $(x + \widetilde{q}m)/R$ is the correct result, or m must be subtracted to obtain the correct result.

4.2.2   Implementation

We now briefly consider some aspects of implementation. In general, the algorithm requires the relatively costly computation of m^{-1} (Chap. 6), but if several reductions are to be carried out, then the one-time cost of computing the value can be amortized. Nevertheless, we will see that the cost need not necessarily be a concern. In a serial-sequential implementation of the basic algorithm only the least significant digit of m^{-1} is actually required; in particular, for binary computation the required bit will be immediately available, without any real computation.

Also, the division by R need not be a real division, and this is easily arranged. Suppose data are represented in radix r, that arithmetic is carried out in radix r or a power of radix r, and that $R = r^n$. Then computing the quotient and remainder with respect to R are simple operations: the remainder is just the least significant n digits, and the other digits comprise the quotient (Fig. 4.4).

Example 4.7   Let $x = 34567, m = 121, r = 10$, and $R = 10^3$. Then

$$R^{-1} = 87$$
$$m^{-1} = 281$$
$$\widetilde{m} = 719$$
$$\widetilde{q} = (34\mathbf{567} \bmod 1000 * 719) \bmod 1000$$
$$\quad = (567 * 719) \bmod 1000 \qquad \text{take least significant 3 digits of } x$$
$$\quad = 407\mathbf{673} \bmod 1000 = 673 \quad \text{take least significant 3 digits of product}$$
$$y = (34567 + 673 * 121)/1000$$
$$\quad = 116\mathbf{000}/1000 = 116 \qquad \text{drop least significant 3 digits}$$

$$x R^{-1} \bmod m = 34567 * 87 \bmod 121 = 116.$$

$\square$

A straightforward architecture for the implementation of Montgomery reduction—as given in Eqs. 4.11–4.15 and with $R = 2^n$—is shown in Fig. 4.5. (LSDs/MSDs denotes least/most significant digits.) This corresponds to the Barrett-

Fig. 4.4 Sequential-parallel
Montgomery-reduction unit

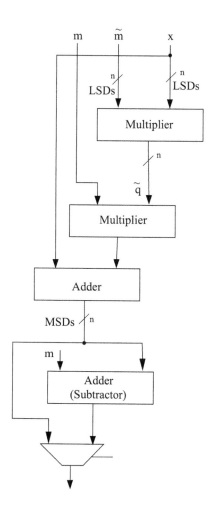

reduction architecture[5] of Fig. 4.2 and would be suitable for an implementation with
"pre-built" units. Otherwise, a much faster implementation can be obtained on a
similar basis as that for the Barrett-reduction unit of Fig. 4.3; the corresponding
version here is shown in Fig. 4.5. Such an architecture will be best for an
implementation in which a multiplier is a high-performance one, with a high
degree of parallelism. (We shall discuss alternatives more cost-effective alternatives
multiplication.)

With the choice $R = r^n$ and the various values represented in radix r, the
algorithm can be expressed in a simpler form that is well suited to sequential
implementation.

---

[5]We leave it to the reader to devise a version that corresponds to Fig. 4.1.

Fig. 4.5 High-performance
Montgomery-reduction unit

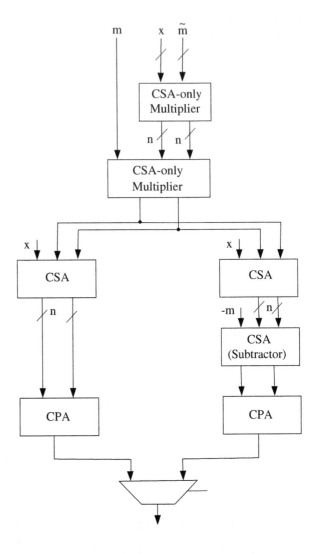

The basic idea[6] is to compute $\widetilde{q}$ one digit at a time and have running multiplications with m and additions to x. The details are as follows.

The value

$$x + \widetilde{q}m = x + \sum_{i=0}^{n-1} \widetilde{q}_i r^i m$$

---

[6]For the reader with the requisite background: the algorithm is essentially Hensel Division [7].

may be computed as

$$U_{i+1} = U_i + \tilde{q}_i r^i m \qquad i = 0, 1, 2, \ldots n - 1, \text{ with } U_0 = x$$

(In what follows, $U_{i,j}$ will denote bit j of U_i.)

The digit-serial algorithm for the computation of an approximation z to $xR^{-1} \bmod m$ is thenv

$$\tilde{m} = -m^{-1} \tag{4.17}$$

$$U_0 = x \tag{4.18}$$

$$\tilde{q}_i = U_{i,i} \tilde{m}_0 \bmod r \tag{4.19}$$

$$U_{i+1} = U_i + \tilde{q}_i m r^i \qquad i = 0, 1, 2, \ldots n - 1 \tag{4.20}$$

$$z = \frac{U_n}{r^n} \tag{4.21}$$

A correction, which consists of a subtraction of m, then follows if $z \geq m$. The final result is $y = xR^{-1} \bmod m$.

The following example corresponds to Example 4.7.

Example 4.8 Let $x = 34567$, $m = 121$, $r = 10$, and $R = 10^3$. Then:

$$R^{-1} = 87$$

$$m^{-1} = 281$$

$$\tilde{m} = 719$$

$$\tilde{m}_0 = 9$$

$$U_0 = 34567 \qquad U_{0,0} = 7$$

$$q_0 = 7 * 9 \bmod 10 = 3$$
$$U_1 = 34567 + 3 * 121 * 10^0 = 34930 \qquad U_{1,1} = 3$$

$$q_1 = 3 * 9 \bmod 10 = 7$$
$$U_2 = 34930 + 7 * 121 * 10^1 = 43400 \qquad U_{2,2} = 4$$

$$q_2 = 4 * 9 \bmod 10 = 6$$
$$U_3 = 43400 + 6 * 121 * 10^2 = 116000$$

$$z = 116000/10^3 = 116$$
$$y = 116$$

□

The correctness of the new algorithm can be established on the same basis as for the original algorithm—here, that U_n is exactly divisible by r^n, that z is an approximation of $xR^{-1} \bmod m$, and that any required correction of z is at most a single subtraction of m. Some of the details are as follows.

First, observe that in the example above the representation of U_i has i trailing 0s, which correspond to a multiplicative factor of r^i on the value represented by the other digits. Therefore, U_i is divisible by r^i. (The general case is easily proved by induction.) And so U_n is divisible by $R = r^n$.

Second, observe that the third step of the algorithm consists of adding multiples of m to a value that is initially x. That is

$$U_n = x + jm \qquad \text{for some integer } j \tag{4.22}$$

So, from

$$z = \frac{U_n}{r^n} = \frac{x + jm}{r^n}$$

and $R = r^n$, a line of reasoning similar to that in part (b) of the correctness proof of the original algorithm gives us

$$z \equiv xR^{-1} \pmod{m}$$

Lastly

$$U_n = x + \sum_{i=0}^{n-1} \tilde{q}_i m r^i$$

$$< mR + r^n m$$

$$= 2r^n m$$

Therefore

$$\frac{U_n}{r^n} < 2m$$

and at most one subtraction, of m, may be required to bring z into the appropriate range.

The second observation above suggests that a better algorithm than that of Eqs. 4.17–4.21 is possible: The computation of U_{i+1} is so as to ensure that at the end of the iterations U_n is divisible by r^n. Since U_{i+1} is divisible by r^{i+1}, the same effect (as division) on the final value can be obtained by simply nominally dividing U_{i+1} by r in each iteration. With that, what was bit i of U_i is now the least significant bit of U_i, and the revised algorithm for the computation of $y = xR^{-1} \bmod m$ is

$$\tilde{m} = -m^{-1} \tag{4.23}$$

$$U_0 = x \tag{4.24}$$

$$\tilde{q}_i = U_{i,0}\tilde{m}_0 \mod r \tag{4.25}$$

$$U_{i+1} = \frac{U_i + \tilde{q}_i m}{r} \qquad i = 0, 2, \ldots n - 1 \tag{4.26}$$

$$y = \begin{cases} U_n & \text{if } U_n < m \\ U_n - m & \text{otherwise} \end{cases} \tag{4.27}$$

The values of U_i are much smaller here than in the algorithm of Eqs. 4.17–4.21, which has significant implications for hardware implementation.

Example 4.9 (See Example 4.8.) Let $x = 34567$, $m = 121$, $r = 10$, and $R = 10^3$. Then:

$$R^{-1} = 87$$

$$m^{-1} = 281$$

$$\tilde{m} = 719$$

$$\tilde{m}_0 = 9$$

$$U_0 = 34567$$

$$q_0 = 7 * 9 \mod 10 = 3$$

$$U_1 = (34567 + 3 * 121)/10 = 3493$$

$$q_1 = 3 * 9 \mod 10 = 7$$

$$U_2 = (3493 + 7 * 121)/10 = 4340$$

$$q_2 = 4 * 9 \mod 10 = 6$$

$$U_3 = (434 + 6 * 121)/10 = 116$$

$$y = 116$$

□

With $r = 2$, the algorithm of Eqs. 4.23–4.27 can be simplified: Since one of the conditions at the start of this section is that $\gcd(m, R) = 1$, m must be odd; that is $m_0 = 1$. And

Fig. 4.6 Serial Montgomery
reduction unit (binary)

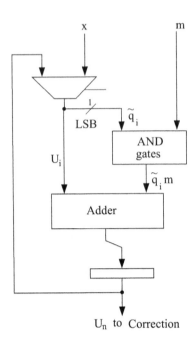

$$\tilde{m}_0 \bmod 2 = -m_0^{-1} \bmod 2$$
$$= (2 - m_0)^{-1} \bmod 2$$
$$= 1$$

Therefore, in Eq. 4.25:

$$\tilde{q}_i = U_{i,0}\tilde{m}_0 \bmod 2$$
$$= [(U_{i,0} \bmod 2)(\tilde{m}_0 \bmod 2)] \bmod 2$$
$$= U_{i,0}$$

That is, $\tilde{q}_i$ is just the least significant bit of U_i. Combining this with the fact that the nominal division by 2 in Eq. 4.26 can be implemented as just a wired shift that drops the least significant bit of the dividend, we have the architecture of Fig. 4.6. The Correction part is as in Fig. 4.4.

A faster arrangement than that of Fig. 4.6 can be obtained by replacing the carry-propagate Adder with a carry-save adder, as shown in Fig. 4.7. The Assimilation and Correction here is as in Fig. 4.5: a CPA (for $y = \tilde{y}$) and a CSA-CPA (for $y = \tilde{y} - m$).

With the conditions $x < mR$ and $m < R$, if we assume that m is represented in n bits and take $R = 2^n$, then $x < 2^{2n}$; then we may compare Montgomery reduction with Barrett reduction on the basis of the remarks made at the end of Sect. 4.1 (and Figs. 4.2 and 4.4). Montgomery reduction requires:

Fig. 4.7 High-performance
Montgomery reduction unit

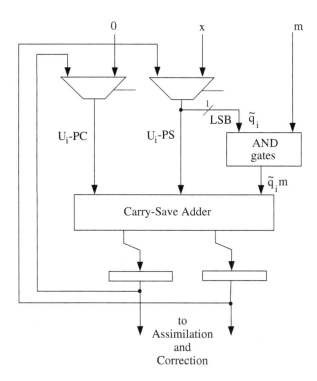

- an n-bit$\times n$-bit multiplication to compute $\widetilde{q}$,
- an n-bit$\times n$-bit multiplication to compute $\widetilde{q}m$,
- a $2n$-bit addition to compute $\widetilde{y}$,
- an $(n + 1)$-bit subtraction for the correction of $\widetilde{y}$, and
- the "precomputation" of $\widetilde{m}$.

The "precomputation" here is (in general) more costly than that in Barrett reduction, but the totality of the other operations cost slightly less than the totality of the other operations in Barrett reduction.

4.3   Lookup-Table Reduction

We next describe a method of reduction that is based on the straightforward use of additions and subtractions and which, depending on the circumstances, can be implemented more efficiently than general direct division. The method is especially useful for certain special moduli, for which it can be greatly simplified (Sect. 4.4).

Given a modulus m and x represented in p bits—i.e., $x = \sum_{i=0}^{p-1} x_i 2^i$—we have

$$x \bmod m = \left[\sum_{i=0}^{p-1} \left(x_i 2^i \right) \bmod m \right] \bmod m \qquad (4.28)$$

Since x_i is 0 or 1, the value $x \bmod m$ may be computed by computing the values $2^i \bmod m$ wherever $x_i = 1$ in the term $x_i 2^i$ and then adding up these modulo m.

Example 4.10 Let $x = 10111001 1000_2 = 2968$ and $m = 17$. Then

$$x = \left(2^{11} \bmod 17 + 2^9 \bmod 17 + 2^8 \bmod 17 + 2^{11} \bmod 17 + 2^7 \bmod 17 \right.$$

$$\left. + 2^4 \bmod 17 + 2^3 \bmod 17 \right) \bmod 17$$

$$= (8 + 2 + 1 + 9 + 16 + 8) \bmod 17$$

$$= 10$$

□

Such a method will be inefficient for anything but small values of p, given that the computation of $2^i \bmod m$ is, in general, not a trivial operation. The values $2^i \bmod m$ may be "pre-computed" and stored for later "lookup" as necessary; but, except for small values of p, the number of additions will be excessively large. A much better approach is to apply the basic idea to "blocks" of bits instead of individual bits.

Let us suppose that p in Eq. 4.28 is a multiple of n. (If that is not the case, then we can obtain an appropriate value by extending the representation of x with 0s at the most significant end.) And suppose the representation of x is split into j blocks of k bits each:

$$\mathbf{x}_{j-1} = x_{kj-1} x_{kj-2} \cdots x_{k(j-1)}$$

$$\vdots$$

$$\mathbf{x}_1 = x_{2k-1} x_{2k-2} \cdots x_k$$

$$\mathbf{x}_0 = x_{k-1} x_{k-2} \cdots x_0$$

Then

$$x \bmod m = \left(\mathbf{x}_{j-1} 2^{k(j-1)} \bmod m + \mathbf{x}_{j-2} 2^{k(j-2)} \bmod m + \cdots \right. \qquad (4.29)$$

$$\left. + \mathbf{x}_1 2^k \bmod m + \mathbf{x}_0 2^0 \bmod m \right) \bmod m$$

So the computation is that of the values $\mathbf{x}_i 2^{ik} \bmod m$ ($i = 0, 1, 2, \ldots, j-1$) and their addition modulo m.

Fig. 4.8 Look-up table
reduction unit

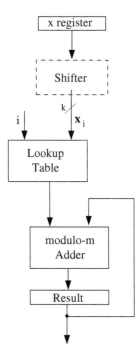

Example 4.11 (See Example 4.10.) Let $x = 101110011000_2 = 2968\ (p = 12)$ and $m = 17$. With three blocks $(j = 3, k = 4)$, we have $\mathbf{x}_2 = 1011_2 = 11, \mathbf{x}_1 = 1001_2 = 9, \mathbf{x}_0 = 1000_2 = 8$. So:

$$2968 \bmod 5 = \left(11 * 2^8 \bmod 17 + 9 * 2^4 \bmod 17 + 8 \bmod 17\right) \bmod 17$$

$$= (11 + 8 + 8) \bmod 17$$

$$= 10$$

□

Figure 4.8 shows the architecture of a serial "block reduction" unit, of which an implementation operates as follows. The Result register is initialized to zero. Assuming j blocks of k bits each, in each of j cycles the cycle-index, i, and k bits of x are used to form an address that is used to access the LUT; the next value to be added is read out; a modular addition takes place; and the contents of the x shift register are shifted by k bit-positions.

There are tradeoffs to be made in the choice of parameters for the implementation. Let n be the precision of x. Then[7] $j = n/k$, and the number of LUT entries is $2^k * (n/k)$. Minimizing that number requires a small block size, but a small block

---

[7]We assume that n is divisible by k; otherwise, it is extended with 0s to make it so.

size means more addition cycles. Thus the tradeoffs involve the size of the LUT, the time to read out an entry from the LUT, and the time for a single modular addition. The time to shift the contents of the operand register should also be taken into account, as an ordinary shift register might be excessively slow. (The last of these can be dealt with by using an ordinary register and a multiplexer whose inputs are the various components of x and whose output is the block of bits for a given cycle.)

4.4   Special-Moduli Reduction

The reduction method described in Sect. 4.3 (Eqs. 4.28 and 4.29) is especially useful for certain moduli, for which simplifications are possible that eliminate the actual powers of two, with the effect that the reduction is simplified to a set of straightforward additions and some smaller subreductions. We next describe such simplifications—for just the moduli $2^n \pm 1$, but it will be readily apparent that these are just particular instances of reduction with respect to $2^n \pm c$, for some positive integers n and c.

The other special moduli that are discussed are those that are significant because of their inclusion in certain standards [5, 6]. The methods used for reduction with these moduli may be regarded as generalizations of those used for the moduli $2^n - c$, and the moduli have been chosen specifically to make reduction easy.

Modulus m $= 2^n - 1$

Since

$$2^n \bmod (2^n - 1) = \left[(2^n - 1) + 1\right] \bmod (2^n - 1) = 1 \qquad (4.30)$$

we have

$$2^{in} \bmod (2^n - 1) = \left[2^n \bmod (2^n - 1)\right]^i \bmod (2^n - 1)$$

$$= 1^i \bmod (2^n - 1)$$

$$= 1$$

Therefore

$$\left(x_i 2^{jn}\right) \bmod (2^n - 1) = x_i \bmod (2^n - 1)$$

and Eq. 4.29, with $k = n$, becomes

$$x \bmod (2^n - 1) = \left[x_{j-1} 2^{n(j-1)} \bmod (2^n - 1) + x_{j-2} 2^{n(j-2)} \bmod (2^n - 1) + \cdots \right.$$

$$\left. + x_1 2^n \bmod (2^n - 1) + x_0 \bmod (2^n - 1)\right] \bmod (2^n - 1)$$

$$= \left[x_{j-1} \bmod \left(2^n - 1 \right) + x_{j-2} \bmod \left(2^n - 1 \right) + \cdots \right.$$

$$\left. + x_1 \bmod \left(2^n - 1 \right) + x_0 \bmod \left(2^n - 1 \right) \right] \bmod \left(2^n - 1 \right)$$

$$= \left(x_{j-1} + x_{j-1} + \cdots x_0 \right) \bmod \left(2^n - s1 \right) \tag{4.31}$$

Example 4.12 (See Examples 4.10 and 4.11.) Let $x = 10111001100_2 = 2968$ ($p = 12$) and $m = 1111_2 = 15$ ($n = 4$). With $j = 3$, we have $x_2 = 1011_2 = 11$, $x_1 = 1001_2 = 9$, and $x_0 = 1000_2 = 8$. So:

$$2968 \bmod 15 = (11 + 9 + 8) \bmod 15$$

$$= 13$$

□

Modulus m = $2^n + 1$

Since

$$2^n \bmod \left(2^n + 1 \right) = \left[\left(2^n + 1 \right) - 1 \right] \bmod \left(2^n + 1 \right) = -1 \tag{4.32}$$

we have

$$2^{in} \bmod \left(2^n + 1 \right) = \left[2^n \bmod \left(2^n + 1 \right) \right]^i \bmod \left(2^n + 1 \right)$$

$$= (-1)^i \bmod \left(2^n + 1 \right)$$

$$= \begin{cases} 1 & \text{if } i \text{ is even} \\ -1 & \text{otherwise} \end{cases}$$

So, from Eq. 4.29, with $k = n$, if j is odd, then:

$$x \bmod \left(2^n + 1 \right) = \left[x_{j-1} 2^{n(j-1)} \bmod \left(2^n + 1 \right) - x_{j-2} 2^{n(j-2)} \bmod \left(2^n + 1 \right) + \cdots \right.$$

$$\left. - x_1 2 * n \bmod \left(2^n + 1 \right) + x_0 \bmod \left(2^n + 1 \right) \bmod \left(2^n - 1 \right) \right]$$

$$= \left[x_{j-1} \bmod \left(2^n + 1 \right) - x_{j-2} \bmod \left(2^n + 1 \right) + \cdots \right.$$

$$\left. - x_1 \bmod \left(2^n + 1 \right) + x_0 \bmod \left(2^n + 1 \right) \bmod \left(2^n - 1 \right) \right]$$

$$= \left(x_{j-1} - x_{j-1} + \cdots - x_1 + x_0 \right) \bmod \left(2^n - 1 \right) \tag{4.33}$$

And if j is even, then:

$$x \bmod \left(2^n + 1 \right) = \left[-x_{j-1} \bmod \left(2^n + 1 \right) + x_{j-2} \bmod \left(2^n + 1 \right) \cdots - \right.$$

$$\left. + x_1 \bmod \left(2^n + 1 \right) - x_0 \bmod \left(2^n + 1 \right) \right] \bmod \left(2^n - 1 \right)$$

$$= \left(-x_{j-1} + x_{j-1} - \cdots - x_1 + x_0 \right) \bmod \left(2^n - 1 \right) \tag{4.34}$$

Example 4.13 Let $x = 1011100110000_2 = 2968$ ($p = 12$) and $m = 10001_2 = 17$ ($n = 4$). With $j = 3$, we have $\mathbf{x}_2 = 1011_2 = 11, \mathbf{x}_1 = 1001_2 = 9$, and $\mathbf{x}_0 = 1000_2 = 8$. So:

$$2968 \bmod 17 = (11 - 9 + 8) \bmod 17$$

$$= 10$$

Let $x = 1101101110011000_2 = 56216$ ($p = 12$) and $m = 10001_2 = 17$ ($n = 4$). With $j = 4$, we have $\mathbf{x}_3 = 1100_2 = 13, \mathbf{x}_2 = 1011_2 = 11, \mathbf{x}_1 = 1001_2 = 9$, and $\mathbf{x}_0 = 1000_2 = 8$. So:

$$56216 \bmod 17 = (-13 + 11 - 9 + 8) \bmod 17$$

$$= -3 \bmod 17$$

$$= 14$$

$\square$

Implementation

We discuss implementations for only the modulus $2^n - 1$. Implementations for the modulus $2^n + 1$ will have some broad similarities with those for the modulus $2^n - 1$, but there is one major difference: as we shall see, the latter can easily use conventional adders; for the former, the most straightforward implementations will be based on generalized modulo-m adders, although modulo-$(2^n + 1)$ subtractors can be used for the subtractions. This is because of the relative difficulty of modulo-$(2^n + 1)$ addition (Sect. 5.1.2).

In modulo-$(2^n - 1)$ addition with n-bit operands, it is easy to determine if the result of an (intermediate) addition is less than the modulus or greater than the modulus. In the former case there will be no carry-out from the addition, and the result is correct modulo $2^n - 1$; in the latter case there will be a carry-out, and discarding that carry and adding a 1 will give the correct result modulo $2^n - 1$. The case where the intermediate result is equal to the modulus—$11 \cdots 1$ in binary—is slightly more complicated: the case must be detected separately, and then adding a 1 and discarding the carry out will leave the correct result modulo $2^n - 1$.

The formal justification for the preceding remarks is given in Sect. 5.1.2, on the design of modulo-$(2^n - 1)$ adders. There is, however, one difference between modular addition as discussed there and the situation here. In an "ordinary" modular addition, each of the operands will be less than the modulus; here, it is possible to have some $\mathbf{x}_j = 2^n - 1$. This presents no difficulty as long as the result in a sequence of additions is handled properly, which is quite straightforward: the procedure above is carried out for each addition, with one more addition after the nominal last one. (The justification for this is given in Sect. 5.2, in the context of repeated additions in multiplication.)

Suppose the only adders available are carry-propagate adders (CPAs) of the types described in Sect. 1.1. Then a straightforward implementation is one that uses one CPA for the addition of each x_j. The carry-out from each adder is fed as a carry-in to the next adder in sequence, to effect the procedure described above. This can be done until the last stage, at which point there is no "next adder"; an additional adder (an incrementor) is therefore required after the last stage. There is also the possibility that the result of the $(k-1)$st adder might be $2^n - 1$. A straightforward way to deal with this is to also subtract $2^n - 1$ and then choose between the original result and the result of the subtraction.[8] With two's-complement representation, the subtraction is as the addition of the two's complement of the subtrahend, which is the ones' complement (shown below as $\overline{2^n - 1}$) and a 1 (a carry-in to the adder). If the subtraction produces a carry, then discarding that carry leaves the correct result; otherwise, the undiminished value is the correct one.[9] Note that the adder for the subtraction will be of $(n+1)$-bit precision in order to accommodate operand signs. Table 4.1 shows an example computation, for $k = 4$. The computation is that of

$$63199 \bmod 15 = 1111011011011111_2 \bmod 1111_2$$
$$= (1111_2 + 0110_2 + 1101_2 + 1111_2) \bmod 1111_2 \qquad (k = 4)$$
$$= (15 + 6 + 13 + 15) \bmod 15$$
$$= 4$$

There is no carry out of CPA5, so the output of CPA4 is the correct result. On the other hand, had the computation been that of, say, $(8 + 7 + 9 + 6) \bmod 15$, there would have been a carry out of CPA5, and discarding that carry would leave the correct result of 0.

A less costly implementation than that suggested by Table 4.1 would consist for a single CPA used, in several cycles at add each x_j, one at a time. The last addition may be carried out in the same CPA, although this will require multiplexing in every cycle, or in an extra CPA outside the "loop."

The faster alternative to the arrangements above is to use carry-save adders (CSA). A (partial) carry out of one CSA gets feed, end-around, into the next CSA or into the final assimilation CPA. An example architecture is shown in Fig. 4.9, for $k = 4$.

Other Special Moduli

From operands a and b and modulus m, modular multiplication is the computation of $ab \bmod m$. A direct way to carry out the computation is to multiply a and

---

[8]A different, less costly but not necessarily much faster, way is to detect the pattern $11 \cdots 1$ and replace it with $00 \cdots 0$.

[9]The justification for this follows given in Sect. 5.1.2 for addition modulo $2^n - 1$.

Table 4.1 Example modulo-$(2^n - 1)$ reduction

CPA 1:	15	=	1 1 1 1	
	6	=	0 1 1 0	
			1	
	(1)	←	0 0 1 1	
CPA 2:			0 1 0 1	
	13	=	1 1 0 1	
			1	EAC from CPA1
	(1)	←	0 0 1 1	
CPA 3:			0 0 1 1	
	15	=	1 1 1 1	
			1	EAC from CPA2
	(1)	←	0 0 1 1	
CPA 4:			0 0 1 1	
			0 0 0 0	
			1	EAC from CPA3
	4	=	0 1 0 0	

				CPA 5:	00011	
					10000	$2^n - 1$
					1	
					10101	

Fig. 4.9 Fast modulo-$(2^n - 1)$-reduction unit

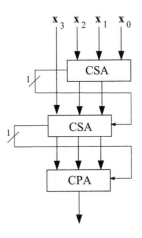

b and then reduce the product relative to m. If each operand is less than m, and therefore $ab < m^2$, and the modulus is of a special form—a *generalized Mersenne number*[10]—then the reduction can be made very efficient, by, essentially,

---

[10]A *Mersenne prime* is one of the form $2^k - 1$, for some positive integer k. A *Mersenne number* has the same form but is not necessarily prime. And a *generalized Mersenne number* has the form $2^k + c_{n-1}2^{k-1} + \cdots + c_1 2^1 + c_0$, where the c_i are integers.

generalizing the methods above for the moduli $2^n \pm 1$. Accordingly, the moduli used in certain standards are chosen to be of such a form [5, 6]. The aforementioned standards include suggestions for how to efficiently carry out reductions with respect to the given moduli, and we next describe two of these. In what follows, the computation is of $x \bmod m$, with $x < m^2$.

Modulus m $= 2^{192} - 2^{64} - 1$

x is a 384-bit number that may be expressed as

$$x = x_5 2^{320} + x_4 2^{256} + x_3 2^{192} + x_2 2^{128} + x_1 2^{64} + x_0 \tag{4.35}$$

where each x_i is a 64-bit integer.

Then

$$x \bmod m = (t + s_1 + s_2 + s_3) \bmod m \tag{4.36}$$

where t and the s_i are 192-bit terms that are obtained from the concatenation (denoted $\|$) of 64-bit values:

$$
\begin{aligned}
t &= x_2 \parallel x_1 \parallel x_0 \\
s_1 &= 0 \parallel x_3 \parallel x_3 \\
s_2 &= x_4 \parallel x_4 \parallel 0 \\
s_3 &= x_5 \parallel x_5 \parallel x_5
\end{aligned}
$$

The t and s_i values are obtained as follows.

The last three terms of Eq. 4.35 give a 128-bit value that is evidently less than m, whence the term t.

For the term $x_3 2^{192}$:

$$2^{192} \bmod m = 2^{192} - \left(2^{192} - 2^{64} - 1 \right)$$

$$= 2^{64} + 1$$

which gives $s_1 = x_3 2^{64} + x_3$.

For the term $x_4 2^{256}$:

$$2^{256} \bmod m = \left(2^{64} 2^{192} \right) \bmod m$$

$$= 2^{64} \left(2^{64} + 1 \right)$$

$$= 2^{128} + 2^{64}$$

which gives $s_2 = x_4 2^{128} + x_4 2^{64}$.

And for the term $\mathbf{x}_5 2^{320}$:

$$2^{320} \bmod m = \left(2^{256} 2^{64}\right) \bmod m$$

$$= \left[\left(2^{128} + 2^{64}\right) 2^{64}\right] \bmod m$$

$$= \left(2^{192} + 2^{128}\right) \bmod m$$

$$= 2^{128} + 2^{64} + 1$$

which gives $\mathbf{s}_3 = \mathbf{x}_5 2^{128} + \mathbf{x}_5 2^{64} + \mathbf{x}_5$.

As with implementations for the other reduction units above, one for Eq. 4.36 may be use only ordinary carry-propagate adders, or such carry-propagate adders in combination with carry-save adders, or one or more modulo-m carry-propagate adders. And the arrangement may be serial, or sequential, or parallel, or a "hybrid". We give two and leave it to the reader to devise others.

A simple architecture that uses a single modulo-m adder is shown in Fig. 4.10. The Concatenation Logic consists of appropriate wiring to produce the terms $\mathbf{t}, \mathbf{s}_1, \mathbf{s}_2$, and $\mathbf{s}_3$ from the primary operand. The modulo-m adder, whose design is described in Chap. 5, adds up these terms one at a time. (The Result register is initialized to zero.) An architecture for a faster implementation is shown in Fig. 4.11. As usual, the $-m$ is added as a ones' complement and a 1, and likewise for the $-2m$.

Fig. 4.10
Modulo-$(2^{192} - 2^{64} - 1)$
reduction unit

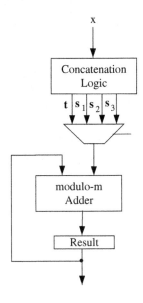

Fig. 4.11 Fast modulo-$(2^{192} - 2^{64} - 1)$ reduction unit

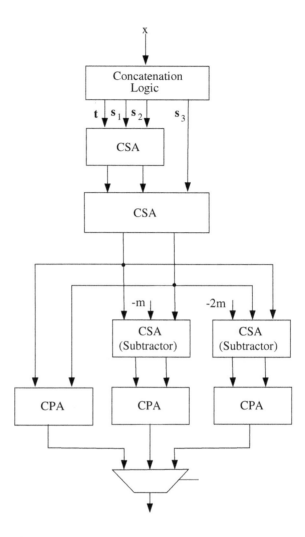

Modulus $\mathbf{m} = 2^{224} - 2^{96} + 1$

The method used for this modulus is the same as for the modulus $2^{192} - 2^{64} - 1$, but with the argument partitioned into 32-bit pieces.

x may be expressed as

$$x = \mathbf{x}_{13}2^{416} + \mathbf{x}_{12}2^{384} + \mathbf{x}_{11}2^{352} + \mathbf{x}_{10}2^{320} + \mathbf{x}_9 2^{288} + \mathbf{x}_8 2^{256} + \mathbf{x}_7 2^{224}$$
$$+ \mathbf{x}_6 2^{192} + \mathbf{x}_5 2^{160} + \mathbf{x}_4 2^{128} + \mathbf{x}_3 2^{96} + \mathbf{x}_2 2^{64} + \mathbf{x}_1 2^{32} + \mathbf{x}_0$$

where each $\mathbf{x}_i$ is a 32-bit integer.

And

$$x \bmod m = (\mathbf{t} + \mathbf{s}_1 + \mathbf{s}_2 - \mathbf{d}_1 - \mathbf{d}_2) \bmod m \tag{4.37}$$

where the $\mathbf{t}$, $\mathbf{s}_i$, and $\mathbf{d}_i$ are 224-bit terms:

$$
\begin{aligned}
\mathbf{t} &= \mathbf{x}_6 \ \| \ \mathbf{x}_5 \ \| \ \mathbf{x}_4 \ \| \ \mathbf{x}_3 \ \| \ \mathbf{x}_2 \ \| \ \mathbf{x}_1 \ \| \ \mathbf{x}_0 \\
\mathbf{s}_1 &= \mathbf{x}_{10} \ \| \ \mathbf{x}_9 \ \| \ \mathbf{x}_8 \ \| \ \mathbf{x}_7 \ \| \ \mathbf{0} \ \| \ \mathbf{0} \ \| \ \mathbf{0} \\
\mathbf{s}_2 &= \mathbf{0} \ \ \| \ \mathbf{x}_{13} \ \| \ \mathbf{x}_{12} \ \| \ \mathbf{x}_{11} \ \| \ \mathbf{0} \ \| \ \mathbf{0} \ \| \ \mathbf{0} \\
\mathbf{d}_1 &= \mathbf{x}_{13} \ \| \ \mathbf{x}_{12} \ \| \ \mathbf{x}_{11} \ \| \ \mathbf{x}_{10} \ \| \ \mathbf{x}_9 \ \| \ \mathbf{x}_8 \ \| \ \mathbf{x}_7 \\
\mathbf{d}_2 &= \mathbf{0} \ \ | \ \ \mathbf{0} \ \| \ \mathbf{0} \ \| \ \mathbf{0} \ \| \ \mathbf{x}_{13} \ \| \ \mathbf{x}_{12} \ \| \ \mathbf{x}_{11}
\end{aligned}
$$

Equation 4.37 may be implemented in the general form exemplified by Figs. 4.10 and 4.11.

Other Moduli

The other moduli in the standards are

- $2^{256} - 2^{224} + 2^{192} + 2^{96} - 1$
- $2^{384} - 2^{128} - 2^{96} + 2^{32} - 1$
- $2^{521} - 1$

Reduction for the first two of these can be done in a manner similar to that for $2^{224} - 2^{96} + 1$ and $2^{224} - 2^{96} + 1$, and the third has the form $2^n - 1$, which has been dealt with above.

Other similar or related methods for reduction are given in [4].

References

1. P. Barrett. 1987. Implementing the Rivest Shamir and Adleman public key encryption algorithm on a standard digital signal processor. In: *Advances in Cryptology*, Lecture Notes in Computer Science (Springer, Germany), Vol. 263, pp. 311–323.
2. P. L. Montgomery. 1985. Modular multiplication without trial division. *Mathematics of Computation*, 44(170):519–521.
3. M. Knezevic, F. Vercauteren, and I. Verbauwhede. 2010. Faster interleaved modular multiplication based Barrett and Montgomery reduction methods. *IEEE Transactions on Computers*, 59(12):1715–1721.
4. W. Hasenplaugh, G. Gaubatz, and V. Gopal. 2007. Fast modular reduction. *Proceedings, 18th Symposium on Computer Arithmetic*, pp. 225–229.
5. National Institute of Standards and Technology. 1999. Recommended Elliptic Curves for Federal Government Use. Gaithersburg, Maryland, USA.
6. American National Standards Institute. 1999. ANSI X9.62: Public Key Cryptography for the Financial Services Industry: the Elliptic Curve Digital Signature Algorithm (ECDSA). Washington, D.C., USA.
7. M. Shand and J. Vuillemin. 1993. Fast implementations of RSA cryptography. *Proceedings, 11th International Symposium on Computer Arithmetic*, pp. 252–259.

8. M. J. Flynn and S. F. Oberman. 2001. *Advanced Computer Arithmetic Design*. Wiley-Interscience, New York, USA.
9. N. Ohkubo, M. Suzuki, T. Shinbo, T. Yamanaka, A. Shimizu, K. Sasaki, and Y. Nakagome. 1995. A 4.4ns CMOS 54*54-b multiplier using pass-transistor multiplexor. *IEEE Journal of Solid-State Circuits*, 30(3):251–257.
10. J.-F. Dhem. 1998. Design of an Efficient Public-Key Cryptographic Library for RISC-based Smart Cards. Ph. D. Thesis, Catholic University of Louvain, Belgium.

Chapter 5
Modular Addition and Multiplication

Abstract This chapter consists of two sections that cover algorithms and hardware architectures for modular addition and multiplication: $(x + y)$ mod m and xy mod m. Subtraction and division are also included—as the addition of an inverse and as multiplication by an inverse. The underlying algorithms and hardware structures are those of Chap. 1, modified for modular arithmetic. For both operations we shall consider generic algorithms and hardware structures for arbitrary moduli and also those for special moduli.

A primary difference between ordinary modular arithmetic and the modular arithmetic of cryptography is in the high precisions used in the latter, with operands represented in hundreds or even thousands of bits. One implication of this difference is that some of the algorithms and hardware designs for the former are not always appropriate for the latter, and this is especially so for multiplication. We will not make specific remarks on high-precision operations. The discussions in Sects. 1.1.5 and 1.2.4 largely carry over to the present context, and there is little to add.

The difference between ordinary addition and modular addition is not large; that is not so with multiplication. If the operands for a modular addition are within the correct range, then ensuring that the result too is within range is a relatively simple task. On the other hand, with modular multiplication ensuring that a result is within range generally requires modular reduction, which is generally almost equivalent to division. Thus the first part of the chapter is a short and straightforward one, whereas the second is not.

5.1 Addition

The first subsection is on "generic" modular adders, i.e., those for an arbitrary modulus. The second subsection is on adders for special moduli, of the form $2^n \pm 1$. And the third section consists of a few remarks on subtraction.

© Springer Nature Switzerland AG 2020

A. R. Omondi, *Cryptography Arithmetic*, Advances in Information Security 77,
https://doi.org/10.1007/978-3-030-34142-8_5

5.1.1 Generic Structures

The result of adding, modulo m, two numbers x and y, where $0 \le x, y < m$, is given by

$$(x + y) \bmod m = \begin{cases} x + y & \text{if } x + y < m \\ x + y - m & \text{otherwise} \end{cases} \qquad (5.1)$$

This equation can be implemented directly or indirectly, in a variety of ways.

Let us suppose that we have an ordinary two-operand adder (Sect. 1.1), i.e., an adder that computes modulo-2^k sums for k-bit operands. (We shall refer to such an adder as a "basic adder.") Then a simple, direct algorithm to effect Eq. 5.1 is as follows.

(i) x and y are added to yield an intermediate sum, s'.
(ii) s' is compared with m.
(iii) If $s' < m$, then s' is the correct result.
(iv) If $s' \ge m$, then m is subtracted from s', to yield a new value, s'', which is then the correct result.

The procedure just described suggests three sequential additions: the primary one; one for the "corrective" subtraction, which would be effected as the addition of the negation of the subtrahend; and one for the comparison, which would consist of a subtraction followed by sign determination. In practice the latter two additions can, as we show below, be combined into a single one, in the following way.

For the subtraction, we shall assume that the negation of the subtrahend is in two's-complement representation. If x, y, and m are each represented in n bits, then the arithmetic will be in $n + 1$ bits, the additional bit being for sign.[1] We shall therefore assume $(n + 1)$-bit representations for the actual arithmetic, even with all operands and results representable in n bits each.

When interpreted as an unsigned number, the two's-complement representation of negative m, which we shall denote $\tilde{m}$, corresponds to the numerical value $2^{n+1} - m$ (Section 1.1.6). With that interpretation

$$x + y + \tilde{m} = (x + y - m) + 2^{n+1}$$

If $x + y \ge m$, then $x + y + \tilde{m} \ge 2^{n+1}$, and the 2^{n+1} in the equation represents a carry out of the most significant bit-position; discarding that carry is equivalent to subtracting 2^{n+1} and leaves the correct result of $x + y - m$. On the other hand, if $x + y < m$, then $x + y + \tilde{m} < 2^{n+1}$, and there is no carry-out. Therefore, the carry-out from the second addition is equivalent to the result of the nominal comparison.

---

[1] Strictly, the first addition may be in n bits.

Table 5.1 Examples of modular addition

$n = 4,\ m = 12$					
5	=	00101	8	=	01000
+4	=	00100	+7	=	00111
9	=	01001	5	=	01111
		01001			01111
$+\widetilde{m}$	=	10100	$+\widetilde{m}$	=	10100
		11101	3	=	00011

As usual, the addition of $\widetilde{m}$ will be as the addition of the ones' complement and a 1 (injected as a carry-in to the adder).

Examples are given in Table 5.1. In the first case the second addition does not produce a carry, so the result (i.e., 9) from the first addition is correct. In the other case there is a carry, which is discarded, so the correct result (i.e., 3) is that from second addition is the correct one.

A straightforward arrangement to implement the preceding algorithm is one that uses a single adder, in three steps. First, x and y are added, and the result s' is stored. Next, $\widetilde{m}$ is added to s', to obtain s''. Finally, s' or s'' is selected as the result, according to the sign of the latter. The corresponding hardware architecture is shown in Fig. 5.1. The unit operates in two cycles, one for the computation of $s' = x + y$ and one for the computation of $s'' = x + y + \widetilde{m}$. Note that a carry-out cannot occur in the first cycle because $x + y < 2^{n+1}$. A *nominal* overflow can occur, but it does not matter because the result is taken as an unsigned number.

An alternative to the organization of Fig. 5.1 is one that uses two basic adders arranged in sequence, as shown in Fig. 5.2. This new arrangement is more costly, but it can be faster on three grounds. First, the register delay in Fig. 5.1 is eliminated. Second, the new arrangement is more amenable to pipelining. And third, depending on the basic-adder design, the two adders in Fig. 5.2 can largely operate concurrently. The explanation for the last point is as follows.

Suppose, for example, that the two base adders are ripple adders. The result, s', from the first adder will be available one bit at a time, starting from the least significant end. Therefore, the second addition, to compute s'', can start as soon as the first result bit from the first adder is available, and the computation of the first result bit from the second adder can be overlapped with that of second bit from the first adder. Extending this reasoning, we see that overlap is possible in the computation of the other sum bits. Similar reasoning may be applied to other adder designs. For example, where each adder is partitioned into blocks, with all the sum bits from a block available at the same time, the overlap can be at the level of blocks.

Fig. 5.1 Modulo-m adder,
single basic adder

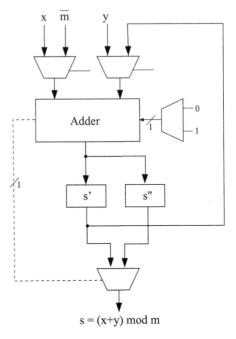

$$s = (x+y) \bmod m$$

Fig. 5.2 Modulo-m adder,
two basic adders in sequence

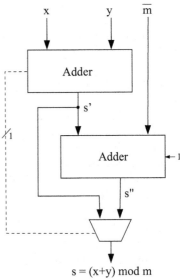

$$s = (x+y) \bmod m$$

Two base adders can also be used in a different arrangement that is, in principle, faster than that of Fig. 5.2, by concurrently computing both s' and s'' and then selecting one of the two. The computation of s'' requires that three operands be reduced to one, which is easily done by using a combination of a carry-save adder (CSA) and a carry-propagate adder (CPA), as shown in Fig. 5.3. Although two CPAs

Fig. 5.3 Modulo-*m* adder, two concurrent basic adders

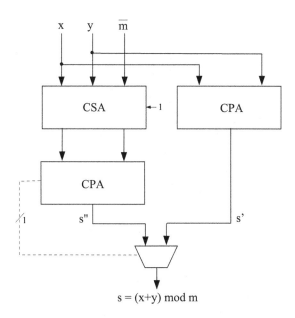

$$s = (x+y) \bmod m$$

are shown, in practice the underlying design of a CPA may permit some sharing of logic, so the total cost need not be twice that of a single adder.

The essential idea in the aforementioned sharing will be found in the design of a carry-select adder, for example. Such an adder basically computes preliminary two sums, $x + y$ and $x + y + 1$, one of which is then selected as the final result; but the detailed design is such that there is little replication of logic in the computation of the two intermediate results. One can envisage an extension of this arrangement into a more general one for the computation of $x + y$ and $x + y + \widetilde{m}$.

5.1.2 Special-Moduli

Moduli of the form $2^n \pm 1$ find much use in modular arithmetic, and those that are prime are especially useful in cryptography. As we show below, modulo-$(2^n - 1)$ addition is almost exactly ones'-complement addition and is therefore much easier to implement than general modular addition. Modulo-$(2^n + 1)$ addition is somewhat more complex than modulo-$(2^n - 1)$ addition. We start with the latter.

Suppose the operands in addition are x and y, with $0 \leq x, y < 2^n - 1$ and therefore representable in n bits each. We may distinguish between three cases in the addition:

(i) $0 \leq x + y < 2^n - 1$
(ii) $x + y = 2^n - 1$
(iii) $x + y > 2^n - 1$, i.e., $x + y = 2^n + w, \ w \geq 0$

In (i), the value $x + y$ is the correct modulo-$(2^n - 1)$ result. On the other hand, in each of (ii) and (ii), a nominal subtraction of $2^n - 1$ is required to obtain the correct result, which is zero in (ii) and $w + 1$ in (iii). It is easy to determine when (iii) is the case: the 2^n is reflected as the carry-out from the addition. But there is no carry-out in either (i) or (ii). To distinguish between the two cases, the value $2^n - 1$ must be detected explicitly. How that is done is explained below.

Correction in case (iii) is straightforward. Subtracting $2^n - 1$ is equivalent to subtracting 2^n and adding 1, which can be done by discarding the carry out (i.e., subtracting 2^n) and adding 1. The carry-out also indicates the need to add the 1, so the desired effect can be achieved by adding the carry-out to the intermediate sum. Such "end-around-carry" addition is exactly what happens in ones'-complement addition (Sect. 1.2.1).

Case (ii) highlights the difference between ones'-complement addition and addition modulo $2^n - 1$. In ones'-complement notation, there are two representations for zero: $00 \cdots 0$ and $11 \cdots 1$. On the other hand, in modular arithmetic there is only one binary representation for zero, i.e., $00 \cdots 0$, and $11 \cdots 1$ represents $2^n - 1$. Thus, although no correction is required for an intermediate result of $11 \cdots 1$ in ones'-complement arithmetic, one is necessary in modular arithmetic. That correction is easily done by adding a 1 and ignoring the carry-out.

Let s' denote the intermediate sum $x + y$ in (i)–(iii), and x_i, y_i, and s'_i denote bits i of x, y, and s'. Then case (ii) holds when $\prod_{i=0}^{n-1} s_i' = 1$, which is possible only if

$$P = \prod_{i=0}^{n-1} x_i \oplus y_i = 1$$

Note that P is just the block-propagate signal P_0^{n-1} in carry-lookahead and parallel-prefix adders (Eqs. 1.14 and 1.17); so its production in such adders will not require any extra logic or time.

On the basis of the preceding remarks, the computation of $s = (x + y)$ mod $(2^n - 1)$ has three main parts:

(i) Add x and y, to obtain the intermediate sum s' and a carry-out c_{n-1}.
(ii) Form the signal $z = P + c_{n-1}$.
(iii) If $z = 0$, then the result s is s'; otherwise s is $s' + 1$.

Some examples are given in Table 5.2.

The hardware implementation of a modulo-$(2^n - 1)$ adder consists of a straightforward modification to a ones'-complement adder (Figs. 1.12 and 1.13 in Sect. 1.2.1): the end-around-carry signal in such an adder is replaced with the z of (ii) above. A second addition will be required in most types of adders, but not with a conditional-sum adder or a parallel-prefix adder. At the penultimate stage of a conditional sum-adder, two intermediate sums that differ by one are available, with a selection made in the last stage. That selection may be made according to z, but the P signal must still be generated before that. In a parallel-prefix adder, the addition of a 1, conditional on z, is done by simply having an extra level of prefix

Table 5.2 Addition modulo $(2^n - 1)$

$n = 4$, $m = 15$					
$4 =$	0100		$11 =$		1011
$+7 =$	0111		$+9 =$		1001
$11 =$	1011		$1 \leftarrow$		0100
			$c_{n-1} =$		1
			$5 =$		0101
$9 =$	1001				
$+6 =$	0110				
	1111				
$P =$	1				
$0 =$	0000	discard carry			

operator, and the P signal will be the last block-propagate signal from the prefix tree.

Addition modulo $2^n + 1$ is not as easy as addition modulo $2^n - 1$. In both cases, when the intermediate result $s' = x + y$ is equal to or exceeds the modulus, it is necessary to subtract 2^n and to also subtract 1 in the former case and subtract -1 (i.e., add 1) in the latter case. In implementation, subtracting 1 is more difficult than adding 1 (as we explain below).

With the modulus $2^n + 1$, $0 \leq x, y \leq 2^n$; so the operands and modulus will be represented in $n + 1$ bits, and the arithmetic will be of $n + 2$ bits (with one bit for sign). Let the binary representation of s' be $s'_{n+1} s'_n s'_{n-1} \cdots s_0$, with a carry-out c_{n+1} from the addition that produces s'. Then a correction is required if either $s'_{n+1} = 1$ and at least one of the other bits is a 1 (i.e., if the logical OR of the other bits is 1) or if $c_{n+1} = 1$; in both cases $s' \geq 2^n + 1$, with $x = y = 2^n$ in the second case. To obtain the correct result, it is necessary to subtract $2^n + 1$. That can be done by subtracting 1 and then setting bit n of the subtraction-result to 0, the latter being equivalent to subtracting 2^n. The order of the two corrective actions is important, because in the second case $s_n = 0$. Note that a carry produced in the subtractive addition is discarded, according to the rules given in Sect. 1.1.6 (for operands of unlike sign).

Adding 1 (as in the modulo $2^n - 1$ case) is easy, since it can be done by injecting a carry into the least significant position of an adder. On the other hand, subtracting 1 (as in the modulo $2^n + 1$ case) requires the addition of $11 \cdots 1$ (assuming two's-complement representation) and is more difficult.

Examples are given in Table 5.3. Subtraction is as the addition of a two's complement of the subtrahend.

Table 5.3 Addition modulo $(2^n + 1)$

$n = 4, \; m = 17$

5	=	000101		12	=	001100
+4	=	000100		+13	=	001101
9	=	001001		25	=	011001

		011001	
−1	=	111111	
		011000	discard carry
8	=	01000	set bit n to 0

16	=	010000
+16	=	010000
32	=	100000

		100000	
−1	=	111111	
		011111	
15	=	01111	set bit n to 0

A special representation has been proposed to help solve the aforementioned subtraction problem. In *diminished-one* representation, the binary representation of a number z is taken to be what would ordinarily be the representation of the number $z - 1$ [1–4]. The requirement for a subtraction of 1 is thus converted to that of an addition of 1. Zero requires special handling in such a system: it is represented by the binary pattern that would normally be used to represent 2^n— i.e., the pattern $100 \cdots 00$—and in the implementation of the arithmetic it is treated as an exceptional case.

Consider non-modular diminished-ones addition. If the operands are $x' = x - 1$ and $y' = y - 1$, adding them yields $s' = x + y - 2$. To get the correct diminished value for $x - y$ a 1 should be added to s'.

Now consider addition modulo $2^n + 1$ with diminished-one representation, excluding the special case of a zero operand, which would be handled in an obvious way. Let the binary representation of s' above be $s'_n s'_{n-1} \cdots s'_0$. If $x + y < 2^n + 1$, then $s' < 2^n - 1$; so $s_n = 0$, and a 1 should be added to get the correct result. On the other hand, if $x + y \geq 2^n + 1$, then $s_n = 1$; to get the correct modular result we should add 1, as before, and also subtract $2^n + 1$. Subtracting 2^n is accomplished by setting $s'_n = 0$. No additional action is required, as adding 1 and then subtracting 1 is equivalent to not doing anything.

Hardware for a diminished-ones modulo-$(2^n + 1)$ addition will be faster and less costly than for a generic modular adder (Figs. 5.1, 5.2, and 5.3). But, on the whole, diminished-one representations and arithmetic are of dubious practical worth. That is because conversion to and from the representation require full carry-propagate additions and subtractions, which are worthwhile only if there are numerous computations to be carried out and the intermediate operands are retained in the same form so that the "one-off" cost of conversions is therefore amortized. Such a situation does not often arise; so, despite a fair amount of study, diminished-ones representation finds little real use. We will therefore not consider the representation any further in the context of addition and refer the interested reader to the published literature [2, 3].

Given the preceding remarks, we may conclude that there is probably little value in devising addition units specifically for the modulus $2^n + 1$. Modulo-$(2^n + 1)$ multiplication might, in certain circumstances, be considered an exceptional case, which is considered below.

5.1.3 Subtraction

Given x and y such that $0 \le x, y < m$, the definition of residue subtraction suggests that the computation of $(x - y) \bmod m$ be carried out by first computing the additive inverse of y (i.e., $m - y$) and then adding x (modulo m). This would involve three additions: one to compute $m - y$, one to add x, and one (a subtraction) to correct the initial result that exceeds the modulus. A much better method is as to compute it as

$$(x - y) \bmod m = \begin{cases} x - y & \text{if } x - y \ge 0 \\ x - y + m & \text{otherwise} \end{cases} \qquad (5.2)$$

This equation is similar to Eq. 5.1, but with the arithmetic operations changed to the converse. The condition that separates the two cases too is changed: the test for $x - y \ge 0$ is easier than that for $x + y > m$, as the latter requires a nominal subtraction, and this difference can have practical implications in the implementation.

If x, y, and m are each represented in n bits, then the arithmetic will be in $n + 1$ bits, with one bit for sign. The subtraction is carried out by adding to x the negation (i.e., two's complement) of the subtrahend. If $\tilde{y}$ is that negation, then the subtraction is equivalent to the addition of the unsigned number $2^{n+1} - y$. That is, the intermediate result s' of operating on the operands is

$$s' = x - y \qquad \text{(signed)}$$
$$= x + \tilde{y}$$
$$= 2^{n+1} + (x - y) \qquad \text{(unsigned)}$$

Table 5.4 Modular subtraction

$n = 4$, $m = 12$						
9 =	01001		4	=	00100	
−4 =	11100		−9	=	10111	
5 =	00101	discard carry	−5	=	11011	
					11011	
			+m	=	01100	
			7	=	00111	discard carry

If $x \geq y$, then $s' \geq 2^{n+1}$; the 2^{n+1} in the expression for s' indicates a carry out from the addition, and discarding that carry leaves the correct result of $x - y$. On the other hand, the absence of a carry out indicates that $x < y$, and so m should be added to s'. This addition will produce a carry, since $x - y + m \geq 0$, and discarding that carry leaves the correct result of $x - y + m$. The algorithm for $s = (x - y) \bmod m$:

(i) x and $-y$ are added to yield an intermediate sum s' with carry-out c_n.
(ii) If $c_n = 1$, then $s = s'$ (after discarding the carry-out).
(iii) Otherwise, add m to s' to obtain s, with the carry-out discarded.

Examples are shown in Table 5.4.

Modulo-m subtractors will be similar to modulo-m adders, such as those of Figs. 5.1, 5.2, and 5.3. Thus, for example, simple changes to the inputs and outputs in Fig. 5.1 give the design of Fig. 5.4. The subtraction takes place in two cycles. The first cycle consists of the subtraction $s' = x - y$, as the addition of the two's complement of the subtrahend, which in turn is the addition of the ones' complement and a 1 that is included as a carry into the adder. The second cycle consists of the addition of $s'' = s' + m$. One of s' and s'' is then chosen as the result. The modifications required of Figs. 5.2 and 5.3 are similarly straightforward, as is the design of a combined modulo-m adder-subtractor.

Special Moduli

Subtraction with the moduli $2^n \pm 1$ follows the general form above for an arbitrary modulus but with some simplifications. For the modulus $2^n - 1$, the negation of the subtrahend is the ones'-complement representation, and for $2^n + 1$ it is the two's complement. The required precisions are $n + 1$ bits in the former case and $n + 2$ in the latter. The following is a discussion of modulo-$(2^n - 1)$ subtraction (Table 5.5).

Fig. 5.4 Modulo-*m*
subtractor, single basic adder

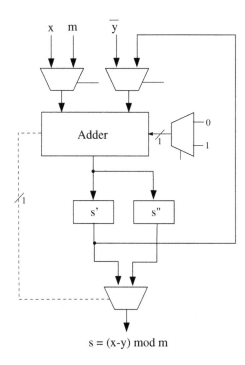

$$s = (x\text{-}y) \bmod m$$

Table 5.5 Modulo-$(2^n - 1)$ subtraction

$n = 4, \ m = 15$						
9	=	01001		4	=	00100
−4	=	11011		−4	=	11011
	1 ⟵	00100				11111
c_n	=	1		P	=	1
5 =		00101		0	=	00000

4	=	00100
−9	=	10110
−5	=	11010
10	=	01010

set $s'_n = 0$

When interpreted as the $(n + 1)$-bit representation of an unsigned number, the
ones'-complement representation of $-y$ is that of $\widetilde{y} = 2^{n+1} - y - 1$ (Sect. 1.1.6). So

$$x - y = x + \widetilde{y}$$
$$= 2^{n+1} + [x - (y + 1)]$$

Three cases may be distinguished of the value of $s' = x + \tilde{y}$:

First, if $x \geq y + 1$, then the 2^{n+1} represents a carry out, c_n, from the addition. Discarding that carry leaves $x - y - 1$, and adding a 1 to that yields the correct result; the latter is just the addition of an end-around carry.

Second, if $x = y$, then $s' = 2^{n+1} - 1$, and it is necessary to subtract $2^{n+1} - 1$; i.e., subtract 2^{n+1} and add 1. Adding 1 to s' produces a carry, and discarding that carry effects the subtraction of 2^{n+1}. This case occurs if $s'_n s'_{n-1} \cdots s'_0 = 11 \cdots$, which corresponds to the carry-propagation signal (Sect. 1.1.1)

$$P = P_0^n = \prod_{i=0}^{n} x_i \oplus y_i$$

Third, if $x < y$, then there is no carry out. In this case $2^n - 1$ should be added to s'; i.e., 2^n should be added and 1 subtracted. Since s' is negative, $s'_n = 1$. So, adding 2^n would change s' to 0 and generate a carry that should be added as an end-around carry. Adding that carry and subtracting 1 is equivalent to doing nothing. Therefore, in this case the correct result may be obtained by simply setting $s'_n = 0$.

An algorithm for subtraction modulo $2^n + 1$ can be devised in a manner similar to that done above for addition modulo $2^n + 1$. Nevertheless, given the comments above on addition modulo $2^n + 1$, we will not bother with subtraction and leave it to the interested reader to look into the details.

5.2 Multiplication

The most straightforward method for the computation of $xy \bmod m$ is to compute the ordinary product xy and then carry out a modular reduction. That is not always the best for cryptography applications, as the intermediate values can be very large, but it may nevertheless be considered for an implementation with "pre-built" units for the two basic functions. As both ordinary multiplication and reduction have been discussed in Chaps. 1 and 4, there is little to add to the matter. We shall therefore focus on those methods that include reduction as an intrinsic part of the computation of the modular product.

Ordinary multiplication consists of the addition of multiples of the multiplicand, with the multiples determined by the digits of the multiplier. In sequential multiplication, the multiples are added one at a time to running partial product that starts at zero and ends up as the sought product. In modular multiplication, reductions may be carried out on the partial products, on the basis that

$$(a + b) \bmod m = (a + b \bmod m) \bmod m \qquad (5.3)$$

Partial products are reduced as they are formed; so a corresponds to a partial product and b corresponds to a multiplicand multiple.

Let x and y be the two operands, each less than the modulus m and representable in n radix-r digits and with a product z: $x = \sum_{i=0}^{n-1} x_i r^i$ and $y = \sum_{i=0}^{n-1} y_i r^i$, and $z = \sum_{i=0}^{n-1} z_i r^i$, where x_i, y_i and z_i are digits i of x, y, and z. And let Z_i denote the partial product in iteration i. Two algorithms for the sequential computation of $z = xy$ are (Section 1.2.1)

$$Z_0 = 0 \qquad (5.4)$$

$$Z_{i+1} = Z_i + r^i y_i x \qquad i = 0, 1, 2, \ldots, n - 1 \qquad (5.5)$$

$$z = Z_n \qquad (5.6)$$

for a right-to-left scan of the multiplier digits and

$$Z_0 = 0 \qquad (5.7)$$

$$Z_{i+1} = r Z_i + y_{n-i-1} x \qquad i = 0, 1, 2, \ldots, n - 1 \qquad (5.8)$$

$$z = Z_n \qquad (5.9)$$

for left-to-right scan.

Ordinary sequential multiplication uses the first algorithm because the r^i factor in the multiplicand multiple—a factor that in implementation implies a left shift of i digits positions—is removed by shifting the partial product one place to the right in each iteration instead of shifting the multiple i places to left in each iteration. Each digit that is shifted out is a digit of the final product, so the additions are of n digits each. On the other hand, the additions with the second algorithm must be of $2n$ digits, $2n$ being the precision of the largest multiplicand multiple.

If the two multiplication algorithms above are modified *directly* for modular arithmetic, then the situation of the preceding paragraph is reversed: the second algorithm is the better one. The implementation "optimization" mentioned above of the first algorithm is not directly possible in modular arithmetic, as all digits of a partial product must be included in a reduction; so, in general, the partial products will be of $2n$ digits, in contrast with $n + 1$ digits in the second algorithm. The former is therefore no better than an algorithm that consists of first computing xy and then reducing that. (The two examples in Tables 5.6 and 5.7 show this.) Nevertheless, we shall see that the first algorithm can still be applied in an efficient way in modular multiplication.

Table 5.6 Modular multiplication, right-to-left multiplier scan

$x = 456$, $y = 123$, $m = 511$, $r = 10$, $n = 3$				
i	$y_i x$	$y_i x r^i$	U_i	Z_i
0	1368	1368	368	0
1	912	9120	9466	346
2	456	45,600	45,868	268
3	–	–	–	389
$xy \bmod m = 389$				

Table 5.7 Modular multiplication, left-to-right multiplier scan

$x = 456$, $y = 123$, $m = 511$, $r = 10$, $n = 3$				
i	$y_{n-i-1} x$	$r Z_i$	U_i	Z_i
0	456	0	456	0
1	912	4560	5472	456
2	1368	3620	4988	362
3	–	–	–	389
$xy \bmod m = 389$				

5.2.1   Multiplication with Direct Reduction

A straightforward algorithm for the computation of $z = xy \bmod m$ based on Eqs. 5.4–5.6 is

$$Z_0 = 0 \tag{5.10}$$

$$U_i = Z_i + r^i y_i x \qquad i = 0, 1, 2, \ldots, n - 1 \tag{5.11}$$

$$Z_{i+1} = U_i \bmod m \tag{5.12}$$

$$z = Z_n \tag{5.13}$$

of which an example is given in Table 5.6.

And an algorithm based on Eqs. 5.7–5.9 is

$$Z_0 = 0 \tag{5.14}$$

$$U_i = r Z_i + y_{n-i-1} x \qquad i = 0, 1, 2, \ldots, n - 1 \tag{5.15}$$

$$Z_{i+1} = U_i \bmod m \tag{5.16}$$

$$z = Z_n \tag{5.17}$$

with an example given in Table 5.7.

A comparison of the magnitudes of the U_i values in Table 5.6 and the U_i values in Table 5.7 confirms the statements made above, to the effect that in the straightforward versions the second algorithm is the better of the two. The following discussion is therefore limited to that algorithm.

The computation of U_i in Eq. 5.15 is largely straightforward. It consists of the computation of a small multiple of x, a left shift to effect the multiplication of Z_i by r, and an addition. The complexity of the computation of $y_{n-i-1}x$ depends on the value of r, but in practice it is unlikely to be high. If $r = 2$, then the multiple is just 0 or x; and, for hardware implementation, other practical values of r will typically be restricted to 4 or 8, for easy multiplier recoding. On the other hand, the computation of $U_i \bmod m$ is fundamentally difficult, as it essentially implies (from the definition of a residue) a nominal division.

Now,

$$U_i = rZ_i + y_{n-i-1}x$$
$$\leq r(m-1) + (r-1)(m-1)$$
$$\leq (2r-1)(m-1)$$

So, a subtraction of a multiple of m, up to $(2r-2)m$, may be required to reduce U_i to Z_{i+1} such that $0 \leq Z_{i+1} \leq m-1$. The difficulty is in determining the correct multiple to be subtracted—the classic problem in integer division. If r is sufficiently small, then the number of possible multiples will be small enough that all multiples may be checked efficiently against Z_{i+1}—one at a time or concurrently—and the correct one chosen. With $r = 2$, Eq. 5.16 is now

$$Z_{i+1} = \begin{cases} U_i & \text{if } U_i < m \\ U_i - m & m \leq U_i < 2m \\ U_i - 2m & \text{otherwise} \end{cases}$$

which can be implemented with one adder used twice, as in Fig. 4.1. Nevertheless, for hardware implementation, the fundamental problem remains: exact results are required of the comparisons, which implies the use of carry-propagate adders. Therefore, the basic algorithm and obvious architecture are best suited to an implementation with "pre-built" units (especially adders), and such an implementation will be much slower than would be the case if carry-save adders could be used.

As in ordinary multiplication and division, the use of a carry-propagate adder (CPA) in the main loop will give an implementation that is quite slow. A faster implementation requires the use of a carry-save adder (CSA), but that raises the basic difficulty in division: in order to know when a subtraction is exactly necessary, a partial-carry and partial-sum must be assimilated, which requires the use of a CPA. As in division, the solution here is to use approximation. We next describe such a method [16].

Let C_i denote the partial carry in iteration i and S_i denote the partial sum in iteration i. And assume the radix is two. Then the starting algorithm, obtained directly from that of Eqs. 5.14–5.17, is

$$(C_0, S_0) = (0, 0)$$

$$(C_{i+1},\ S_{i+1}) = 2C_i + 2S_i + xy_{n-i-1} \qquad i = 0, 1, 2, \ldots, n-1$$

$$(\widetilde{C}_{i+1},\ \widetilde{S}_{i+1}) = C_i + S_i - m$$

$$(C_{i+1},\ S_{i+1}) = (\widetilde{C}_{i+1}\ \widetilde{S}_{i+1}) \text{ if SIGN}(\widetilde{C}_{i+1},\ \widetilde{S}_{i+1}) \geq 0$$

$$z = C_n + S_n$$

where the computations in the right-hand sides of the middle two equations are carry-save additions, SIGN is the exact sign of the result of assimilating the partial carry and partial sum, and the last addition is an assimilating carry-propagate addition. The algorithm requires exact determination of the sign. A better algorithm is obtained as follows.

Define the function T on as k-bit integer u as

$$T(u) = u - u \bmod 2^t \qquad 0 \leq t \leq k-1$$

T replaces the least significant t bits of u with 0s, so

$$T(u) \leq u < T(u) + 2^t$$

Suppose the reduction of (C_i, S_i) is then by carrying out q times the two steps:

$$(\widetilde{C}_{i+1}, \widetilde{S}_{i+1}) = C_i + S_i - m \tag{5.18}$$

$$(C_{i+1}, S_{i+1}) = (\widetilde{C}_{i+1}, \widetilde{S}_{i+1}) \text{ if } T(\widetilde{C}_{i+1}) + T(\widetilde{S}_{i+1}) \geq 0 \tag{5.19}$$

In the second step, the assimilation addition (or equivalent), which gives a sign approximation, involves only the most significant $n - t$ bits of the operands.

Let (C_J, S_J) denote the values at the start of the q instances of Eqs. 4.18–4.19 and (C_K, S_K) denote the ending values. Then:

$$0 \leq C_J + S_J < (q+1)m + 2^t$$

$$0 \leq C_K + S_K < m + 2^t$$

where the additions of C and S are assimilation ones. If the estimated value in Eq. 4.19 is positive in all q instances, then qm is subtracted from the starting values, so

$$C_K + S_K = C_J + S_J - qm < m + 2^t$$

And if the value is negative in any step, then it stays negative to the end, and the condition

$$T(\widetilde{C}_{i+1}) + T(\widetilde{S}_{i+1}) < 0$$

in the last instance of Eq. 4.19 implies

$$T(\widetilde{C}_{i+1}) + T(\widetilde{S}_{i+1}) \leq -2^t$$

since $T(\cdots)$ is always a multiple of 2^t. Therefore:

$$T(\widetilde{C}_{i+1}) + T(\widetilde{S}_{i+1}) < \widetilde{C}_{i+1} + \widetilde{S}_{i+1} < T(\widetilde{C}) + T(\widetilde{S}_{i+1}) + 2^{t+1}$$

and

$$\widetilde{C}_{i+1} + \widetilde{S}_{i+1} < 2^{t+1} - 2^t = 2^t$$

In Eq. 4.18 $C + S$ is reduced by m, and in the last instance there is no reduction, because the estimated value is negative. So

$$C_K + S_K = \widetilde{C}_K + \widetilde{S}_K + m$$

which implies

$$C_K + S_K < m + 2^t$$

In the initial algorithm m is subtracted in each iteration, which would correspond to nq subtractions with the replacement of Eqs. 4.18–4.19. This can be improved upon by instead subtracting $2^{k-j}m$ in iteration j, where $q + 1 \leq 2k$ and $k = 1, 2, \ldots, k$. For example, if $q = 3$ and $k = 2$, then what would have been three subtractions of m are replaced with two subtractions—one of $2m$ and one of m.

Putting together all of the above, the final algorithm is

$$(C_0, S_0) = (0, 0)$$

$$(C_i, S_i) = 2C_{i-1} + 2S_{i-1} + xy_{n-i} \qquad i = 1, 2, \ldots, n$$

$$(\widetilde{C}_i, \widetilde{S}_i) = C_i + S_i - 2m$$

$$(C_i, S_i) = (\widetilde{C}_i, \widetilde{S}_i) \text{ if } T(\widetilde{C}_{i+1}) + T(\widetilde{S}_{i+1}) \geq 0$$

$$(\widetilde{C}_i, \widetilde{S}_i) = C_i + S_i - m$$

$$(C_i, S_i) = (\widetilde{C}_i, \widetilde{S}_i) \text{ if } T(\widetilde{C}_i) + T(\widetilde{S}_i) \geq 0$$

The final assimilation step is omitted, as it may require an additional correction, which is discussed below.

The value of t determines the precision and accuracy of the estimation and, therefore, of the complexity of the control logic and the required correction in the assimilation of C_n and S_n. After the second-last step of the algorithm:

$$C_i + S_i < m + 2^t$$

So, after the next shift-add

$$0 \le C_{i+1} + S_{i+1} < 3m + 2^{t+1}$$

With $q = 3$:

$$3m + 2^{t+1} \le (q+1)m + 2^t = 4m + 2^t$$

which implies $2^t \le m$ or $t \le n - 1$.
    Therefore:

$$0 \le C_{i+1} + S_{i+1} < 3m + 2^{t+1} \le 3m + 2^n \le 2^{n+2}$$

And after the step in which $2m$ is subtracted:

$$-2^{n+1} \le -2m \le C_{i+1} + S_{i+1} < N + 2^n < 2^{n+1}$$

If $t = n - 1$, then the sign estimation is computed from the five most significant bits of the partial carry and partial sum, and the result will be in the range $[0, 2m)$. So, the final assimilation step is

$$z = \begin{cases} C_n + S_n & \text{if } C_n + S_n < m \\ C_n + S_n - m & \text{otherwise} \end{cases}$$

This is just a case of Eq. 1.1 and may be implemented as described.
    An example application of the algorithm is given in Table 5.8.
    Figure 5.5 shows a straightforward architecture for the algorithm. The assimilation-and-correction part is omitted; an arrangement similar to that Fig. 4.5 (bottom half) will do for that. The y register is a left-shift register; and the PC-PS registers are initialized to zeros. Multiplication by two is as a wired left shift, and $-m$ and $-2m$ are each added as the ones' complement and a 1. A cheaper but slower implementation would avoid the duplication of logic and have just one carry-save adder, one logic estimation unit, and one carry-propagate adder—with all used twice.
    Modifying direct-reduction algorithms, such as those above, for efficient high-radix computation is difficult; the modular reduction will require something close to actual division. We next give a brief description of such an algorithm.
    If q_i is the quotient from the division of U_i by m, then we may express the algorithm of Eq. 5.14–5.17 as

$$Z_0 = 0$$
$$U_i = rZ_i + y_{n-i-1}x \qquad i = 0, 1, 2, \dots, n - 1$$
$$q_i = U_i \div m$$

Table 5.8 Fast direct-reduction multiplication

$x = 48 = 110000_2$, $y = 47 = 101111_2$, $m = 50 = 110010_2$, $n = 6$.

i	C_i	S_i	$\tilde{C}_i$	$\tilde{S}_i$	$T(\tilde{C}_i) + T(\tilde{S}_i)$
0	000000000	000000000	–	–	–
1	000000000	000110000	–	–	–
	000000000	000110000	001000000	110101100	111000000
	000000000	000110000	000000000	111111110	111100000
2	000000000	001100000	–	–	–
	000000000	001100000	000000000	111111100	111100000
	010000000	110101110	010000000	110101110	000100000
3	000100000	001101100	–	–	–
	001011000	111010000	001011000	111010000	000000000
	001011000	111010000	110110000	111100000	
4	101100000	100100000	–	–	–
	001000000	111011100	001000000	111011100	000000000
	001000000	111011100	110011000	001010010	111000000
5	101100000	100001000	–	–	–
	101100000	100001000	000010000	111110100	111100000
	010010000	110100110	010010000	110100110	000100000
6	00100000	001011100	–	–	–
	010111000	110000000	010111000	110000000	001000000
	010111000	110000000	100010000	011110110	111100000

$(C_6, S_6) = (010111000, 110000000) = (184, -128)$, $C_6 + S_6 - 50 = 6 = 48 * 47 \bmod 50$

$$Z_{i+1} = U_i - q_i m$$

$$z = Z_n$$

Implementing this algorithm requires division—a costly operation—in each iteration, as well as the two multiplications (by x and m) that are not simple operations if $r > 2$ and the range of q_i is unrestricted. In [17] these difficulties are dealt with as follows.

Instead of the exact q_i, an approximation $\tilde{q}_i$ is used. This approximation is obtained by approximately dividing U_i by $2^k m$, where $r = 2^k$; the subtraction $U_i - q_i m$ is then changed to $U_i - \tilde{q}_i 2^{2k} m$. Restrictions on the possible values of k and $\tilde{q}_i$ ensure that each Z_i is less than $2m$ and, therefore, that a single subtraction of m is sufficient to correct Z_n.

The method used to compute $\tilde{q}_i$ is similar to that in SRT division (Sect. 1.3)—it involves, essentially, the comparison of a few leading bits of the partial dividend with a few leading bits of the divisor—but is more complex, as it requires the parallel computation of all the possible values of $U_i - q_i 2^k m$. The arrangement and various restrictions also ensure that it is relatively easy to compute $\tilde{q}_i m$.

Fig. 5.5 Fast direct-reduction modular multiplier

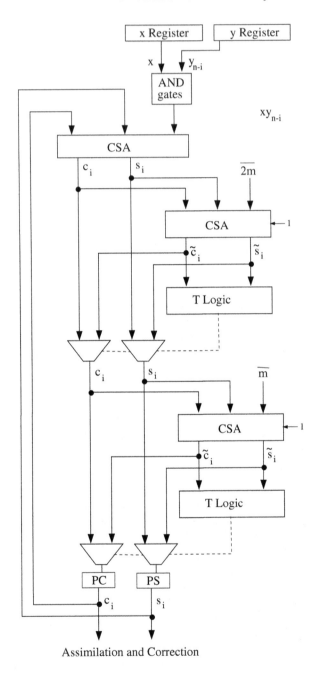

Assimilation and Correction

To facilitate the efficient computation of $y_{n-i-1}x$, y_{n-i-1} is recoded into the radix-4 digit set $\{\bar{1}, 0, 1, 2\}$. The multiplication is then carried out as the addition of easily computed, weighted powers of four.

The corresponding architecture employs carry-save adders in the primary arithmetic and so an implementation ought to be fast. Nevertheless, both the algorithm and the architecture are sufficiently complex that their real merits over, say, a fast radix-2 implementation are not easily ascertained.

5.2.2 Multiplication with Barrett Reduction

Barrett reduction (Sect. 4.1) consists of the computation of an approximation to the quotient from an integer division, the computation of an approximate remainder from that approximate quotient, and then, if necessary, a correction of that remainder. The radix-r computations for $a = u \bmod m$, with $u < m^2$, are

$$\widetilde{m} = \lfloor r^{2n}/m \rfloor \tag{5.20}$$

$$\widetilde{q} = \left\lfloor \frac{\lfloor u/r^{n-1} \rfloor * \widetilde{m}}{r^{n+1}} \right\rfloor \tag{5.21}$$

$$\widetilde{a} = u - \widetilde{q}m \tag{5.22}$$

$$a = \begin{cases} \widetilde{a} & \text{if } \widetilde{a} < m \\ \widetilde{a} - m & \text{otherwise} \end{cases} \tag{5.23}$$

To compute $xy \bmod m$ with this algorithm it suffices to add the computation $u = xy$. The algorithm can then be combined easily with the multiplication algorithm of Eqs. 5.14–5.17 to produce an algorithm that sequentially computes the $xy \bmod m$. As with the algorithm of Sect. 5.2.1, the essence of such an algorithm is to reduce the partial products as they are generated.

If $0 \le x, y < m$, $y = \sum_{i=0}^{n-1} y_i r^i$, and $2^{n-1} \le m < 2^n$, then the combined multiplication-reduction algorithm for the computation of $z = xy \bmod m$ is

$$Z_0 = 0 \tag{5.24}$$

$$\widetilde{m} = \lfloor r^{2n}/m \rfloor \tag{5.25}$$

$$U_i = rZ_i + y_{n-i-1}x \qquad i = 0, 1, 2, \ldots, n-1 \tag{5.26}$$

$$\widetilde{q}_i = \left\lfloor \frac{\lfloor U_i/r^{n-1} \rfloor * \widetilde{m}}{r^{n+1}} \right\rfloor \tag{5.27}$$

$$Z_{i+1} = U_i - \widetilde{q}_i m \tag{5.28}$$

$$z = \begin{cases} Z_n & \text{if } Z_n < m \\ Z_n - m & \text{if } m \le Z_n < 2m \\ Z_n - 2m & \text{otherwise} \end{cases} \tag{5.29}$$

Table 5.9 Modular multiplication with Barrett reduction

$x = 456$, $y = 123$, $m = 511$, $r = 10$, $n = 3$, $\widetilde{m} = 1956$.

i	$y_{n-i-1}x$	rZ_i	U_i	$\widetilde{q}_i$	Z_i
0	456	0	456	0	0
1	912	4560	5472	10	456
2	1368	3620	4988	9	362
3	–	–	–	–	389

$xy \bmod m = 389$

$x = 2345$, $y = 1234$, $m = 3457$, $r = 10$, $n = 4$, $\widetilde{m} = 28{,}926$.

i	$y_{n-i-1}x$	rZ_i	U_i	$\widetilde{q}_i$	Z_i
0	2345	0	2345	0	0
1	4690	23,450	28,140	8	2345
2	7035	4840	11,875	3	484
3	9380	15,040	24,420		1504
4	–	–	–	–	3678

$xy \bmod m = 3678 - 3457 = 221$

An example computation is given in Table 5.9, which corresponds to Tables 5.6 and 5.7. It is straightforward to devise a corresponding architecture, with two full multipliers, two carry-propagate adders, a multiplexer, etc.

A variant of the preceding algorithm, based on the "extension" described at the end of Sect. 4.1, and corresponding architectures for implementation will be found in [6]. Both the basic algorithm and the variant have an obvious shortcoming: the repeated multiplications, to compute $\widetilde{q}_i$ and Z_{i+1}, are not conducive to high-performance implementation, even with $r = 2$. The use of carry-save adders and recoding will help, but only partially. Nevertheless, they do have some merit—if cost, not performance, is the primary consideration. That is because the alternatives would require larger adders and multipliers. To see this, consider the first example in Table 5.9, and suppose the alternative consists of a straightforward multiplication followed by a reduction (e.g., Barrett reduction). The product of 456 and 123 is 56,088, which is far larger than any value in the table.

Residue number systems (Sect. 1.1.4) appear attractive for the high precisions in cryptography, since what would otherwise be high-precision operations are replaced with numerous low-precision ones, and there has been some work on adapting the Barrett algorithm to such representations. A fundamental problem in such an approach is the efficient computation of the approximate quotient of Eq. 5.27; a partially satisfactory solution for that is given in [15]. But the problem of conversions remains.

5.2.3 Multiplication with Montgomery Reduction

With $u < mR$ and $\gcd(m, R) = 1$, Montgomery reduction (Sect. 4.2) computes $z = uR^{-1} \bmod m$, where R^{-1} is the multiplicative inverse of R with respect to m, through these steps:

$$\widetilde{m} = -\left(m^{-1} \bmod R\right) \bmod R \tag{5.30}$$

$$\widetilde{q} = u\widetilde{m} \bmod R \tag{5.31}$$

$$= (u \bmod R)\widetilde{m} \bmod R \tag{5.32}$$

$$\widetilde{z} = \frac{u + \widetilde{q}m}{R} \tag{5.33}$$

$$z = \begin{cases} \widetilde{z} & \text{if } \widetilde{z} < m \\ \widetilde{z} - m & \text{otherwise} \end{cases} \tag{5.34}$$

z is the result, $uR^{-1} \bmod m$.

With $R = r^n$, a serial-sequential version of the algorithm is

$$\widetilde{m} = -m^{-1} \tag{5.35}$$

$$U_0 = u \tag{5.36}$$

$$\widetilde{q}_i = U_i\widetilde{m} \bmod r \tag{5.37}$$

$$= [(U_i \bmod r)(\widetilde{m} \bmod r)] \bmod r \tag{5.38}$$

$$U_{i+1} = \frac{U_i + \widetilde{q}_i m}{r} \qquad i = 0, 1, 2, \dots, n-1 \tag{5.39}$$

$$z = \begin{cases} U_n & \text{if } Z_n < m \\ U_n - m & \text{otherwise} \end{cases} \tag{5.40}$$

The Montgomery reduction algorithm can be used as the basis of one that computes $xy \bmod m$ in two multiply-reduce steps, as follows.

Suppose we have a single algorithm that takes the operands, $a, b, m,$ and R and computes $abR^{-1} \bmod m$; that is, multiplication is intrinsically part of the reduction process. Then $xy \bmod m$ can be computed with two applications of the algorithm:

(i) Apply the algorithm with $a = x$ and $b = y$ and obtain $z = xyR^{-1} \bmod m$.
(ii) Apply the algorithm with $a = z$ and $b = \widetilde{R}$, where $\widetilde{R}$ is the "pre-computed;;" value $R^2 \bmod m$.

The result from (ii) is $(xy) \bmod m$:

$$z\tilde{R}R^{-1} \bmod m = \left(xyR^{-1}\right)\left(R^2 \bmod m\right)R^{-1} \bmod m$$

$$= xyR^{-1}R^2R^{-1} \bmod m$$

$$= xy \bmod m$$

The single algorithm is obtained by combining the steps of the basic multiplication algorithm (Eqs. 5.4–5.6), with operands a and $b = \sum_{i=0}^{n-1} b_i r^i$, and the basic Montgomery-reduction multiplication algorithm for the computation of $z = abR^{-1} \bmod m$, with $R = r^n$ is

$$Z_0 = 0 \tag{5.41}$$

$$U_i = Z_i + b_i a \qquad\qquad i = 0, 1, 2, \ldots, n-1 \tag{5.42}$$

$$\tilde{q}_i = [(U_i \bmod r)(\tilde{m} \bmod r)] \bmod r \tag{5.43}$$

$$Z_{i+1} = \frac{U_i + \tilde{q}_i m}{r} \tag{5.44}$$

$$z = \begin{cases} Z_n & \text{if } Z_n < m \\ Z_n - m & \text{otherwise} \end{cases} \tag{5.45}$$

A single subtraction at the end of the iterative process suffices because $0 \leq Z_i < 2m - 1$ for all i. This is evident for Z_0, and, by induction, if it is so for Z_k, then

$$Z_{k+1} = \frac{Z_i + b_i a + \tilde{q}_i m}{r}$$

$$< \frac{[(2m-1) + (r-1)(m-1) + (r-1)m]}{r}$$

$$= 2m - 1$$

Note that with r as the representation radix, if the implementation radix is r or a power or r, then reduction modulo r is a trivial operation: it consists of simply taking the least significant digit of the operand. Therefore, Eq. 5.43 may be replaced with

$$\tilde{q}_i = u_0 \tilde{m}_0 \bmod r \tag{5.46}$$

where u_0 is the least significant digit of U_i.

An example computation is shown in Table 5.10, which corresponds to Tables 5.6, 5.7, and 5.9.

Table 5.10 Multiplication with interleaved Montgomery reduction

(a) First multiplication-reduction

$a = x = 456, \ b = y = 123, \ m = 511, \ R = 10^3$

$R^{-1} = 209, \ m^{-1} = 591, \ \widetilde{m} = 409$

i	$b_i a$	U_i	u_0	$u_0 \widetilde{m}_0$	$\widetilde{q}_i$	$U_i + \widetilde{q} m$	Z_i
0	1368	1368	8	72	2	2390	0
1	912	1151	1	9	9	5750	239
2	456	1031	1	9	9	5630	575
	–	–	–	–	–	–	563

$z = xyR^{-1} \bmod m = 563 - 511 = 52$

(b) Second multiplication-reduction

$a = z = 52, \ b = R^2 \bmod m = 484$

i	$b_i a$	U_i	u_0	$u_0 \widetilde{m}_0$	$\widetilde{q}_i$	$U_i + \widetilde{q} m$	Z_i
0	208	208	8	72	2	1230	0
1	416	539	9	81	1	1050	123
2	208	313	3	27	7	3890	105
	–	–	–	–	–	–	389

$z = abR^{-1} \bmod m = xy \bmod m = 389$

In the multiplication shown in Table 5.10, the role of the second multiplication-reduction is essentially to remove the R^{-1} factor and thus obtain $xy \bmod m$ from $xyR^{-1} \bmod m$. This is clearly seen if the factor of $R^2 \bmod m$ is "split" into one $R \bmod m$ for each operand; that is, each operand is multiplied by that factor before proceeding with any multiplication-reduction: Suppose that instead of x and y, the operands in the algorithm are $xR \bmod m$ and $yR \bmod m$. Then one multiplication-reduction yields

$$z' = (xR \bmod m)(yR \bmod m)R^{-1} \bmod m \qquad (5.47)$$

$$= xRyRR^{-1} \bmod m$$

$$= xyR \bmod m$$

from which another, similar, multiplication-reduction with 1 as the other operand, yields

$$z = (z' * 1)R^{-1} \bmod m \qquad (5.48)$$

$$= (xyR \bmod m)R^{-1} \bmod m \qquad (5.49)$$

$$= (xyR)R^{-1} \bmod m \qquad (5.50)$$

$$= xy \bmod m \qquad (5.51)$$

If only a single modular multiplication is required, then the second multiplication-reduction can be done away with, by modifying only one of the operands in Eq. 5.47, say x to $xR \bmod m$:

$$z = (xR \bmod m)yR^{-1} \bmod m \qquad (5.52)$$

$$= xRyR^{-1} \bmod m$$

$$= xy \bmod m$$

But, as we explain below, it is the version with two modified operands that is of special interest.

The forms $xR \bmod m$ and $yR \bmod m$ are the *Montgomery residues* of x and y with respect to m and R. The set $\{kR \bmod m : 0 \le k \le m - 1\}$ is a complete set of residues with respect to m, so it is permissible to work with this set instead of the "original" residues. Computing a Montgomery residue is, however, not an easy operation. One method that is routinely suggested in the standard literature is a Montgomery reduction in which $xR \bmod m$ is obtained from the reduction of the product of x and $R^2 \bmod m$:

$$x \left(R^2 \bmod m \right) R^{-1} \bmod m = xR \bmod m$$

This computation may be carried out by first computing $xR^2 \bmod m$ and then applying the Montgomery-reduction algorithm of Sect. 4.2 or by combining the multiplication and reduction, as in the algorithm of Sect. 5.2.3. Regardless of which of the two methods is used, $R^2 \bmod m$ or $xR^2 \bmod m$ must be computed by means other than Montgomery multiplication, and neither will be a simple computation. A careful consideration of what is involved also shows that for hardware implementation the aforementioned methods are probably no better than simply computing the product xR and then directly reducing that modulo m. Nevertheless, as we next explain, the one-time cost of computing Montgomery residues can be well worthwhile, depending on the circumstances.

Equation 5.47 shows that the Montgomery multiplication of two Montgomery residues yields a Montgomery residue, and this is what is special about the algorithm. If there is a sequence of modular multiplications to be carried out—as in modular exponentiation—then all the intermediate values may held as Montgomery residues, with only one special, extra multiplication (Eqs. 5.48–5.51) for the conversion of the final result. The one-time costs of converting to and from Montgomery residues are then amortized. The key point, therefore, is that Montgomery multiplication is quite efficient if the cost of computing Montgomery residues is excluded. (We shall return to this in Chap. 6.)

In summary, if we wish to compute $xy \bmod m$, then we may convert x and y to the corresponding Montgomery residues and then use the algorithm of Eqs. 5.41–5.45 in two steps, the second of which converts from a Montgomery residue:

- In the first step $a = xR \bmod m$ and $b = yR \bmod m$; the result is $z = xyR \bmod m$.
- In the second step $a = z$ and $b = 1$.

Table 5.11 Montgomery multiplication

(a) First multiplication-reduction

$x = 456,\ y = 123,\ m = 511\ R = 10^3,\ R^{-1} = 209,\ m^{-1} = 591,\ \widetilde{m} = 409$

$a = xR \bmod m = 360\ b = yR \bmod m = 188$

i	$b_i a$	U_i	u_0	$u_0 \widetilde{m}_0$	$\widetilde{q}_i$	$U_i + \widetilde{q} m$	Z_i
0	2880	2880	0	0	0	2880	0
1	2880	3168	8	72	2	4190	288
2	360	779	9	81	1	1290	419
–	–	–	–	–	–	–	129

$z = abR^{-1} \bmod m = 129 = xyR \bmod m$

(b) Second multiplication-reduction

$a = z = (xyR) \bmod m = 129,\ b = 001$

i	$b_i a$	U_i	u_0	$u_0 \widetilde{m}_0$	$\widetilde{q}_i$	$U_i + \widetilde{q} m$	Z_i
0	129	129	9	81	1	640	0
1	0	64	4	16	6	3130	64
2	0	313	3	27	7	3890	313
–	–	–	–	–	–	–	389

$z = abR^{-1} \bmod m = xy \bmod m = 389$

An example is shown in Table 5.11. (In practice, there will most likely be many computations between initial and final conversions.)

Assuming the availability of only carry-propagate adders, an architecture for Montgomery multiplication is shown in Fig. 5.6, for the computation of $xyR^{-1} \bmod m$, with $r = 2$. With this radix, Eq. 5.46 becomes $\widetilde{q}_i = u_0$. (The explanation for this is given in Sect. 4.2.) The Z register is initialized to zero. The y register is a right-shift register that shifts once in each cycle, so that its least significant bit (y_0) is bit y_i of the operand. The division by r is a wired shift that drops the least significant bit. The Correction part is as in Fig. 4.4.

A faster multiplier than that of Fig. 5.6 can be obtained by replacing the carry-propagate adders with carry-save adders, as shown in Fig. 5.7. Assimilation and Correction are as in Fig. 4.5.

The standard algorithmic technique for the design of high-performance multipliers is the use of a large radix, through multi-bit, fixed-length scanning of the multiplier operand—k bits at a time for radix 2^k—combined with multiplier recoding (Sect. 1.2.2). The use of a large radix here is problematic: the computations of xy_i and $\widetilde{q}_i m$ (Fig. 5.7) are not simple operations if $k > 1$.

Let us briefly consider a straightforward modification of the architecture of Fig. 5.7, for radix-4 computation; that is, with the multiplier scanned two bits at a time, a digit y_i'. The y shift register, carry-save-adders (CSAs), Z registers, are appropriately modified. The AND gates that compute xy_i are replaced with a Multiple-Formation unit that produces one of $0, x, 2x$, and $3x$, according to the value of y_i'. A simple way to do this is the use of two multiplexers: one with inputs $0, x$, and $2x$ (obtained through a wired shift), and the other with inputs 0 and x; so the outputs of the multiplexers are 0 and 0 ($0 + 0 = 0$), or x and 0 ($x + 0 = x$), or

Fig. 5.6 Montgomery
multiplier

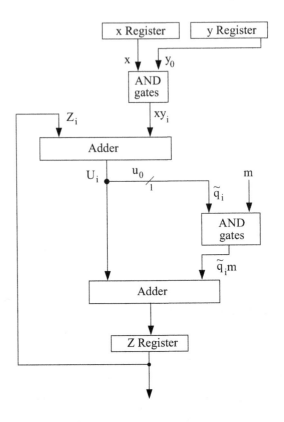

$2x$ and 0 ($2x + 0 = 2x$), or $2x$ and x ($2x + x = 3x$). The two multiplexer outputs together with what were the two CSA inputs now go into a 4:2 compressor.[2] The logic for $\tilde{q}_i m$ is similarly replaced. Lastly, the logic to compute $\tilde{q}_i$ will now be a small, two-gate-level circuit that carries out a very small multiplication. Recoding may also be used. The modified multiplier will have a cycle that is slightly longer than that of the original, but there is a trade-off in the reduced number of cycles. The details would have to be worked out to determine any worthwhile benefit.

The alternative to fixed-length scanning is variable length scanning, which permits skipping past a string of any number of contiguous 0s or contiguous 1s in the multiplier operand without performing an arithmetic operation (for a string of 0s) or performing just two (for a string of 1s). The effect is to, essentially, recode the multiplier into a variable-length string. The two operations for a string of 1s are an addition and a subtraction (Eq. 1.33, Sect. 1.1.2):

$$\left(2^j + 2^{j-1} + \cdots + 2^i\right) x = \left(2^{j+1} - 2^i\right) x$$

---

[2]Recall that such a compressor can be built to have a delay that is slightly larger than that through a 3:2 compressor.

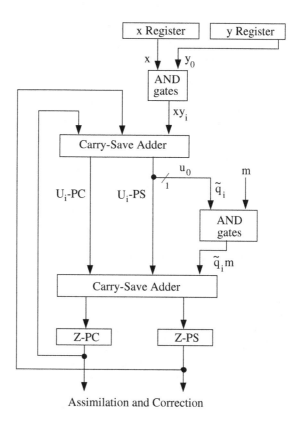

Fig. 5.7 Fast Montgomery multiplier

Assimilation and Correction

So, implementing the technique requires a barrel shifter for y, another barrel shifter to compute $2^{j+1}x$ (by shifting x), and a third to compute $2^i x$ (by shifting x). The operational time of the multiplier will be variable, according to the operands at hand, which is why (in addition to cost) the technique is not normally used. Nevertheless, an implementation on such a basis has been devised for Montgomery multiplication [12], and we next give a very simplified description of that.

In [12], for the computation of xy_i, the multiplier y is first recoded into the digit-set $\{\bar{1}, 0, 1\}$; this can be done with a simple circuit. Thus, in principle, the multiple to be produced is some power of two or zero, according to the sign and position of a digit. In order to determine the latter, another circuit is used to produce an "expanded" form of the recoded multiplier operand. The expansion consists of partitioning the operand into groups of bits, (y_i', z_i), where y_i' is a digit of the recoded operand and z_i is the number of 0s in the group. This expanded form is subsequently used in a circuit that includes a barrel shifter to produce the corresponding power of two. Relative to the architecture of Fig. 5.7, a single CSA is still used at the top level, but the third input now comes from a more complex circuit. The circuit used for the computation of $\tilde{q}_i m$ is more complex. In essence, it should consist of a barrel shifter for $2^{j+1}m$ and a barrel shifter for $2^i m$, with $j+1$

and i supplied by the barrel shifter for y. The two values produced are then added together, in a 4:2 compressor, with what are the other inputs to the second-level CSA in Fig. 5.7. In practice, the implementation includes a lookup table that produces the equivalent of $j + 1$ and i and whose inputs are some low-order bits of U_i (produced by a small carry-propagate adder that assimilates the corresponding bits of U_i-PC and U_i-PS), the output of the y barrel shifter, and some low-order bits of m.

The Montgomery-multiplication industry has been a substantial one, and the published literature is extensive. The work includes high-radix recoding, carry-save-adder design, and low-level techniques to improve performance or power consumption; examples will be found in [7, 8, 10–13]. As with the Barrett reduction algorithm, the use of residue number systems has been investigated [14]; but here too the fundamental problems are yet to be fully solved.

5.2.4 Multiplication with Special Moduli

Because reduction modulo $2^n \pm 1$ is relatively simple (Sect. 4.4), the computation of $xy \bmod (2^n \pm 1)$ may reasonably be carried out first computing xy and then reducing that. Such an approach is especially well suited to cases in which one has to work with "pre-built" units.

With the modulus $2^n - 1$, the operands will be of n bits each and the product xy will be of $2n$ bits. Suppose the product is split into two n-bit parts, $\mathbf{x}_h$ and $\mathbf{x}_l$:

$$xy = 2^n \mathbf{x}_h + \mathbf{x}_l$$

Since $2^n \bmod (2^n - 1) = [(2^n - 1) + 1] \bmod (2^n - 1) = 1$,

$$xy \bmod (2^n - 1) = (\mathbf{x}_h + \mathbf{x}_l) \bmod (2^n - 1) \tag{5.53}$$

This can readily be implemented with a single modulo-$(2^n - 1)$ adder. And for very large operands, xy may be split into more than two pieces that are then added in a tree of adders (consisting of carry-save adders and carry-propagate adders).

Similarly, with the modulus $2^n + 1$, the product will be of $2n + 1$ bits, and splitting this into a one-bit and two n-bit parts:

$$xy = 2^{2n} \mathbf{x}_h + 2^n \mathbf{x}_m + \mathbf{x}_l$$

Since $2^n \bmod 2^n + 1 = [(2^n + 1) - 1] \bmod (2^n + 1) = -1$,

$$xy \bmod (2^n + 1) = (\mathbf{x}_h - \mathbf{x}_m + \mathbf{x}_l) \bmod (2^n + 1)$$

If diminished-one representation is used with the modulus $2^n + 1$, then the value 2^n cannot be a direct operand, as its usual representation is used to represent zero, which is handled as an exceptional case. Therefore, xy will be of $2n$ bits that may be split into two n-bit parts, $\mathbf{x}_h$ and $\mathbf{x}_l$:

$$xy \bmod \left(2^n + 1\right) = (\mathbf{x}_l - \mathbf{x}_h) \bmod (2^n - 1) \tag{5.54}$$

This equation also holds if normal representation is used but the cases of one or both operands being 2^n are excluded and treated as special cases. That is what we will assume in what follows.

The alternative to computing an entire product and then reducing it is to reduce each partial product as it is generated and added. The additions of partial products may be done in carry-save adders, with the final partial-carry and partial-sum from these adders assimilated in a modulo-$(2^n \pm 1)$ carry-propagate adder. We shall assume that each operand in both cases is less than the modulus.

The algorithms given are based on the basic radix-2 multiplication algorithm of Eqs. 5.10–5.13, modified for modular arithmetic[3]

$$Z_0 = 0$$

$$Z_{i+1} = \left(Z_i + 2^i y_i x\right) \bmod m \qquad i = 0, 1, 2, \ldots, n-1$$

$$= \left(Z_i \bmod m + 2^i y_i x \bmod m\right) \bmod m$$

$$= \left(Z_i + 2^i y_i x \bmod m\right) \bmod m \qquad \text{since } Z_i < m \tag{5.55}$$

$$z = Z_n$$

The unreduced multiplicand multiple in Eq. 5.55 is

$$M_i = 2^i y_i x$$

$$= 2^i y_i \sum_{j=0}^{n-1} 2^j x_j$$

$$= 2^i y_i \left(2^{n-1} x_{n-1} + 2^{n-2} x_{n-2} + \cdots + 2^{n-i} x_{n-i} + 2^{n-i-1} x_{n-i-1} + \cdots + x_0\right)$$

$$= y_i \left(2^{i-1} 2^n x_{n-1} + 2^{i-2} 2^n x_{n-2} + \cdots + 2^n x_{n-i}\right)$$

---

[3]In ordinary multiplication the 2^i factor is taken care of by shifting the partial product instead of the multiplicand multiple. That is possible because the lower order i bits of the ith partial product are not included in the corresponding addition. On the other hand, with modular multiplication—specifically the required reductions—all bits of a partial product must be included in the arithmetic; therefore, $2^i x y_i$ is taken in its entirety.

$$+2^{n-1}x_{n-i-1} + 2^{n-2}x_{n-i-2} + \cdots + 2^i x_0\Big)$$

$$= y_i \left[\left(2^{n-1}x_{n-i-1} + 2^{n-2}x_{n-i-2} + \cdots + 2^i x_n \right) \right.$$

$$\left. +2^n \left(2^{i-1}x_{n-1} + 2^{i-2}x_{n-2} + \cdots + x_{n-1} \right) \right] \quad (5.56)$$

Modulus $2^n - 1$

From Eqs. 5.53 and 5.56:

$$M_i \bmod (2^n - 1) = y_i \left(2^{n-1}x_{n-i-1} + 2^{n-2}x_{n-i-2} + \cdots + 2^i x_0 \right.$$

$$\left. +2^{i-1}x_{n-1} + 2^{i-2}x_{n-2} + \cdots + x_{n-i} \right) \bmod (2^n - 1)$$

Observe that the expression in the brackets corresponds to the binary pattern $x_{n-i-1}x_{n-i-2} \cdots x_0 \, x_{n-1}x_{n-2} \cdots x_{n-i}$, which is just an i-place cyclic right shift of $x_{n-1}x_{n-1} \cdots x_0$. And because $x < 2^n - 1$, there must be some x_k, $0 \le k \le n - 1$, such that $x_k = 0$. Therefore, the value represented by $x_{n-i-1}x_{n-i-2} \cdots x_0 x_{n-1}x_{n-2} \cdots x_{n-i}$ must also be less than $2^n - 1$. And since y_i is 0 or 1:

$$M_i \bmod (2^n - 1) = y_i \left(2^{n-1}x_{n-i-1} + 2^{n-2}x_{n-i-2} + \cdots + 2^i x_0 \right.$$

$$\left. +2^{i-1}x_{n-1} + 2^{i-2}x_{n-2} + \cdots + x_{n-i} \right) \quad (5.57)$$

An example computation is shown in Table 5.12; $|\cdots| \bmod m$ denotes $\cdots$ $\bmod m$. Note that M_i is not actually computed but is included solely for clarity the bits of $\mathbf{x}_h$ are the bits shifted in at the low end (shown in bold font) to form $M_i \bmod m$. The latter is, of course, a simple computation that involves addition of a 1 for and end-around-carry or to take into account the value $2^n - 1$ (Sect. 5.1.2). The relevant details in multiplication are given below.

The multiplication algorithm may be implemented with a single modulo-$(2^n - 1)$ adder used repeatedly, in a design that corresponds to that of Fig. 1.15. A much faster implementation for sequential multiplication would use a carry-save adder (CSA) in the loop and a modulo-$(2^n - 1)$ carry-propagate adder (CPA) to assimilate the final outputs of the CSA, in a design that corresponds to that of Fig. 1.16. An architecture for the latter is shown in Fig. 5.8. The y register is a right-shit register that shifts by one bit-position in each cycle. The multiplicand multiples, M_i, are formed by cyclically the contents of the x register shifted by one bit-position in each cycle and then producing zero or the contents of that register, according to the value of y_0.

The reduction of a partial product as it is formed requires an addition modulo-$(2^n - 1)$, and such an addition involves the addition of an "end-around-carry." With the arrangement of Fig. 5.8 this takes place at two places. First, during the "looping"

Table 5.12 Example of modulo-$(2^n - 1)$ multiplication

$x = 53 = 110101_2$, $y = 43 = 101011$, $m = 63 = 111111_2$, $n = 6$					
		M_i			
i	y_i	$\mathbf{x}_h$	$\mathbf{x}_l$	M_i mod 63	Z_i
0	1		110101	110101	000000
1	1	1	101010	101011	$(000000 + 110101)$ mod $m = 110101$
2	0	00	000000	000000	$(110101 + 101011)$ mod $m = 100001$
3	1	110	101000	101110	$(100001 + 000000)$ mod $m = 100001$
4	0	0000	000000	000000	$(100001 + 101110)$ mod $m = 010000$
5	1	11010	100000	111010	$(010000 + 000000)$ mod $m = 010000$
–	–	–	–	–	$(010000 + 111010)$ mod $m = 001011$
$53 * 43$ mod $63 = 001011_2 = 11$					

Fig. 5.8 Sequential modulo-$(2^n - 1)$ multiplier

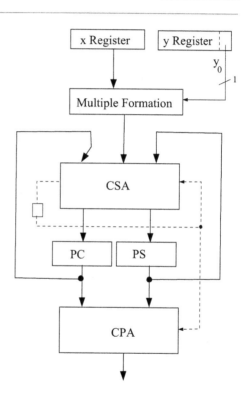

the reduction is effected by turning the partial-carry output from the most significant bit of the CSA output in one cycle into the least significant bit of the CSA for the addition in the next cycle; in effect, we have a modulo-$(2^n - 1)$ CSA. Second, in the last cycle the partial-carry output bit is fed, as a carry-in, into the modulo-$(2^n - 1)$ CPA. We next explain one aspect of the design that might not be immediately apparent.

In modulo-$(2^n - 1)$ addition as described in Sect. 5.1.2, it is necessary to detect the value $2^n - 1$, which is represented by $11 \cdots 1$, and then make an adjustment in order to get the correct result. That is not necessary here, because each successive CSA makes any required correction and the final CPA makes any final correction required. To see this, ignore for the moment the fact that the intermediate results are in partial-carry/partial-sum form, or imagine that each CSA has been replaced by an ordinary adder (i.e., one with no end-around carry). Suppose, then, that the partial product is $2^n - 1$ and the multiplicand multiple is a. Now, $(2^n - 1) + a = 2^n + (a - 1)$. If $a > 0$, then the 2^n is a carry from the addition, and the addition of a 1 (the carry) at the next addition makes the necessary correction; otherwise, i.e., if $a = 0$, then the $2^n - 1$ is carried forward. At the last stage the modulo-$(2^n - 1)$ CPA contains logic that makes any necessary final correction.

With a parallel CSA array, similar to that of Fig. 1.19, the multiplicand multiples will be formed by appropriate wiring, and end-around-carries will be passed from one CSA to the next and finally to the CPA.

High-radix multiplier recoding, the primary technique used to speed up ordinary multiplication, can be used here too. The essential idea in recoding is exactly as described in Sect. 1.2.2, but with two points to note here. First, as indicated above, here it is the "full" shifted multiplicand multiple, $2^i x y_i$, that is considered—but not actually computed—instead of assuming relative shifts to account for the 2^i. Second, modulo-$(2^n - 1)$ arithmetic is essentially ones'-complement arithmetic, so subtractions are effected by adding the ones' complement of the subtrahend. Putting all that together with Eq. 5.54, the table for radix-4 recoding is shown in Table 5.13, which corresponds to Table 1.8a. (The index i is incremented by two for each multiplicand multiple.) The values possible values added are ± 0, $\pm [2^i x \bmod (2^n - 1)]$, and $\pm [2(2^i x) \bmod (2^n - 1)]$, which correspond to ± 0, $\pm x$, and $\pm 2x$ in Table 1.8a.

Modulus $2^n + 1$

Table 5.13 Radix-4 recoding table for modulo-$(2^n - 1)$ multiplier

$y_{i+1} y_i$	y_{i-1}	Action
00	0	Add $00 \cdots 0$
00	1	Add $x_{n-i-1} x_{n-i-2} \cdots x_0 x_{n-1} \cdots x_{n-i}$
01	0	Add $x_{n-i-1} x_{n-i-2} \cdots x_0 y_{n-1} \cdots x_{n-i}$
01	1	Add $x_{n-i-2} x_{n-i-3} \cdots x_0 x_{n-1} \cdots x_{n-i-1}$
10	0	Subtract $x_{n-i-2} x_{n-i-3} \cdots x_0 x_{n-1} \cdots x_{n-i-1}$
		(Add $\overline{x}_{n-i-2} \overline{x}_{n-i-3} \cdots \overline{x}_0 \overline{x}_{n-1} \cdots \overline{x}_{n-i-1}$)
10	1	Subtract $x_{n-i-1} x_{n-i-2} \cdots x_0 x_{n-1} \cdots x_{n-i}$
		(Add $\overline{x}_{n-i-1} \overline{x}_{n-i-2} \cdots \overline{x}_0 \overline{x}_{n-1} \cdots \overline{x}_{n-i}$)
11	0	Subtract $x_{n-i-1} x_{n-i-1} \cdots x_0 x_{n-1} \cdots x_{n-i}$
		(Add $\overline{x}_{n-i-1} \overline{x}_{n-i-1} \cdots \overline{x}_0 \overline{x}_{n-1} \cdots \overline{x}_{n-i}$)
11	1	Subtract $00 \cdots 000 \cdots 0$
		(Add $11 \cdots 1$)

Multiplication modulo $2^n + 1$ is more difficult than that modulo $2^n - 1$, and a multiplier for the former will not be as efficient as that for the latter. The basic difficulty is evident from a comparison of Eqs. 5.53 and 5.54. The addition of Eq. 5.53 translates into a simple cyclic shift because one operand has i trailing 0s and the other has $n - i$ leading 0s. On the other hand, the operation in Eq. 5.54 is a subtraction, and in complement representation leading 0s in the subtrahend turn into 1s.

Much work has been done on the design of modulo-(2^n+1) multipliers, although mostly for diminished-one representation, e.g., [3–5, 9]. The following discussion, based on [1], is for a system with a normal representation but with the value zero not used for an operand and 2^n represented by $00 \cdots 0$. The design can be modified easily to allow zero and instead use a special flag for 2^n as well as for diminished-one representation.

Given two operands, x and y, with neither equal to 2^n—special cases that are dealt with separately—the modulo-$(2^n + 1)$ product is

$$xy \bmod (2^n + 1) = \left(\sum_{i=0}^{n-1} 2^i y_i x \right) \bmod (2^n + 1)$$

Each partial product, $M_i' \triangleq 2^i x y_i$, is nominally represented in $2n$ bits, and we may split the representation of $2^i x$, which is

$$\overbrace{00\cdots0}^{n-i \ 0s} x_{n-1} x_{n-2} \cdots x_0 \overbrace{00\cdots0}^{i \ 0s} v$$

into two n-bit pieces, U and L:

$$L = x_{n-i-1} x_{n-i-2} \cdots x_0 \overbrace{00\cdots0}^{i \ 0s}$$

$$U = \overbrace{00\cdots0}^{n-i \ 0s} x_{n-1} x_{n-2} \cdots x_{n-i}$$

Then $2^i x = 2^n U + L$, and, with Eq. 5.54:

$$M_i' \bmod \left(2^n + 1\right) = (y_i L - y_i U) \bmod (2^n + 1)$$

Now, from Sect. 1.1.6, the numeric value, $\bar{z}$, of the 1s complement of a number U is

$$\bar{z} = 2^n - 1 - z$$

so

$$-z \bmod (2^n + 1) = \left(-2^n + 1 + \bar{z}\right) \bmod (2^n + 1)$$

$$= \left[-(2^n + 1) + 2 + \overline{z}\right] \bmod (2^n + 1)$$

$$= (\overline{z} + 2) \bmod (2^n + 1)$$

Therefore

$$M_i' \bmod (2^n + 1) = [y_i (L - U)] \bmod (2^n + 1)$$

$$\left[y_i (L + \overline{U} + 2)\right] \bmod (2^n + 1)$$

That is

$$xy \bmod (2^n + 1) = \sum_{i=0}^{n-1} y_i (x_{n-i-1} x_{n-i-2} \cdots x_0 \overbrace{00 \cdots 0}^{i \text{ 0s}} \tag{5.58}$$

$$+ \overbrace{11 \cdots 1}^{n-i \text{ 1s}} \overline{x}_{n-1} \overline{x}_{n-2} \cdots \overline{x}_{n-i} + 2) \bmod (2^n + 1)$$

For simplification and correctness, two extra terms are added to the right-hand side of this equation. The first is a term that corresponds to $\overline{y}_i (-0 \cdots 00 \cdots 0) = \overline{y}_i (1 \cdots 11 \cdots 1) + 1$ and which allows some factoring. The second is an extra 1, because a normal adder performs modulo-2^n operation, and

$$(a + b + 1) \bmod (2^n + 1) = (a + b + \overline{c}) \bmod 2^n$$

where c is the carry-out from the addition of a and b.

After some simplification, Eq. 5.58 yields

$$xy \bmod (2^n + 1) = \sum_{i=0}^{n-1} [y_i (x_{n-i-1} x_{n-i-2} \cdots x_0 \overline{x}_{n-1} \overline{x}_{n-2} \cdots \overline{x}_{n-i})$$

$$\overline{x}_i (\overbrace{00 \cdots 0}^{n-i \text{ 0s}} \overbrace{11 \cdots 1}^{i \text{ 1s}}) + 1 + 2] \bmod (2^n + 1)$$

$$\overset{\triangle}{=} \sum_{i=0}^{n-1} [(M_i + 1) + 2]1 \bmod (2^n + 1) \tag{5.59}$$

So, for fast multiplication, the process consists of addition of the terms of Eq. 5.59 in a structure of modulo-$(2^n + 1)$ CSAs and an assimilating modulo-$(2^n + 1)$ CPA, with each CSA having the extra precision to include the 2. A modulo-$(2^n + 1)$ CSA is similar to a modulo-$(2^n - 1)$, except that the end-around carry is inverted.

We now consider the special cases and the required corrections. Suppose x is 2^n but y is not. Then

Fig. 5.9 Modulo-$(2^n + 1)$ multiplier

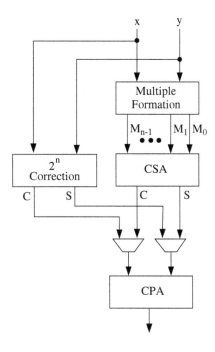

$$xy \bmod (2^n + 1) = 2^n y \bmod (2^n + 1)$$

$$= -y \bmod (2^n + 1)$$

$$= (\bar{y} + 2) \bmod (2^n + 1)$$

Similarly, if y is 2^n but x is not, then $xy \bmod (2^n + 1) = (\bar{x} + 2) \bmod (2^n + 1)$. And if both are equal to 2^n, then $xy \bmod (2^n + 1) = 1$.

The general form of a modulo-$(2^n + 1)$ multiplier is shown in Fig. 5.9. The 2^n Correction Unit: Since the assimilating carry-propagate adder adds in a 1 (C_{in} above), in this case the inputs to the adder that are required to get the correct results (i.e., $\bar{y} + 2$, or $\bar{x} + 2$, or 1) are $\bar{y} + 1$, or $\bar{x} + 1$, or 0. That is,

$$(PC, PS) = \begin{cases} (\bar{y}, 1) & \text{if } x = 2^n \\ (\bar{x}, 1) & \text{if } y = 2^n \\ (0, 0) & \text{if } x = y = 2^n \end{cases}$$

Recoding here too is somewhat more complicated than with modulus $2^n - 1$. Table 5.14 gives one part of the actions for radix-4 recoding and corresponds to Table 5.12. (The index i is incremented by two for each multiplicand multiple.) Also, each multiplicand multiple now has added a two-bit correction term $T = t_{i+1} t_i$:

Table 5.14 Radix-4 recoding table for modulo-$(2^n + 1)$ multiplier

$y_{i+1}y_i$	y_{i-1}	Action
00	0	Add $00 \cdots 011 \cdots 1$
00	1	Add $x_{n-i-1}x_{n-i-2} \cdots x_0 \overline{x}_{n-1} \cdots \overline{x}_{n-i}$
01	0	Add $x_{n-i-1}x_{n-i-2} \cdots x_0 \overline{x}_{n-1} \cdots \overline{x}_{n-i}$
01	1	Add $x_{n-i-2}x_{n-i-3} \cdots x_0 \overline{x}_{n-1} \cdots \overline{x}_{n-i-1}$
10	0	Subtract $x_{n-i-2}x_{n-i-3} \cdots x_0 x_{n-1} \cdots x_{n-i-1}$
		(Add $\overline{x}_{n-i-2}\overline{x}_{n-i-3} \cdots \overline{x}_0 x_{n-1} \cdots x_{n-i-1}$)
10	1	Subtract $x_{n-i-1}x_{n-i-2} \cdots x_0 \overline{x}_{n-1} \cdots \overline{x}_{n-i}$
		(Add $\overline{x}_{n-i-1}\overline{x}_{n-i-2} \cdots \overline{x}_0 x_{n-1} x_{n-i}$)
11	0	Subtract $x_{n-i-1}x_{n-i-2} \cdots x_0 \overline{x}_{n-1} \cdots \overline{x}_{n-i}$
		(Add $\overline{x}_{n-i-1}\overline{x}_{n-i-2} \cdots \overline{x}_0 x_{n-1} \cdots x_{n-i}$)
11	1	Subtract $00 \cdots 000 \cdots 0$
		(Add $11 \cdots 100 \cdots 0$)

$$t_0 = \overline{y}_1 \overline{y}_0 \overline{y}_{n-1} + y_1 \overline{y}_0 y_{n-1} + y_0 \overline{y}_{n-1}$$

$$t_1 = \overline{y}_1 + \overline{y}_0 \overline{y}_{n-1}$$

$$t_i = y_{i+1}\overline{y}_i + y_{i+1}y_{i-1} + \overline{y}_i y_{i-1}$$

$$t_i = \overline{y}_{i+1}$$

And the 2 in Eq. 5.59 is replaced with 1.

References

1. R. Zimmerman. 1999. Efficient VLSI implementation of modulo $(2^n \pm 1)$ addition and multiplication. *Proceedings, International Symposium on Computer Arithmetic*, pp. 158–167.
2. H. T. Vergos and C. Efstathiou. 2008. A unifying approach for weighted and diminished-one modulo $2^n + 1$ addition. *IEEE Transactions on Circuits and Systems II*, 55:1041–1045.
3. L. Sousa and R. Chaves. 2005. A universal architecture for designing efficient modulo $2^n + a$ multipliers. *IEEE Transactions on Circuits and Systems I*, 52(6):1166–1178.
4. R. Muralidharan and C.-H. Chang. 2012. Area-power efficient modulo $2^n - 1$ and modulo $2^n + 1$ multipliers for $\{2^n - 1, 2^n, 2^n + 1\}$ based RNS. *IEEE Transactions on Circuits and Systems I*, 59(10):2263–2274.
5. J. W. Chen, R. H. Yao, and W. J. Wu. 2011. Efficient modulo $2^n + 1$ multipliers. *IEEE Transactions on Very Large Scale Integration (VLSI) Systems*, 19(12):2149–2157.
6. M. Knezevic, F. Vercauteren, and I. Verbauwhede. 2010. Faster interleaved modular multiplication based on Barrett and Montgomery reduction methods. *IEEE Transactions on Computers*, 59(12):1715–17121.
7. H. Orup. 1995. Simplifying quotient determination in high-radix modular multiplication. In: *Proceedings, 12th Symposium on Computer Arithmetic*, pp. 193–199.
8. P. Kornerup. 1993. High-radix modular multiplication for cryptosystems. In: *Proceedings, 12th Symposium on Computer Arithmetic*, pp. 277–283.

9. A. V. Curiger, H. Bonennenberg, and H. Kaeslin. 1991. Regular VLSI architectures for multiplication modulo $2^n + 1$. *IEEE Journal of Solid-State Circuits*, 26(7):990–994.

10. M.-D. Shieh, J.-H. Chen, W.-C. Lin, and H.-H. Wu. 2009. A new algorithm for high-speed modular multiplication design. *IEEE Transactions on Circuits and Systems I*, 56(9): 2009–2019.

11. S.-R. Kuang, J. -P. Wang, K.-C. Chang, and H.-W. Hsu. 2013. Energy-efficient high-throughput Montgomery modular multipliers for RSA cryptosystems. *IEEE Transactions on Very Large Scale Integration (VLSI) Systems*, 21(11):1999–2009.

12. A. Reza and P. Keshavarzi. 2015. High-throughput modular multiplication and exponentiation algorithms using multibit-scan—multibit-shift technique. *IEEE Transactions on Very Large Scale Integration (VLSI) Systems*, 23(9): 1710–1719.

13. S.-R. Kung, K.-Y. Wu, and R.-Y. Lu. 2016. Low-cost high-performance VLSI architecture for Montgomery modular multiplication. *IEEE Transactions on Very Large Scale Integration (VLSI) Systems*, 24(2):434–443.

14. J.-C. Bajard, L.-S. Didier, and P. Kornerup. 1998. An RNS Montgomery modular multiplication algorithm. *IEEE Transactions on Computers*, 47(7):766–776.

15. H. K. Garg and H. Xiao. 2016. New residue based Barrett algorithms: modular integer computations. *IEEE Access*, 4:4882–4890.

16. C. K. Koc. 1995. RSA Hardware Implementation. Report, RSA Laboratories. Redwood City, California, USA.

17. H. Orup and P. Kornerup. 1991. A high-radix hardware algorithm for calculating the exponential M^E modulo N. *Proceedings, 10th IEEE Symposium on Computer Arithmetic*, pp. 51–56.

Chapter 6
Modular Exponentiation, Inversion, and Division

Abstract Modular exponentiation is the computation of x^e mod m, and multiplicative modular inversion is the computation of y such that $x * y$ mod $m = 1$. This chapter consists of two sections, one each on the two operations. Modular division is included implicitly in the second, as in practice it is effected as multiplication by an inverse.

Compared with ordinary modular arithmetic, there are two aspects worth noting here, one minor and one major. The minor one is that exponentiation is not a particularly important operation in ordinary modular arithmetic; the major one is that operands in cryptography are usually of very high precision. The latter has significant practical implications.

6.1 Exponentiation

The computation of $y = x^e$ mod m may be done directly and easily by first computing x^e through repeated multiplications and then carrying out a modular reduction:

$$u = x^e \tag{6.1}$$

$$y = u \ \text{mod} \ m \tag{6.2}$$

This will work well for small or moderately sized numbers, but for the large ones that are typical in cryptography the intermediate value x^e can be extremely large, and the method is unlikely to be practical. For high-precision operands, the intermediate values should be kept small, and this can be done by interleaving modular reductions with the multiplications. Such reductions are based on the fact that

$$ab \ \text{mod} \ m = a(b \ \text{mod} \ m) \ \text{mod} \ m \tag{6.3}$$

© Springer Nature Switzerland AG 2020
A. R. Omondi, *Cryptography Arithmetic*, Advances in Information Security 77,
https://doi.org/10.1007/978-3-030-34142-8_6

and this can be applied recursively to the computation of $y = x^e \bmod m$:

$$y = x \left(x^{e-1} \bmod m \right) \bmod m \tag{6.4}$$

Whether the computation consists of "plain" exponentiation followed by reduction (Eqs. 6.1–6.2) or of multiplications and reductions interleaved (Eq. 6.4) the starting point is the underlying method for the computation of an ordinary exponential. We shall therefore first consider how to efficiently compute x^e and then how to include the modular reductions.

The simplest way to compute $y = x^e$ is the standard paper-and-pencil one of multiplication at a time: start with $y = x$ and multiply $e-1$ times by x. This requires $e - 1$ multiplications. We next give a much more efficient method, which requires at most $\lceil \log_2 e \rceil$ multiplications and a similar number of squaring operations.

Suppose the binary representation of e is $e_{n-1} \cdots e_3 e_2 e_1 e_0$, where $e_i = 0$ or $e_i = 1$; $i = 0, 1, 2, \ldots, n - 1$. That is, $e = \sum_{i=0}^{n-1} e_i 2^i$. Then

$$x^e = x^{e_{n-1} 2^{n-1} + \cdots + e_3 2^3 + e_2 2^2 + e_1 2^1 + e_0 2^0}$$

$$= x^{e_{n-1} 2^{n-1}} \cdots x^{e_3 2^3} x^{e_2 2^2} x^{e_1 2^1} x^{e_0 2^0}$$

$$= \left(x^{2^{n-1}} \right)^{e_{n-1}} \cdots \left(x^8 \right)^{e_3} \left(x^4 \right)^{e_2} \left(x^2 \right)^{e_1} (x)^{e_0} \tag{6.5}$$

For example, the binary representation of decimal 25 is 11001, and

$$x^{25} = x^{16} x^8 x^0 x^0 x^1$$

In Eq. 6.5 we observe that:

- The bits $e_0, e_1, e_2, e_3 \ldots$ correspond, in that order, to the powers x^1, x^2, x^4, x^8, and so forth; and these powers may be computed by starting with x and repeatedly squaring.
- The effect of e_i is that the corresponding power of x is included in the product if $e_i = 1$, but not if $e_i = 0$.

A basic method for the computation of the exponential may therefore consist of the computation of a sequence Z_i of squares and a corresponding sequence Y_i of an accumulation of squares, the latter according to the bits in the binary representation of the exponent. This is a well-known algorithm that is usually referred to as the *square-and-multiply* algorithm [1]. For the computation of $y = x^e$:

$$Y_0 = 1 \tag{6.6}$$

$$Z_0 = x \tag{6.7}$$

$$Y_{i+1} = \begin{cases} Y_i Z_i & \text{if } e_i = 1 \\ Y_i & \text{otherwise} \end{cases} \quad i = 0, 1, 2, \ldots n - 1 \tag{6.8}$$

Table 6.1 Example computation of x^e, right-to-left exponent scan

$e = 25 = 11001_2, \ n = 5$			
i	e_i	Z_i	Y_i
0	1	x	1
1	0	x^2	x
2	0	x^4	x
3	1	x^8	x
4	1	x^{16}	$x * x^8 = x^9$
5	–	–	$x^9 * x^{16} = x^{25}$

$$Z_{i+1} = Z_i^2 \tag{6.9}$$

$$y = Y_n \tag{6.10}$$

As an example, the computation of $x^{25} = x^{16} * x^8 * 1 * 1 * x^1$ is as shown in Table 6.1.

We can also compute x^e in a square-and-multiply algorithm with a left-to-right scan of the exponent, by rewriting Eq. 6.5 into

$$x^e = ((((\cdots (x^{e_{n-1}})^2 x^{e_{n-2}})^2 \cdots x^{e_3})^2 x^{e_2})^2 x^{e_1})^2 x^{e_0}$$

in which we observe that if the exponent is scanned from e_{n-1} to e_0, then at each step in the computation we should square the result up to that point and multiply that by the contribution of the current bit of the exponent. The said contribution is

$$x^{e_i} = \begin{cases} x & \text{if } e_i = 1 \\ 1 & \text{otherwise} \end{cases}$$

We thus have this algorithm:

$$Z_n = 1 \tag{6.11}$$

$$Y_{i-1} = \begin{cases} x Z_i & \text{if } e_i = 1 \\ Z_i & \text{otherwise} \end{cases} \qquad i = n, n-1, \dots, 1 \tag{6.12}$$

$$Z_{i-1} = Y_i^2 \tag{6.13}$$

$$y = Y_0 \tag{6.14}$$

As an example,

$$x^{25} = \left(\left(\left((x)^2 * x \right)^2 * 1 \right)^2 * 1 \right)^2 * x$$

and the computation is as shown in Table 6.2.

Table 6.2 Example
computation of x^e,
left-to-right exponent scan

$e = 25 = 11001_2$, $n = 5$			
i	e_i	Z_i	Y_i
5	–	1	–
4	1	x^2	$x * 1 = x$
3	1	x^6	$x * x^2 = x^3$
2	0	x^{12}	x^6
1	0	x^{24}	x^{12}
0	1	–	$x * x^{24} = x^{25}$

A major difference between the algorithms of Eqs. 6.6–6.10 and that of Eqs. 6.11–6.14 is that in the former the core computations (Eqs. 6.8–6.9) can be carried out in parallel, which is not possible with the second algorithm (Eqs. 6.12–6.13). Therefore, the former can yield a faster, if more costly, implementation.

The binary square-and-multiply idea can be extended to one in which the exponent is represented in a radix larger than two, as is done in high-radix multiplication (Sect. 1.2.2), thus improving performance by proportionally reducing the number of iterations. We next describe this for the algorithm of Eqs. 6.11–6.14.

Suppose n, the number of bits for the exponent representation is a multiple of k, with $k \geq 2$. (If that is not the case, then it can be made so by appending 0s at the most significant end of the representation.) Then for radix-2^k computation, which is taking two bits at a time of the exponent, the multiplication in Eq. 6.12, i.e.,

$$Y_i = x^{e_i} Z_i \qquad i = n - 1, n - 2, \ldots, 1, 0$$

becomes

$$Y_i = x^{E_i} Z_i \qquad i = m - 1, m - 2, \ldots, 1, 0; \; m = n/k$$

where E_i is the radix-2^k digit with the binary representation $e_{k(i+1)-1} \cdots e_{ki+2}$ $e_{ki+1}e_{ki}$. And Eq. 6.13, i.e.,

$$Z_{i-1} = Y_{i-1}^2 \qquad i = n - 1, n - 2, \ldots, 1, 0$$

becomes

$$Z_{i-1} = Y_{i-1}^{2^k} \qquad i = m - 1, m - 2, \ldots, 1, 0$$

So, for example, the radix-4 algorithm is

$$Z_{n-1} = 1 \tag{6.15}$$

$$Y_i = x^{e_{2i+1}e_{2i}} Z_i \qquad i = m - 1, m - 2, \ldots, 1, 0; \; m = n/2 \tag{6.16}$$

$$Z_{i-1} = Y_{i-1}^4 \tag{6.17}$$

Table 6.3 Example radix-4
computation of x^e

$e = 25 = 011001_2$, $n = 5$, $k = 2$			
i	$e_{2i+1}e_{2i}$	Z_i	Y_i
2	01	1	$x^1 * 1 = x$
1	10	x^4	$x^2 * x^4 = x^6$
0	01	$(x^6)^4 = x^{24}$	$x * x^{24} = x^{25}$

of which an example computation (which corresponds to those of Tables 6.1 and 6.2)
is given in Table 6.3.

There are numerous other versions of the basic algorithm that greatly reduce the
number of multiplications required [1]. Nevertheless, almost all these versions have
complexities that make them ill-suited to hardware implementation, and we will not
consider them.

Exponentiation with Direct Reduction

We will assume that the computation is of $y = x^e \bmod m$ with $x < m$. That is
acceptable because if we start with $u > m$ and wish to compute $u^e \bmod m$, then

$$u^e \bmod m = (u \bmod m)^e \bmod m$$

$$= x^e \bmod m \qquad \text{where } x = u \bmod m$$

The two "ordinary-exponentiation" algorithms above can be modified easily
into algorithms for modular exponentiation, by replacing all multiplications with
modular multiplications. Thus, from Eqs. 6.6–6.10 we obtain

$$Y_0 = 1 \tag{6.18}$$

$$Z_0 = x \tag{6.19}$$

$$Y_{i+1} = \begin{cases} Y_i Z_i \bmod m & \text{if } e_i = 1 \quad i = 0, 1, 2, \ldots n - 1 \\ Y_i & \text{otherwise} \end{cases} \tag{6.20}$$

$$Z_{i+1} = Z_i^2 \bmod m \tag{6.21}$$

$$y = Y_n \tag{6.22}$$

of which an example computation is shown in Table 6.4.

And from the algorithm of Eqs. 6.11–6.14 we obtain

$$Z_n = 1 \tag{6.23}$$

$$Y_{i-1} = \begin{cases} x Z_i \bmod m & \text{if } e_{i-1} = 1 \quad i = n, n - 1, \ldots, 1 \\ Z_i & \text{otherwise} \end{cases} \tag{6.24}$$

Table 6.4 Example computation of $x^e \bmod m$, right-to-left exponent scan

$x = 14,\ e = 25 = 11001_2,\ m = 23$			
i	e_i	Z_i	Y_i
0	1	14	1
1	0	12	14
2	0	6	14
3	1	13	14
4	1	8	21
5	–	–	7
$14^{25} \bmod 23 = 7$			

Table 6.5 Example computation of $x^e \bmod m$, left-to-right exponent scan

$x = 14,\ e = 25 = 11001_2,\ m = 23$			
i	e_i	Z_i	Y_i
5	–	1	–
4	1	11	14
3	1	3	7
2	0	9	3
1	0	12	9
0	1	–	7
$14^{25} \bmod 23 = 7$			

$$Z_{i-1} = Y_i^2 \bmod m \tag{6.25}$$

$$y = Y_0 \tag{6.26}$$

of which an example computation is given in Table 6.5.

An architecture derived directly from the second algorithm is shown in Fig. 6.1. (We leave it to the reader to devise an architecture without the duplication of units.) The contents of the e register are shifted left by one bit-position in each cycle; so e_{n-1} in iteration i is the i-th bit of the exponent to be processed. The rest of the diagram is self-explanatory, although it should be noted that squaring can be performed faster than general multiplication (Sect. 1.2.6).

As given, the architecture is evidently ill-suited for a high-performance implementation, and the requirement for modular reductions in every iteration mean than modifications intended to achieve high performance will almost definitely be problematic. Therefore, in the first instance, the architecture might be considered if only "pre-built" units are available, or if the modulus is such that fast reduction is possible. It should, however, be noted that the combination of Multiplier and Reduction units is just a modular multiplier. So, the architecture could be the basis of an implementation with a fast, direct-reduction multiplier, such as that of Fig. 5.5, Sect. 5.2.1. Nevertheless, even such a multiplier cannot be used directly: the outputs of each multiplier (Fig. 6.1) must be kept in partial-carry/partial-sum form at the end of the multiplication iterations, and that requires several changes, including additional carry-save adders in the critical path. It should also be noted that the multiplier of Fig. 5.5 cannot be adapted easily for high-radix computation.

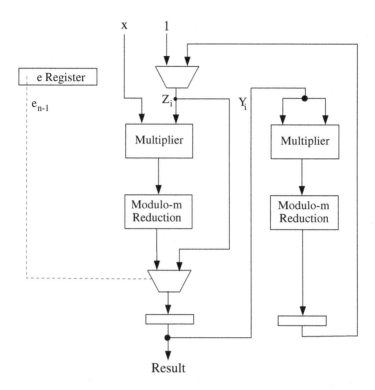

x 1

Fig. 6.1 Modular exponentiation unit

A high-radix, direct-reduction method for exponentiation is described in [5]. The method is based on high-radix, direct-reduction multiplication, with approximate division used in the reduction. (A brief description of that multiplication algorithm is given at the end of Sect. 5.2.1.) The superiority of the method over a fast binary one is unclear.

The fundamental difficulties in relation to the architecture of Fig. 6.1 (and similar architectures) arise from the basic algorithm and are mostly resolved in Montgomery exponentiation, in which the intermediate results are in a more helpful form.

Exponentiation with Montgomery Reduction

The algorithms of Eqs. 6.6–6.10 and Eqs. 6.11–6.14 require modular reductions in each iteration, and, for a general modulus, those will be costly operations. More efficient algorithms can be obtained by instead modifying the basic square-and-multiply algorithms to use Montgomery reductions and keep intermediate results in Montgomery-residue form. In general, the initial conversion of the operands into Montgomery-residue form will be costly operations, but they have to be done only once.

Given an operand x, a modulus m, and a suitable value R, the Montgomery reduction algorithm computes the value $xR^{-1} \bmod m$, from which $x \bmod m$ can be obtained by another Montgomery reduction (Eqs. 4.11–4.15, Sect. 4.2). The Montgomery reduction process can be combined with multiplication (Eqs. 5.41–5.45). We will use $\otimes$ to denote such multiplication-reduction. That is:

$$x \otimes y = xyR^{-1} \bmod m$$

The Montgomery residue of x, denoted $\tilde{x}$ below, is $xR \bmod m$. The Montgomery multiplication of two such residues produces another similar residue:

$$\tilde{x} \otimes \tilde{y} = xR \bmod m \otimes yR \bmod m$$
$$= (xR \bmod m)(yR \bmod m)R^{-1} \bmod m$$
$$= xyR^2 R^{-1} \bmod m$$
$$= xyR \bmod m$$
$$= \widetilde{xy}$$

The relatively high costs of computing the Montgomery residues mean that Montgomery multiplication is ill-suited to a single modular multiplication. But if used in modular exponentiation, then these costly computations are carried out only once for a sequence of multiplications, and the method is therefore worthwhile, since Montgomery reduction (even with multiplication included) is much less costly than general reduction. If the multiplications-and-reductions in Eqs. 6.6–6.10 and Eqs. 6.11–6.14 are replaced with Montgomery multiplications, then, with Montgomery residues as the initial operands, the intermediate results and final result will be Montgomery residues. A result in conventional form is then obtained through another Montgomery multiplication, with 1 as the other operand:

$$zR \bmod m \otimes 1 = zR * 1 * R^{-1} \bmod m$$
$$= z \bmod m$$

So, for the computation of $y = x^e \bmod m$, the algorithm of Eqs. 6.6–6.10 may be replaced with

$$\tilde{x} = xR \bmod m \qquad (6.27)$$
$$\tilde{1} = R \bmod m \qquad (6.28)$$
$$Y_0 = \tilde{1} \qquad (6.29)$$

Table 6.6 Example of Montgomery modular exponentiation

$x = 14$, $e = 25 = 11001_2$, $m = 23$, $R = 32$			
$R^{-1} = 18$, $\widetilde{x} = 11$, $\widetilde{1} = 9$			
i	e_i	Z_i	Y_i
0	1	11	9
1	0	$11^2 * 18 \bmod 23 = 16$	$11 * 9 * 18 \bmod 23 = 11$
2	0	$16^2 * 18 \bmod 23 = 8$	11
3	1	$8^2 * 18 \bmod 23 = 2$	11
4	1	$2^2 * 18 \bmod 23 = 3$	$11 * 2 * 18 \bmod 23 = 5$
5	–	–	$5 * 3 * 18 \bmod 23 = 17$
$y = 14^{25} \bmod 23 = 17 * 1 * 18 \bmod 23 = 7$			

$$Z_0 = \widetilde{x} \tag{6.30}$$

$$Y_{i+1} = \begin{cases} Y_i \otimes Z_i & \text{if } e_i = 1 \quad i = 0, 1, 2, \ldots n - 1 \\ Y_i & \text{otherwise} \end{cases} \tag{6.31}$$

$$Z_{i+1} = Z_i \otimes Z_i \tag{6.32}$$

$$y = Y_n \otimes 1 \tag{6.33}$$

An example computation is given in Table 6.6, which corresponds to Table 6.4.
An algorithm obtained from Eqs. 6.11–6.14 is

$$\widetilde{x} = xR \bmod m \tag{6.34}$$

$$\widetilde{1} = R \bmod m \tag{6.35}$$

$$Z_n = \widetilde{1} \tag{6.36}$$

$$Y_{i-1} = \begin{cases} \widetilde{x} \otimes Z_i & \text{if } e_{i-1} = 1 \quad i = n - 1, n - 2, \ldots, 1 \\ Z_i & \text{otherwise} \end{cases} \tag{6.37}$$

$$Z_{i-1} = Y_i \otimes Y_i \tag{6.38}$$

$$y = Y_0 \otimes 1 \tag{6.39}$$

An example computation is given in Table 6.7, which corresponds to Table 6.5.
    An architecture for the implementation of the second algorithm is shown in
Fig. 6.2. (As with Fig. 6.1, we leave it to the reader to devise a single-multiplier
architecture.) The Montgomery residues $\widetilde{x}$ and $\widetilde{1}$ are assumed to have been computed
beforehand, in the manner described in Sect. 5.2.3. The design of an architecture
for the algorithm of Eqs. 6.27–6.33 is similarly straightforward, with one more
multiplexer, and will have the advantage that in an implementation the two
multipliers can operate in parallel, thus giving better performance.

Table 6.7 Example of Montgomery modular exponentiation

$x = 14$,	$e = 25 = 11001_2$,	$m = 23$,	$R = 32$	
$R^{-1} = 18$,	$\tilde{x} = 11$,	$\tilde{1} = 9$		
i	e_i	Z_i		Y_i
5	–	9		–
4	1	$11^2 * 18 \bmod 23 = 16$		$11 * 9 * 18 \bmod 23 = 11$
3	1	$17^2 * 18 \bmod 23 = 4$		$11 * 16 * 18 \bmod 23 = 17$
2	0	$4^2 * 18 \bmod 23 = 12$		4
1	0	$12^2 * 18 \bmod 23 = 16$		12
0	1	–		$11 * 16 * 18 \bmod 23 = 17$
$y = 14^{25} \bmod 23 = 17 * 1 * 18 \bmod 23 = 7$				

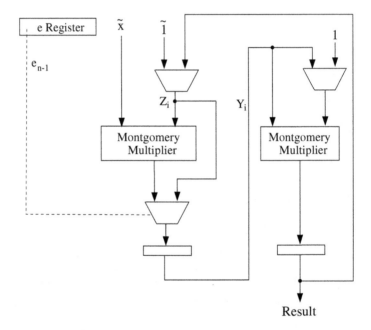

Fig. 6.2 Montgomery modular-exponentiation unit

For high performance, there has been work in developing high-radix Montgomery exponentiation algorithms. Most of the architectures are primarily based on corresponding high-radix Montgomery-multiplication algorithms; examples will be found in [2, 3]. As well, the use of residue number systems has been investigated [4]; the usual problems with such systems remain.

6.2 Inversion and Division

For a given modulus m it is not always the case that inverses with respect to m exist; but they do when m is prime, which is the typical case in cryptography. The following result is an assertion on the existence of modular multiplicative inverses.

Theorem 6.1 *If* $\gcd(a, m) = 1$, *then there exists a unique* x *(the inverse of* a*) such that*

$$ax \equiv 1 \pmod{m}$$

(We will use $|a^{-1}|_m$ or a^{-1} mod m to denote the inverse of a with respect to m and just write a^{-1} if the modulus is evident from the context.)

Fermat's Little Theorem provides a straightforward method for the computation of modular inverses in most cases of interest.

Theorem 6.2 (Fermat's Little Theorem) *If* p *is prime and* $\gcd(a, p) = 1$), *then*

$$a^{p-1} \equiv 1 \pmod{p}$$

Multiplying both sides of this equation by a^{-1}:

$$a^{-1} \equiv a^{p-2} \pmod{p} \tag{6.40}$$

Therefore, inversion can be done easily through exponentiation, which is described above, and nothing more need be said on that. We next describe another method.

The *Euclidean Algorithm* is a well-known method for the computation of the greatest common divisor (gcd) of two integers. The *Extended Euclidean Algorithm* is version that gives the integer solutions, x and y, in the following statement.

Theorem 6.3 (Bezout's Lemma) *If* a *and* b *are nonzero integers, then there exist integers* x *and* y *such that*

$$\gcd(a, b) = ax + by \tag{6.41}$$

Given that the multiplicative inverse of a modulo m exists if and only if $\gcd(a, m) = 1$, Eq. 6.41 indicates that a method for computing the gcd can be used as a basis for one to compute a^{-1} mod m.

If

$$mx + ay = 1 \tag{6.42}$$

then

$$(mx \bmod m + ay \bmod m) \bmod m = 1$$

$$ay \bmod m = 1$$

and therefore

$$y = a^{-1} \bmod m$$

We next describe paper-and-pencil versions of the basic Euclidean Algorithm and the extended version and then look at the binary versions suitable for computer implementations.

The essence of the basic Euclidean Algorithm is the expression for the ordinary integer division of x by y, with quotient q and remainder r:

$$r = x - qy \qquad 0 \le r < y \tag{6.43}$$

The algorithm consists of repeatedly applying an iterative version of Eq. 6.43, along with the observation that if $r = 0$, then $\gcd(x, y) = q$. To compute $\gcd(a, b)$, with $a \ge b > 0$, we repeatedly compute quotients and remainders, Q_i and R_i, as follows.

$$R_0 = a \tag{6.44}$$

$$R_1 = b \tag{6.45}$$

$$Q_i = \lfloor R_{i-1}/R_i \rfloor \qquad i = 1, 2, \ldots, n \tag{6.46}$$

$$R_{i+1} = R_{i-1} - Q_i R_i \tag{6.47}$$

where n is the smallest number such that $R_{n+1} = 0$. At the end, $R_n = \gcd(a, b)$.

Theorem 6.4 *The algorithm of Eqs. 6.44–6.47 computes $R_n = \gcd(a, b)$.*

Example 6.1 Suppose $a = 3054$ and $b = 162$. Then the computation of $\gcd(3054, 162)$ consists of the computations

$$R_0 = 3054$$
$$R_1 = 162$$

$$R_2 = 3054 - 18 * 162 = 138$$
$$Q_2 = \lfloor 3054/162 \rfloor = 18$$

$$R_3 = 162 - 1 * 138 = 24$$
$$Q_3 = \lfloor 162/138 \rfloor = 1$$

$$R_4 = 138 - 5 * 24 = 18$$
$$Q_4 = \lfloor 24/18 \rfloor = 1$$

Table 6.8 Example
application of Extended
Euclidean Algorithm

i	R_i	Q_i	X_i	Y_i
0	161	–	1	0
1	28	5	0	1
2	21	1	1	−5
3	7	3	−1	6
4	0	–	–	–

$$R_5 = 24 - 1 * 18 = 6$$

$$Q_5 = \lfloor 18/3 \rfloor = 6$$

$$R_6 = 18 - 3 * 6 = 0$$

So, $\gcd(3054, 162) = R5 = 6$.   □

The Extended Euclidean Algorithm is an extension of the basic algorithm to compute the solutions x and y in Theorem 6.3 (Eq. 6.41):

$$R_0 = a$$
$$R_1 = b$$
$$X_0 = 1$$
$$X_1 = 0$$
$$Y_0 = 0$$
$$Y_1 = 1$$
if $(R_i > 0)$ **then**
$$\qquad Q_i = \lfloor R_{i-1}/R_i \rfloor$$
$$\qquad R_{i+1} = R_{i-1} - Q_i R_i \qquad\qquad i = 1, 2, 3 \ldots$$
$$\qquad X_{i+1} = X_{i-1} - Q_i X_i$$
$$\qquad Y_{i+1} = Y_{i-1} - Q_i Y_i$$
end if

If on termination $R_{n+1} = 0$, then

$$a X_n + b Y_n = R_n = \gcd(a, b) \tag{6.48}$$

An example computation is given in Table 6.8: $a = 161, b = 28$: $\gcd(a, b) = R_3 = 7$, and $161 X_3 + 28 Y_3 = 161 * (-1) + 28 * 6 = 7$.

For modular inversion, $\gcd(a, b) = 1$; so, from Eq. 6.48:

$$a X_n + b Y_n = 1$$

$$b Y_n \bmod a = 1$$

$$Y_n = b^{-1} \bmod a$$

Table 6.9 Example
computation of multiplicative
inverse

i	R_i	Q_i	Y_i
0	11	–	0
1	8	1	1
2	3	2	−1
3	2	1	3
4	1	1	−4
5	**1**	1	**7**
6	0	–	–

Table 6.10 Example
computation of multiplicative
inverse

i	R_i	Q_i	Y_i
0	100	–	0
1	23	4	1
2	8	2	−4
3	7	1	9
4	1	7	**−13**
5	0	–	–

The values of X_i need not be computed. Thus the algorithm for the computation
of a^{-1} mod m is

$$R_0 = m$$
$$R_1 = a$$
$$Y_0 = 0$$
$$Y_1 = 1$$
if $(R_i > 0)$ **then**
$$\qquad Q_i = \lfloor R_{i-1}/R_i \rfloor$$
$$\qquad R_{i+1} = R_{i-1} - Q_i R_i \qquad\qquad i = 1, 2, 3 \ldots$$
$$\qquad Y_{i+1} = Y_{i-1} - Q_i Y_i$$
end if

An example computation is shown in Table 6.9: $\gcd(41, 7) = R_3 = 1$, and
8^{-1} mod $11 = Y_5 = 7$. The result returned by the algorithm, as given, can
be negative. If a positive result is required—and this is almost always the case—
then a correction should be made by adding the modulus. An example is given in
Table 6.10: 23^{-1} mod $100 = Y_4 = -13$, and adding the modulus gives 87.

Modular division is defined as multiplication by a multiplication inverse and is
usually carried out as such:

$$\frac{a}{b} \bmod m \stackrel{\triangle}{=} a * b^{-1} \bmod m$$

Thus the computation may consist of an application of the Extended Euclidean Algorithm (to compute b^{-1}) followed by a modular multiplication (Sect. 5.2). It is, however, simple to modify the former algorithm so that division is included in the inversion process [7]:

$$R_0 = m$$
$$R_1 = b \qquad\qquad 0 < b < m \text{ and } \gcd(b, m) = 1$$
$$Y_0 = 0$$
$$Y_1 = a \qquad\qquad 0 \leq a < m$$
$$\textbf{if } (R_i > 0) \textbf{ then}$$
$$\qquad Q_i = \lfloor R_{i-1}/R_i \rfloor$$
$$\qquad R_{i+1} = R_{i-1} - Q_i R_i \qquad i = 1, 2, 3 \ldots$$
$$\qquad Y_{i+1} = Y_{i-1} - Q_i Y_i \bmod m$$
$$\textbf{end if}$$

The iterations terminate when $R_{n+1} = 0$, for some n, and then $Y_n = ab^{-1}$ mod m. An example computation is given in Table 6.11: $27/17 \bmod 41 = 4$.

For hardware implementation there are related binary algorithms that are more suitable than those above, as they do not involve division or multiplications [1, 6–10]. In some of these algorithms modular addition or subtraction is the most complex operation, and some do not require even that (since modular additions and subtractions can be replaced with conditionals and ordinary additions and subtractions). Almost all such algorithms are based on the binary gcd algorithm.

The binary gcd algorithm is based on the following facts, for any integers a and b.

- If a and b are both even, then $\gcd(a, b) = 2 \gcd(a/2, b/2)$.
- If a is even and b is odd, then $\gcd(a, b) = \gcd(a/2, b)$.
- If both a and b are odd, then

  - if $a > b$, then $\gcd(a, b) = \gcd(a - b, b)$, and $a - b$ is even;
  - otherwise $\gcd(a, b) = \gcd(b - a, a)$, and $b - a$ is even.

This may be applied recursively in an algorithm for the computation of $a^{-1} \bmod m$, $a < m$. The following is such an algorithm, for m odd. Note that the "do-

Table 6.11 Example of modular division

i	R_i	Q_i	Y_i
0	41	–	0
1	17	2	27
2	7	2	28
3	3	2	12
4	1	3	4
5	0	–	–

Table 6.12 Example of
modular inversion

i	U_i	V_i	R_i	S_i
0	11	8	0	1
1	11	4	0	6
2	11	2	0	3
3	11	1	0	7

Table 6.13 Example of
modular inversion

i	U_i	V_i	R_i	S_i
0	13	5	0	1
1	8	5	12	1
2	4	5	6	1
3	2	5	3	1
4	1	5	8	1

Table 6.14 Example of
modular division

i	U_i	V_i	R_i	S_i
0	13	5	0	7
1	8	5	9	7
2	4	5	11	7
3	2	5	12	7
4	1	5	6	7

nothing" assignments and the use of the variables marked "*" are solely for syntactical correctness. In particular, in hardware implementation such variables would correspond to wires, not registers.

Since $R_i + m$ mod $\equiv R$ (mod m), $R_i/2$ mod m is computed as $R_i/2$ if R_i is even and $(R_i + m)/2$ if R_i is odd. Therefore, m must be odd.

An example computation is given in Table 6.12, which corresponds to Table 6.9: 8^{-1} mod $11 = S_3 = 7$. Another example is given in Table 6.13: 5^{-1} mod $13 = R_4 = 8$.

As with the Extended Euclidean Algorithm, this one too is easily modified for division. To compute ab^{-1} mod m—with $0 \leq a < m, 0 < b < m$, and gcd$(b, m) = 1$—set $S_0 = b$. An example that corresponds to Table 6.13 is shown in Table 6.14: $(4/5)$ mod $13 = 4 * 5^{-1}$ mod $13 = 4 * 8$ mod $13 = 6$.

Part of an architecture for the inversion algorithm is shown in Fig. 6.3. This is the half that is the U_i–R_i datapath; the V_i–S_i half is similar (with appropriate changes of constants and variable names), and its completion is left to the reader. To check whether a value is even or odd, it suffices to examine to the least significant bit of its representation; in Fig. 6.3 this bit is denoted $\widetilde{u}$ (for U_i) and $\widetilde{v}$ (for V_i). A nominal division by two is effected as a wired shift of one bit-position to the right. The computation of $R_i/2$ mod m requires a multiplexer and an adder. Modular adders and subtractors have been described in Chap. 5; so the rest of the diagram is self-explanatory.

One can readily devise alternative binary gcd algorithms and corresponding architectures on the basis of the three main conditions on which the last algorithm is based. (A "direct algorithm will be found in [1].) As an example, repeated

$$U_0 = m$$
$$V_0 = a$$
$$R_0 = 0$$
$$S_0 = 1$$

if $(U_i \neq 1$ and $V_i \neq 1)$ **then**

    if $(U_i$ is even$)$ **then**

$$R_{i+1}^* = R_i/2 \bmod m \qquad\qquad i = 0, 1, 2 \ldots,$$
$$U_{i+1}^* = U_i/2$$
$$S_{i+1}^* = S_i$$
$$V_{i+1}^* = V_i$$

    end if

    if $(V_i$ is even$)$ **then**

$$S_{i+1}^* = S_i/2 \bmod m$$
$$V_{i+1}^* = V_i/2$$
$$R_{i+1}^* = R_i$$
$$U_{i+1}^* = U_i$$

    end if

    if $(U_i$ and V_i are both odd$)$ **then**

        if $(U_i > V_i)$ **then**

$$R_{i+1}^* = (R_i - S_i) \bmod m$$
$$U_{i+1}^* = U_i - V_i$$
$$S_{i+1}^* = S_i$$
$$V_{i+1}^* = V_i$$

        else

$$S_{i+1}^* = (S_i - R_i) \bmod m$$
$$V_{i+1}^* = V_i - U_i$$
$$R_{i+1}^* = R_i$$
$$U_{i+1}^* = U_i$$

        end if

$$R_{i+1} = R_{i+1}^*$$
$$S_{i+1} = S_{i+1}^*$$
$$U_{i+1} = U_{i+1}^*$$
$$V_{i+1} = V_{i+1}^*$$

end if

if $(U_i = 1)$ **then**

    result $= R_i$

else

    result $= S_i$

end if

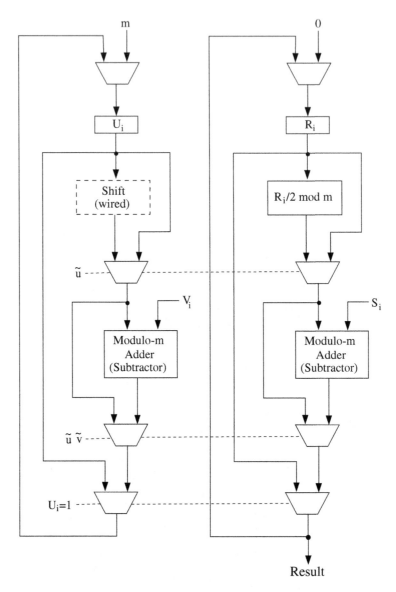

Fig. 6.3 Modular inversion unit

shifting can be accomplished in one "go," in a barrel shifter, instead of (essentially) several cycles through the datapath of Fig. 6.3. Barrel shifter will be costly, but their use will reduce the number of cycles (relative to Fig. 6.3); so there are trade-offs to be considered. The reader will find in [10] a comparative analysis of several binary inversion algorithms, along with comparisons (cost and performance) of corresponding architectures and realizations.

References

1. D. E. Knuth. 1998. *The art of Computer Programming, Vol. 2*. Addison-Wesley, Reading, Massachusetts, U.S.A.
2. N. Nedjah and L. M. Mourelle. 2006. Three hardware architectures for the binary modular exponentiation: sequential, parallel, and systolic. *IEEE Transactions on Circuits and Systems I: Regular Papers*, 53(3):27–633.
3. A. Reza and P. Keshavarzi. 2015. High-throughput modular multiplication and exponentiation algorithms using multibit-scan–multibit-shift technique. *IEEE Transactions on Very Large Scale Integration (VLSI) Systems* , 23(9):1710–1719.
4. F. Gandino, F. Lamberti, G. Paravati, J.-C. Bajard, and P. Montuschi. 2012. An algorithmic and architectural study on Montgomery exponentiation in RNS. *IEEE Transactions on Computers*, 61(8):1071–1083.
5. H. Orup and P. Kornerup. 1991. A high-radix hardware algorithm for calculating the exponential M^E modulo N. *Proceedings, 10th IEEE Symposium on Computer Arithmetic*, pp. 51–56.
6. J. Stein. 1961. Computational problems associated with Racah algebra. *Journal of Computational Physics*, 1:397–405.
7. Takagi, N.: 1998. A VLSI algorithm for modular division based on the binary GCD algorithm. *IEICE Transactions on Fundamentals of Electronics, Communications and Computer Sciences*, E81-A(5):724–728.
8. T. Zhou, X. Wu, G. Bai, and H. Chen. 2002. Fast GF(p) modular inversion algorithm suitable for VLSI implementation. *Electronics Letters*, 38(1):706–707.
9. L. Hars. 2006. Modular inverse algorithms without multiplications for cryptographic applications. *EURASIP Journal on Embedded Systems*, 1:1–13.
10. P. Choi, M. K. Lee, J. T. Kong, and D. K. Kim. 2017. Efficient design and performance analysis of a hardware right-shift binary modular inversion algorithm in GF(p). *Journal of Semiconductor Technology and Science*, 17(3):425–437

Part III

Chapter 7
Mathematical Fundamentals II: Abstract Algebra

Abstract Ordinary arithmetic has the basic operations of addition, subtraction, multiplication, and division defined over the integers and the real numbers. Similar operations can be defined over other mathematical structures—certain subsets of integers, polynomials, matrices, and so forth. This chapter is a short discussion on such generalizations. The first section of the chapter is an introduction to two types of abstract mathematical structures that are especially important in cryptography: *groups* and *fields*. The second section consists of a review of ordinary polynomial arithmetic. The third section draws on the first two sections and covers polynomial arithmetic over certain types of fields. And the last section is on the construction of some fields that are especially important in cryptography.

In the discussions we mention a few mathematical results whose proofs will be found in standard texts on abstract algebra and related subjects [1–3]. For the reader's convenience, most of the proofs are given in Appendix A.

7.1 Groups and Fields

7.1.1 Groups

Definition A **group** is a pair $(S, \circ)$ consisting of a set S and a binary operation $\circ$ such that the following hold.

- (i) For all x and y in S, $x \circ y$ is in S.
- (ii) For all x, y, and z in S, $x \circ (y \circ z) = (x \circ y) \circ z$.
- (iii) There is an identity element e in S such that for all x in S, $x \circ e = e \circ x = x$.
- (iv) For all x in S, there is an element y—called the inverse of x—such that $x \circ y = y \circ x = e$.

(We shall sometimes refer to a "group" when strictly we mean the corresponding set.)

© Springer Nature Switzerland AG 2020
A. R. Omondi, *Cryptography Arithmetic*, Advances in Information Security 77,
https://doi.org/10.1007/978-3-030-34142-8_7

It is straightforward to show that the identity element of a group is unique, and so too is the inverse of each element.

Condition (ii) shows that the positions of the brackets have no "effect." So, we may write $x \circ y \circ z$ for either $x \circ (y \circ z)$ or $(x \circ y) \circ z$, and this may be extended to any number of variables.

Example 7.1 The following are some examples of groups.

- The integers under addition. The identity element is 0, and the inverse of an element is its negation.
- The set of n-by-n matrices of real numbers under matrix addition. The identity element is the matrix of all 0s, and the inverse element is the matrix in which each entry is a negation of the corresponding element in the original matrix.
- The set $\{0, 1\}$ under the logical exclusive-OR operation. The identity element is 0, and the inverse of an element is itself.
- The set of positive rationals under multiplication. The identity element is 1, and the inverse of an element is its reciprocal.
- The set $\{0, 1, 2, \ldots, m - 1\}$ under addition modulo m. The identity element is 0, and the inverse of an element x is $m - x$.
- The set of integers $\{1, 2, 3, \ldots, p - 1\}$ under multiplication modulo a prime p. The identity element is 1, and the inverse of x is $x^{p-2} \bmod p$. (This group is especially important in cryptography.)

On the other hand, the set of all integers under multiplication is not a group, as only 1 and -1 have multiplicative inverses. Nor is the set $0, 1, 2, \ldots, m - 1$ under multiplication modulo m a group: 0 has no inverse, and any other element has an inverse only if it is relatively prime to m.   □

The number of elements in S is known as the *order* of the group, and a group is said to be *finite* if it is of finite order.

A group $(S, \circ)$ is an *Abelian group* (or a *commutative group*) if for all x and y in S, $x \circ y = y \circ x$. The groups in Example 7.1 are also Abelian groups.

If $G = (S, \circ)$ is a group and S' is a subset of S such that $H = (S', \circ)$ is also a group, then H is said to be a *subgroup* of G.

Example 7.1 above shows that the operation $\circ$ may, according to the context, be considered as a generalization of ordinary addition or ordinary multiplication. So, it is common to use $+$ or $*$ (or $\times$ or $\cdot$ or just juxtaposition) for $\circ$, according to whether the operation is considered "additive" or "multiplicative." In the former case, we shall use 0 to denote the identity element and $-x$ to denote the inverse of an element x; and in the latter case we shall use 1 to denote the identity element and $-x^{-1}$ to denote the inverse of an element x. We shall say that a group is *additive* or *multiplicative* according to whether $\circ$ is "additive" or "multiplicative."

With inverses defined, we have straightforward generalizations of ordinary subtraction and division. Subtraction $(-)$ may be defined as the addition of the additive inverse of the subtrahend, and division $(/)$ may be defined as multiplication by the multiplicative inverse of the divisor:

$$x - y \text{ is } x + (-y)$$

$$\frac{x}{y} \text{ is } x * y^{-1}$$

The notational correspondence of ∘ to addition or multiplication may be extended to the case where the operation is applied repeatedly to the same element. Just as in ordinary arithmetic we define multiplication as repeated addition and exponentiation as repeated multiplication, so here too we may define $k * x$ and x^k, for an integer k and group element x, in terms of the group operations $+$ and $*$, as

$$k * x = \overbrace{x + x + \cdots + x}^{k \text{ times}} \tag{7.1}$$

$$x^k = \overbrace{x * x * \cdots * x}^{k \text{ times}} \tag{7.2}$$

according to whether ∘ is considered additive or multiplicative. (Unless confusion is possible, we shall frequently write kx instead of $k * x$.) We stipulate that $0 * x = 0$ and $x^0 = 1$, i.e., the corresponding identity elements.

On the basis of the preceding notational interpretations, we also have the following, for integers k and m integers and group element x.

Additive notation	Multiplicative notation
$-kx = k(-x)$	$x^{-k} = \left(x^{-1}\right)^k$
$kx + mx = (k+m)x$	$x^k * x^m = x^{k+m}$
$k(mx) = (km)x$	$\left(x^k\right)^m = x^{km}$

where $-kx$ is the inverse of kx, and x^{-k} is the inverse of x^k.

An important notion arises from how the "multiplication" and "exponentiation" are related to the identity elements. We shall say that x is of *order* n (or that n is the *order* of x) if x if n is the smallest positive integer such that $nx = 0$ or $x^n = 1$, according to whether the group is additive or multiplicative. If there is no such n, then x is said to be of infinite order.

Continuing with the analogies: In the ordinary arithmetic of real numbers, if $x^y = z$, for some numbers x, y, and z, then we say that y is the *logarithm* of z with respect to the *base* x. Given that the interpretation above of ∘ as addition or multiplication is somewhat arbitrary, we may use that terminology in either case of a relationship to Eq. 7.1 or 7.2. In particular, if $y = kx$, then we shall also refer to k as the logarithm of y with respect to x, even though in ordinary terminology the notation is not suggestive of that.

Definition A group G is said to be **cyclic** if there is an element g in G such that for every element x in G, $x = kg$ or $x = g^k$, for some integer k, according to whether the group operation is considered additive or multiplicative.

In such case, g is known as a *generator* of G, and we say that g *generates* G.

As examples, the set of multiples of 5 is cyclic, with 5 as generator; and the group of positive integers under addition is cyclic, with 1 as generator.

It can be shown that if G is a cyclic group and H is a subgroup of G, then H will also be cyclic; that is, there will be some element of H that generates all its elements.

For cryptography the primary interest is in finite groups, i.e., those of finite order. If g is a generator for such a group and n is the order of the group, then $ng = 0$ or $g^n = 1$ and the group's entire set of elements is the set $\{g, 2g, 3g, \ldots, ng\}$ or $\{g, g^2, g^3, \ldots, g^n\}$, according to whether the group is additive or multiplicative.

Example 7.2 5 is generator for the set $\{0, 1, 2, 3, 4, 5\}$ under addition $(+)$ modulo 6. With the abbreviation of Eq. 7.1:

$$1 * 5 \bmod 6 = 5 \qquad 4 * 5 \bmod 6 = 2$$
$$2 * 5 \bmod 6 = 4 \qquad 5 * 5 \bmod 6 = 1$$
$$3 * 5 \bmod 6 = 3 \qquad 6 * 5 \bmod 6 = 0$$

and so on, in cyclic repetition.

2 is a generator for the set $\{1, 2, 4\}$ under multiplication $(*)$ modulo 7. With the abbreviation of Eq. 7.2:

$$2^1 \bmod 7 = 2 \qquad 2^4 \bmod 7 = 2$$
$$2^2 \bmod 7 = 4 \qquad 2^5 \bmod 7 = 4$$
$$2^3 \bmod 7 = 1 \qquad 3^6 \bmod 7 = 1$$

and so on, in cyclic repetition.

3 is a generator for the set $\{1, 2, 3, 4, 5, 6\}$ under multiplication $(*)$ modulo 7:

$$3^1 \bmod 7 = 3 \qquad 3^4 \bmod 7 = 4$$
$$3^2 \bmod 7 = 2 \qquad 3^5 \bmod 7 = 5$$
$$3^3 \bmod 7 = 6 \qquad 3^6 \bmod 7 = 1$$

and so on, in cyclic repetition.   □

Many cryptosystems are based on the problem of computing logarithms in certain cyclic groups.

Discrete Logarithm Problem

Let G be a cyclic group of finite order n, g be a generator for the group, and x be an element of G. The *discrete logarithm* of x, with respect to base g, is the unique integer k such that $0 \leq k \leq n - 1$ and $x = kg$ or $x = g^k$, according to whether the group is additive or multiplicative. Finding k is known as the *Discrete Logarithm Problem*, and it is a very difficult task if n is large.

7.1.2 Fields

Definition A **field** is a triplet $(S, \circ, \bullet)$ consisting of a set S and two binary operations $\circ$ and $\bullet$ such that the following hold.

  (i) $(S, \circ)$ is an Abelian group.
 (ii) For all x and y in S, $x \bullet y$ is in S.
(iii) If S^* is S with the identity element under $\circ$ excluded, then $(S^*, \bullet)$ is an Abelian group.
 (iv) For all x, y, and z in S, $x \bullet (y \circ z) = (x \bullet y) \circ (x \bullet z)$.

We will refer to $(S, \circ)$ as the *additive group* of F and to $(S^*, \bullet)$ as the *multiplicative group*.

Example 7.3 With addition for $\circ$ and multiplication $\bullet$, the following are fields.

- The set of real numbers under ordinary addition and multiplication.
- The set of complex numbers with complex addition and multiplication.
- The set $\{0, 1, 2, \ldots, p\}$ under addition and multiplication modulo a prime p.
- The set of rational functions, with addition and multiplication defined appropriately.

But the set of all integers under ordinary integer addition and multiplication is not a field because there is no group structure with respect to multiplication: almost all integers do not have multiplicative inverses.   □

The preceding examples show that, as above for groups, we may regard $\circ$ and $\bullet$ as generalizations of ordinary addition and multiplication, refer to them as such, and use $+$ and $*$ (or $\times$ or $\cdot$ or just juxtaposition) for notational convenience. We shall use 0 for the additive identity element, 1 for the multiplicative identity element, $-x$ for the additive inverse of x, and x^{-1} for the multiplicative inverse of x.

Finite fields—i.e., those of finite order—are especially important in cryptography. Such a field is also known as a *Galois field*, and $GF(q)$ will denote a finite field of q elements.

Example 7.4 The following are finite fields.

- $GF(2)$: the set $\{0, 1\}$ with logical XOR as $+$ and logical AND as $*$

- GF(p): the set $\{0, 1, 2, \ldots, p - 1\}$ under addition and multiplication modulo a prime p.

$\square$

A generator for the nonzero elements of GF(q) under multiplication is known as a *primitive element*.[1] As an example, if p is prime, then a primitive root of p is a primitive element of GF(p).

We shall make frequent use of the following two significant results.

Theorem 7.1 *For every element a in GF(q)*

$$a^q = a$$

That is, $a^{q-1} = 1$.

(One may view this as a generalization of Fermat's Little Theorem of Part II of the text.)

Corollary 7.1 *If α is a nonzero element of GF(q), then it is a root of $x^{q-1} - 1$.*

Theorem 7.2 *The nonzero elements of GF(q) form a cyclic group under multiplication.*

It is also a fundamental and important result that

(i) every Galois field has p^m elements for some prime p and positive integer m, and

(ii) for every prime p and positive integer m, there is a Galois field of p^m elements.

There are nominally different ways to represent GF(p^m), for given p and m, but all these will give essentially the same field, even though they might "look" different; that is, the constructions will be isomorphic. So for our purposes it will suffice to consider just a couple of ways to construct the field.

For cryptography, the most important fields are GF(p), with p prime, and GF(2^m), with m a positive integer[2]; the former is known as a *prime field* and the latter as a *binary field*. The construction of GF(p) is straightforward: the elements are represented by the set of integers $\{0, 1, 2, \ldots, p - 1\}$, and the operations $+$ and $*$ are addition and multiplication modulo p. Such a construction will not work for a binary field and, in general, for a non-prime field GF(p^m), p prime and $m > 1$; that can readily be seen by considering, for example, GF(2^3). The construction of GF(p^m) is somewhat more complicated than that of GF(p) and is described in Sect. 7.4.

Hereafter, GF(p) will denote a prime field (i.e., p prime), p will be taken to be prime in GF(p^m), and GF(q) will denote a field with q prime or composite.

---

[1]One may view this as a generalization of the concept of *primitive root* of Part II of the text.

[2]The field GF(3^m) is also much studied but practical uses are relatively rare.

7.2 Ordinary Polynomial Arithmetic

This section is a review of ordinary polynomial arithmetic; the next section is on polynomial arithmetic over finite fields, which is our primary interest for cryptography. Note that unlike the "normal" situation, our main interest in arithmetic here is not in *evaluating* the polynomials, i.e., in computing a number, $f(x)$, given a number x and a polynomial p.

A polynomial $f(x)$ over a field F, in the *indeterminate* x, is given by expression

$$f(x) = f_n x^n + f_{n-1} x^{n-1} + \cdots + f_1 x + f_0$$

where the *coefficients* f_i are elements of F, and n is a nonnegative integer. Examples of typical fields used are the set of integers under the usual addition and multiplication, the set of rational numbers under the usual addition and multiplication, and the set of real numbers under the usual addition and multiplication.

The polynomial in which all the coefficients are 0 is the *zero polynomial*, and a polynomial with $f_n = 1$ is a *monic polynomial*.

The *degree* of a polynomial is the exponent of the leading term—i.e., n in the expression above—provided that $f_n \neq 0$; thus the degree of $f(x) = f_0$ is zero, provided $f_0 \neq 0$. The degree of the zero polynomial is taken to be $-\infty$.

Take two polynomials $a(x) = \sum_{i=0}^{n} a_i x^i$ and $b(x) = \sum_{i=0}^{n} b_i x^i$, and assume, without loss of generality, that $n \geq m$. Then addition and multiplication are given by

$$a(x) + b(x) = \sum_{i=0}^{n} (a_i + b_i) x^i$$

$$a(x)b(x) = \sum_{i=0}^{n+m} \left(\sum_{j=0}^{i} a_k b_{j-i} \right) x^i$$

The case for addition also covers subtraction $(-)$: Let $\overline{b(x)}$ be the polynomial whose coefficients are the additive inverses of the corresponding ones in $b(x)$. Then $a(x) - b(x)$ is defined to be $a(x) + \overline{b(x)}$.

Example 7.5 Suppose $a(x) = 4x^4 + x^3 + x^2 + 1$ and $b(x) = 2x^3 + 3x + 1$. Then

$$a(x) + b(x) = 4x^4 + 3x^3 + x^2 + 3x + 2$$
$$a(x) - b(x) = 4x^4 - x^3 + x^2 - 3x$$
$$a(x) * b(x) = 8x^7 + x^6 + 13x^5 + 7x^4 + 6x^3 + x^2 + 3x + 1$$

□

Fig. 7.1 Multiplication arrays

$$r^5 \quad r^4 \quad r^3 \quad r^2 \quad r^1 \quad r^0$$

$$
\begin{array}{ccc}
 & a_2 & a_1 & a_0 \\
 & b_2 & b_1 & b_0 \\
\hline
 & a_2b_0 & a_1b_0 & a_0b_0 \\
 a_2b_1 & a_1b_1 & a_0b_1 \\
 a_2b_1 & a_1b_1 & a_0b_1 \\
\hline
c_5 \quad c_4 & c_3 & c_2 & c_1 & c_0
\end{array}
$$

(a)

$$x^4 \qquad x^3 \qquad x^2 \qquad x^1 \qquad x^0$$

$$
\begin{array}{ccc}
 & a_2 & a_1 & a_0 \\
 & b_2 & b_1 & b_0 \\
\hline
 & a_2b_0 & a_1b_0 & a_0b_0 \\
 & a_2b_1 & a_1b_1 & a_0b_1 \\
 a_2b_1 & a_1b_1 & a_0b_1 \\
\hline
 a_2b_2x^4 & (a_2b_2+ & (a_1b_0+ & (a_1b_0 & a_0b_0 \\
 & a_1b_2)x^3 & a_1b_1+ & a_0b_1)x \\
 & & a_0b_1)x^2
\end{array}
$$

(b)

We next consider multiplication in slightly more detail. Let $a_n a_{n-1} \cdots a_1 a_0$ be the ordinary radix-r representation of some number a:

$$a = a_n r^n + a_{n-1} r^{n-1} + \cdots a_1 r + a_0 \tag{7.3}$$

and $(f_n f_{n-1} \cdots f_1 f_0)$ be the representation of some polynomial $f(x)$:

$$f(x) = f_n x^n + f_{n-1} x^{n-1} + \cdots f_1 x + f_0 \tag{7.4}$$

There is similarity between Eqs. 7.3 and 7.4. In the interpretation, however, there is a significant difference: a_i is related to r, in that both are numbers and $a < r$, but there is no such relationship between f_i and x. The similarity is reflected in the fact that polynomial addition is carried out in a manner similar to ordinary addition; that is, pairs of "digits" (a_i or f_i) at the same position are added together. And the difference is reflected in the fact that in ordinary addition there are carries from one position to the next—when the radix is exceeded—but there is no corresponding effect in polynomial addition. These difference and similarity carry over to multiplication when that operation is taken as the addition of multiplicand multiples.

Figure 7.1a shows the multiplication array for the product of $a_2 a_1 a_0$ and $b_2 b_1 b_0$ in radix r; the value of f_i ($i = 0, 1, 2, 3, 4, 5$) is determined by the sum of the elements in column i and any carry from the sum in column $i - 1$. Figure 7.1b shows the array for the multiplication of $(a_2a_1a_0)$ and $(b_2b_1b_0)$, polynomials in x; the column sums do not involve any carries. As in Fig. 7.1a, each row of the multiplicand array of Fig. 7.1b corresponds to the product of the multiplicand and a "digit" of the multiplier:

$$(a_2x^2 + a_1x + a_0)(b_2x^2 + b_1x + b_0) = (a_2x^2 + a_1x + a_0)b_0$$
$$+ (a_2x^2 + a_1x + a_0)b_1x$$
$$+ (a_2x^2 + a_1x + a_0)b_2x^2$$

The significance of the preceding remarks is that algorithms for polynomial multiplication may have the same general form as those for ordinary multiplication but without carry-propagation issues.

The division of polynomials is similar to the ordinary paper-and-pencil "long division" of integers [4]. A polynomial $a(x)$ is *divisible* by a polynomial $b(x)$ if there is a polynomial $q(x)$ such that $a(x) = q(x)b(x)$. The *greatest common divisor* (gcd) of two polynomials is the monic polynomial of highest degree that divides both. A polynomial $f(x)$ is said to be *irreducible* if it cannot be expressed as the product of two non-constant polynomial; monic irreducible polynomial is known as a *prime polynomial*. A polynomial might be irreducible over one field but not another; for example $x^2 - 3$ is irreducible over the field of rationals but not over the field of reals, and $x^2 + x + 1$ is irreducible over GF(2) but not over GF(4). We will make much of polynomials that are irreducible over GF(p) but not over GF(p^m).

The ordinary division of an integer a by an integer b yields a quotient q and a remainder r such that $a = qb + r$ and $r < b$. Correspondingly, the division of a polynomial $a(x)$ by a polynomial $b(x)$ yields a quotient polynomial $q(x)$, and remainder polynomial $r(x)$ such that $a(x) = q(x)b(x) + r(x)$, with the degree of $r(x)$ of less than that of $b(x)$.

With the underlying field taken to be set of real numbers under the usual addition and multiplication, a sketch of the polynomial division algorithm is as follows, for a dividend $a(x)$ of degree n and divisor $b(x)$ of degree m.

If $n < m$, then $q(x) = 0$ and $r(x) = a(x)$. Otherwise, we proceed in a manner similar to the paper-and-pencil division of integers: Start with a partial remainder that is initially equal to the dividend, and repeatedly reduce by multiples of the divisor. With each reduction, a part of the quotient is formed, according to the multiplying factor used to form the corresponding divisor multiple. What is left of the partial remainder at the end is the remainder polynomial.

The general algorithm for polynomial division is as follows. Suppose we wish to divide x by y, where

$$a(x) = a_nx^n + a_{n-1}x^{n-1} + \cdots + a_1x + a_0$$
$$b(x) = b_mx^m + b_{m-1}x^{m-1} + \cdots + b_1x + b_0$$

Let N_i denote the ith partial remainder, Q_i denote the ith partial quotient, and $a_j^{(i)}$ denote coefficient j of the ith partial remainder, with $a_j^{(0)} = a_j$. The algorithm is

$$N_0(x) = x$$

$$Q_0(x) = 0$$

$$k = n - m$$

$$q_{i+1}(x) = \frac{a_{n-i}^{(i)}}{b_m} x^{k-i} \qquad\qquad i = 0, 1, 2, \ldots, k$$

$$N_{i+1}(x) = N_i(x) - q_{i+1}(x)b(x)$$

$$q(x) = Q_{k+1}(x)$$

$$r(x) = N_{k+1}(x)$$

Example 7.6 Suppose $a(x) = 4x^4 + 8x^3 - 2x^2 + 7x - 1$ and $b(x) = 2x^2 + 1$. Then the "long division" of $a(x)$ by $b(x)$ is

$$
\begin{array}{r}
2x^2 + 4x - 2 \\
\hline
2x^2 + 1 \,\big|\, 4x^4 + 8x^3 - 2x^2 + 7x - 1 \\
-4x^4 \qquad\quad - 2x^2 \\
\hline
8x^3 - 4x^2 + 7x - 1 \\
-8x^3 \qquad - 4x \\
\hline
-4x^2 + 3x - 1 \\
+4x^2 \qquad + 2 \\
\hline
3x + 1
\end{array}
$$

$$q(x) = 2x^2 + 4x - 2 \text{ and } r(x) = 3x + 1$$

□

Example 7.7 Mechanically translating the long division of Example 7.6 above into the preceding algorithm:

$$b(x) = 2x^2 + 1 \qquad (m = 2)$$

$$\overset{\triangle}{=} b_2 x^2 + b_0$$

$$N_0(x) = 4x^4 + 8x^3 - 2x^2 + 7x - 1$$

$$\overset{\triangle}{=} a_4^{(0)} x^4 + a_3^{(0)} x^3 + a_2^{(0)} x^2 + a_1^{(0)} x + a_0^{(0)} \qquad (n = 4)$$

$$Q_0(x) = 0$$

$$k = 4 - 2 = 2$$

$$q_1(x) = \frac{a_4^{(0)}}{b_2} x^2 = 2x^2 \qquad\qquad (i = 0)$$

$$N_1(x) = (4x^4 + 8x^3 - 2x^2 + 7x + 1) - 2x^2(2x^2 + 1)$$
$$= 8x^3 - 4x^2 + 7x - 1$$
$$\overset{\triangle}{=} a_3^{(1)}x^3 + a_2^{(1)}x^2 + a_1^{(1)}x + a_0^{(1)}$$

$$Q_1(x) = 2x^2$$

$$q_2(x) = \frac{a_3^{(1)}}{b_2} x = 4x \qquad\qquad (i = 1)$$

$$N_2(x) = (8x^3 - 4x^2 + 7x - 1) - 4x(2x^2 + 1)$$
$$= -4x^2 + 3x - 1$$
$$\overset{\triangle}{=} a_2^{(2)}x^2 + a_1^{(2)}x + a_0^{(2)}$$

$$Q_2(x) = 2x^2 + 4x$$

$$q_3(x) = \frac{a_2^{(2)}}{b_2} - 1 = -2 \qquad\qquad (i = 2)$$

$$N_3(x) = (-4x^2 + 3x - 1) + 2(2x^2 + 1)$$
$$= 3x + 1$$
$$Q_3(x) = 2x^2 + 4x - 2$$

We end up with $q(x) = 2x^2 + 4x - 2$ and $r(x) = 3x + 1$.   □

For a computer implementation of the preceding algorithm, two points are worth noting. First, the powers in x are simply "place-holders" and play no role in the arithmetic; that is, the "division array" (Example 7.6) may consist of just coefficients. Second, the computation of q_{i+1} always involves the division of coefficients by b_m, and this can be simplified by reducing it to a multiplication by a pre-computed "reciprocal"—i.e., a multiplicative inverse—or by selecting b_m to be a power of two, the standard operational radix for computer implementation. The division can also be eliminated by working with a monic polynomial, if possible. A third point that is not evident in the preceding: In ordinary division (Sect. 1.3) determining the divisor multiple to be subtracted at each step is not always easy. We will see that there is no such difficulty in polynomial division in cryptography.

7.3 Polynomial Arithmetic Over Finite Fields

The aforementioned division of coefficients shows why they must be elements of a field: if that is not so, then multiplicative inverses might not exist, in which case it would not be possible to divide. Thus, for example, the coefficients may not be taken from the set of all integers, but they may be taken from an appropriate set of modular residues. We shall accordingly speak of a *polynomial defined over a field*.

For cryptography the most important field is GF(p). If the polynomial coefficients are restricted to the elements of GF(p), then addition and multiplication are still straightforward, but the coefficients in the results are now taken modulo p.

Example 7.8 Consider the set of polynomials over GF(5), and take $a(x) = 4x^3 + 4x^2 + 3x + 1$ and $b(x) = 2x^3 + x^2 + x + 1$. Then

$$a(x) + b(x) = 6x^3 + 5x^2 + 4x + 2$$
$$= x^3 + 4x + 2 \quad \text{(coefficients mod 5)}$$

$$a(x) * b(x) = 8x^6 + 12x^5 + 14x^4 + 13x^3 + 8x^2 + 4x + 1$$
$$= 3x^6 + 2x^5 + 4x^4 + 3x^3 + 3x^2 + 4x + 1 \quad \text{(coefficients mod 5)}$$

$\square$

Polynomial subtraction is as the addition of the polynomial whose coefficients are the additive inverses of the coefficients of the subtrahend.

Example 7.9 Take the same polynomials as in Example 7.8: The modulo-5 additive inverses of the coefficients 2 and 1 in $b(x)$ are 3 and 4, respectively. So, the subtraction of $b(x)$ from $a(x)$ is the addition of $\overline{b(x)} = 3x^3 + 4x^2 + 4x + 4$, the additive inverse of $b(x)$:

$$a(x) - b(x) = a(x) + \overline{b(x)}$$
$$= 7x^3 + 8x^2 + 7x + 5$$
$$= 2x^3 + 3x^2 + 2x \quad \text{(coefficients mod 5)}$$

$\square$

Division is more complicated than addition and subtraction, as it also includes the use of multiplicative inverses.

Example 7.10 Suppose we wish to divide $3x^5 + 2x^4 + 2x^3 + x$ by $5x^3 + 4x^2 + x$ over GF(7). In "long-division" form, with subtraction as the addition of the negation (i.e., inverse) of the subtrahend:

$$
\begin{array}{r}
2x^2 + 3x\ + 1 \\[2pt]
\hline
\end{array}
$$

$$
5x^3 + 4x^2 + x \ \overline{\big)\ 3x^5 + 2x^4 + 2x^3 \qquad\quad + x}
$$

$$
\underline{+4x^5 + 6x^4 + 5x^3}
$$

$$
x^4 \qquad\qquad + x
$$

$$
\underline{+6x^4 + 2x^3 + 4x^2}
$$

$$
2x^3 + 4x^2 + x
$$

$$
\underline{+5x^3 + 3x^2 + 6x}
$$

$$
0 \tag{7.5}
$$

□

The explanation for the various steps in this:

- The power of x for the first term of the quotient is $x^5/x^3 = x^2$. For the corresponding coefficient of the quotient, the modulo-7 multiplicative inverse of 5 (the first coefficient of the divisor) is 3. So, to divide 3 by 5 (modulo 7), we multiply 3 by 3 (modulo 7), which gives 2. The first term of the quotient polynomial is therefore $2x^2$.

  The multiple of the divisor to be subtracted is $2x^2(5x^3 + 4x^2 + x) = 3x^5 + x^4 + 2x^3$ (with modulo-7 coefficients), which corresponds to the addition of $4x^5 + 6x^4 + 5x^3$ (in light of subtraction as addition of the inverse). This is the modulo-7 addition of the coefficients shown in the first subtraction.
- For the second term of the quotient, the power in x is $x^4/x^3 = x$, and the coefficient is $1 * 3 = 3$. So the subtraction is that of $3x(5x^3 + 4x^2 + x) = x^4 + 5x^3 + 3x^2$ (with modulo-7 coefficients), which is the addition of $6x^4 + 2x^3 + 4x^4$, whence the second subtraction.
- Lastly, for the third term, the power in x is $x^3/x^3 = 1$, and the coefficient is $5 * 3 \bmod 7 = 1$. The polynomial to be subtracted is $5x^3 + 4x^2 + x$—i.e., $2x^3 + 3x^2 + 6x$—is to be added, whence the third subtraction.

Addition with coefficients in GF(2) is quite simple: the additive inverse of an operand is exactly the same operand, so subtraction is exactly the same as addition, which greatly simplifies division (and multiplication).

Example 7.11 The "long division" of $x^6 + x^5 + x^4 + x^2 + x$ by $x^4 + x + 1$ over GF(2) is

$$
\begin{array}{r}
x^2 + x\ + 1 \\[2pt]
\hline
\end{array}
$$

$$
x^4 + x + 1 \ \overline{\big)\ x^6 + x^5 + x^4 \qquad\ + x^2 + x}
$$

$$
\underline{+x^6 \qquad\qquad\ + x^3 + x^2}
$$

$$
x^5 + x^4 \qquad\qquad + x
$$

$$
\underline{+x^5 \qquad\qquad + x^2 + x}
$$

$$
x^4 + x^3 + x^2
$$

$$
\underline{+x^4 \qquad\qquad + x + 1}
$$

$$
x^3 + x^2 + x + 1
$$

☐

As indicated and shown above, the arithmetic is really on just the coefficients. So, division over GF(2) may be expressed more succinctly by dropping the powers of x and the arithmetic reduced to simple arithmetic on binary strings. And in GF(2), the addition of a bit pair is just their logical exclusive-OR. Thus, for the division of Example 7.11 we would have[3]

$$
\begin{array}{r}
1\,1\,1 \\
1\,0\,0\,1\,1 \, \overline{)\, 1\,1\,1\,0\,1\,1\,0} \\
1\,0\,0\,1\,1\,0\,0 \\
\hline
1\,1\,0\,0\,1\,0 \\
1\,0\,0\,1\,1\,0 \\
\hline
1\,1\,1\,0\,0 \\
1\,0\,0\,1\,1 \\
\hline
1\,1\,1\,1
\end{array}
$$

The significance of this particular example is that the field GF(2^m) is especially important in cryptography, and the field may be constructed in terms of polynomials over GF(2) (Sect. 7.4 and Chap. 10).

The *polynomial reduction* of a polynomial $a(x)$ relative to a modulus polynomial $m(x)$ is defined in a manner similar to that for integers: divide $a(x)$ by $m(x)$ and the remainder polynomial, of degree less than that of $m(x)$. We shall sometimes express this as $a(x)$ mod $m(x)$.

Polynomial congruences too may be defined in a manner similar to congruences for integers (Sect. 2.1). Two polynomials $a(x)$ and $b(x)$ are *congruent* modulo of a polynomial $m(x)$ if they leave the same remainder upon division by $m(x)$. This may be expressed as

$$a(x) \equiv b(x) \pmod{m(x)}$$

Properties similar to those given in Sect. 2.1 for integers also hold here. For polynomials $a(x), b(x), c(x), d(x)$, and $m(x)$:

- $a(x) \equiv b(x) \pmod{m(x)}$ if and only if $a(x)$ and $b(x)$ leave the same remainder on division by $m(x)$
- if $a(x) \equiv b(x) \pmod{m(x)}$, then $b(x) \equiv a(x) \pmod{m(x)}$
- if $a(x) \equiv b(x) \pmod{m(x)}$ and $y \equiv c(x) \pmod{m(x)}$, then $a(x) \equiv c(x) \pmod{m(x)}$
- if $a(x) \equiv b(x) \pmod{m(x)}$ and $c(x) \equiv d(x) \pmod{m(x)}$, then $a(x) + c(x) \equiv b(x) + d(x) \pmod{m(x)}$ and $a(x)c(x) \equiv b(x)d(x) \pmod{m(x)}$

---

[3]Note that in terms of powers of x, the binary strings are to be interpreted from right to left.

- if $a(x) \equiv b(x) \pmod{m(x)}$, then $a(x) + c(x) \equiv b(x) + c(x) \pmod{m(x)}$ and $a(x)c(x) \equiv b(x)c(x) \pmod{m(x)}$
- if $a(x) \equiv b(x) \pmod{m(x)}$, then $a(x)^k \equiv b(x)^k \pmod{m(x)}$ for any positive integer k

And just as in Chap. 2 we have arithmetic for integer residues with respect to integer moduli, so here too we may define arithmetic for polynomial residues with respect to polynomial moduli, by considering the remainders in polynomial division and with the basic operations of polynomial addition and multiplication. We will be especially interested in cases in which the modulus is a prime polynomial: the set of integers $\{0, 1, 2, \ldots, p - 1\}$, with p prime, form a field under residue addition and multiplication and is the standard construction of GF(p); similarly, the set of polynomial residues with respect to a prime polynomial form a field under modular polynomial addition and multiplication, and that field corresponds to GF(p^m) for a modulus polynomial of degree m over GF(p).

7.4   Construction of GF(p^m)

If q is prime, then the field GF(q) can be constructed easily as the set of integers under addition and multiplication modulo q. That is not possible otherwise, because inverses do not always exist with such a construction: consider, for example, a field of order sixteen.

The following is a short discussion on the construction of non-prime finite fields GF(p^m), where p is a prime and $m > 1$; that is, on the representation of the field's elements, which representation determines how the operations of addition and multiplication are defined. For given p and m, there is really only one finite field with p^m elements. Nevertheless, different constructions facilitate in different ways the implementation of the arithmetic operations; we will see this in the context of different *bases* used for representations (Chaps. 10 and 11).

By Corollary 7.1, if a is a nonzero element of GF(p^m), then a is a root of $x^{p^m-1} - 1$. So, a nonzero element of GF(p^m) may be expressed in terms of the roots of an irreducible polynomial that divides $x^{p^m-1} - 1$. For this, a *primitive polynomial* is especially handy, as a root of such a polynomial is also a primitive element.

Definition Let $f(x)$ be a prime polynomial of degree m over GF(p). Then $p(x)$ is a **primitive polynomial** over GF(p) if

- $x^{p^m-1} - 1$ is divisible by $f(x)$, and
- if $x^k - 1$ is divisible by $f(x)$, then $k \geq p^m - 1$.

Example 7.12 For GF(2^3), $f(x) = x^3 + x + 1$ is a primitive polynomial of degree 3 over GF(2). We have $p^m - 1 = 2^3 - 1 = 7$, because over GF(2) none of $x^4 - 1$, $x^5 - 1$, and $x^6 - 1$ is divisible by $f(x)$, but $x^7 - 1$ is.   □

A primitive polynomial exists over every field GF(p) and in every degree. Indeed, there are several such polynomials in each case: for every p and m, there are exactly $\phi(p^m - 1)/m$ primitive polynomials, where ϕ is Euler's totient function. Such a polynomial can be used to construct the field GF(p^m).

The construction of GF(p^m) is based on the following result.

Theorem 7.3 *A root α in GF(p^m) of a primitive polynomial $f(x)$ of degree m over GF(p) is of order $p^m - 1$ and is therefore a primitive element of GF(p^m).*

(We will be interested in only the properties of a root, not its actual value.)

Since α is of order $p^m - 1$, its powers produce $p^m - 1$ distinct elements—$\alpha^1, \alpha^2, \ldots, \alpha^{p^m-1}$—and these may be used to represent the nonzero elements of GF(p^m).

The relationship between α and $f(x)$ also means that each power of α can be expressed as a polynomial[4]: Since α is a root of $f(x) = x^m + f_{m-1}x^{m-1} + \cdots + f_1 + f_0$; so

$$\alpha^m = (-f_{m-1})\alpha^{m-1} + \cdots + (-f_1)\alpha + (-f_0)$$

where $-f_i$ is the additive inverse of f_i modulo p.

So, any $a = \alpha^i$ can be expressed as polynomial $a(\alpha)$ of degree at most $m - 1$:

$$a(\alpha) = a_{m-1}\alpha^{m-1} + a_{m-2}\alpha^{m-1} + \cdots + a_1\alpha + a_0$$

from which one gets an "ordinary" polynomial by replacing α with the indeterminate x, something we will implicitly assume in much of what follows.

Example 7.13 For the construction of GF(2^4), with $f(x) = x^4 + x + 1$, let α be a root of $f(x)$. The nonzero elements of the field are $\alpha^1, \alpha^2, \alpha^3, \ldots, \alpha^{15}$. These can be expressed as polynomials of degree less than four, as follows.

Since

$$\alpha^4 + \alpha + 1 = 0$$

we have

$$\alpha^4 = -\alpha - 1$$

$$= \alpha + 1 \pmod 2$$

---

[4]Replace α with the indeterminate x.

Table 7.1 Polynomial-basis representation of GF(2^4)

Polynomial	Power	Binary
0	0	(0000)
1	α^{15}	(0001)
α	α^1	(0010)
$\alpha + 1$	α^4	(0110)
α^2	α^2	(0100)
$\alpha^2 + 1$	α^8	(0101)
$\alpha^2 + \alpha$	α^5	(0110)
$\alpha^2 + \alpha + 1$	α^{10}	(0111)
α^3	α^3	(1000)
$\alpha^3 + 1$	α^{14}	(1001)
$\alpha^3 + \alpha$	α^9	(1010)
$\alpha^3 + \alpha + 1$	α^7	(1011)
$\alpha^3 + \alpha^2$	α^6	(1100)
$\alpha^3 + \alpha^2 + 1$	α^{13}	(1101)
$\alpha^3 + \alpha^2 + \alpha$	α^{11}	(1110)
$\alpha^3 + \alpha^2 + \alpha + 1$	α^{12}	(1111)

Another element:

$$\alpha^7 = \alpha^4 \alpha^3 = (\alpha + 1)\alpha^3$$
$$= \alpha^4 + \alpha^3$$
$$= \alpha^3 + \alpha + 1$$

And so forth.

The representations of all sixteen elements of GF(2^4) are given in Table 7.1. □

As noted above, in general, for given p and m, there will be several different primitive polynomials that can be used to construct GF(p^m); for example, for GF(2^4) as above, another choice for $f(x)$ is $x^4 + x^3 + 1$. Further, the polynomial used need not be a primitive one; any irreducible polynomial over the same field will do. What is special about a primitive polynomial is that a root is also a primitive element, and so a "powers" construction such as that of Table 7.1 is possible.

Since for a suitable polynomial and root α of that polynomial, every element of GF(p^m) can be expressed uniquely in the form

$$a_{m-1}\alpha^{m-1} + \cdots + a_2\alpha^2 + a_1\alpha + a_0 \qquad a_i \in \text{GF(p)}$$

set $\{1, \alpha, \alpha^2, \ldots, \alpha^{m-1}\}$ therefore constitutes a basis,[5] known as a *polynomial basis*, for $GF(p^m)$. For $GF(2^m)$ and computer representation, each element will be expressed in binary, as $(a_{m-1}a_{m-2}\cdots a_0)$, where a_i is 0 or 1.

With polynomial representations for $GF(p^m)$ as above—i.e., relative to an irreducible polynomial $f(x)$—addition in $GF(p^m)$ may be defined as polynomial addition with the coefficients reduced modulo p, and multiplication in $GF(p^m)$ may be defined as polynomial multiplication with polynomial reduction modulo $f(\alpha)$. Accordingly, $f(\alpha)$ is also known here as a *reduction polynomial*.

To simplify reduction, the polynomial used should have as few nonzero coefficients as possible. Ideally, we want the polynomial to be both sparse and primitive, and some polynomials are both. The polynomials most often used in cryptography are *trinomials* (of three terms) and *pentanomials* (of five terms). Some examples of standard reduction polynomials are as follows [5].

$$x^{163} + x^7 + x^6 + x^3 + 1$$

$$x^{233} + x^{74} + 1$$

$$x^{283} + x^{12} + x^7 + x^5 + 1$$

$$x^{409} + + x^{87} + 1$$

$$x^{571} + x^{10} + x^5 + x^2 + 1$$

Example 7.13 suggests that multiplication can also be carried out very easily in the "powers" representation, by adding exponents modulo $p^m - 1$, which is indeed the case. For example, with Table 7.1:

$$\left(\alpha^3 + \alpha^2 + 1\right)\left(\alpha^2 + 1\right) = \alpha^{13}\alpha^8$$
$$= \alpha^{15}\alpha^6$$
$$= \alpha^6$$
$$= \alpha^3 + \alpha^2$$

Multiplication in this form can be implemented simply by using lookup tables, but the tables are likely to be impracticably large for the large values of m used in cryptography.

Several bases other than polynomial bases exist for $GF(p^m)$, but one type in particular is much used in cryptography: a *normal basis*, from the following result.

---

[5]The terminology comes from linear algebra: a set of k linearly independent vectors is said to be a basis for a k-dimensional vector space if any vector in the space can be expressed as a linear combination of the vectors in the set. The structures here can be shown to be vector spaces in the standard sense.

Table 7.2 Normal-basis
representation of GF(2^4)

(0000) 0	(1000) β
(0001) β^8	(1001) $\beta + \beta^8$
(0010) β^4	(1010) $\beta + \beta^4$
(0011) $\beta^4 + \beta^8$	(1011) $\beta + \beta^4 + \beta^8$
(0100) β^2	(1100) $\beta + \beta^2$
(0101) $\beta^2 + \beta^8$	(1101) $\beta + \beta^2 + \beta^8$
(0110) $\beta^2 + \beta^4$	(1110) $\beta + \beta^2 + \beta^4$
(0111) $\beta^2 + \beta^4 + \beta^8$	(1110) $\beta + \beta^2 + \beta^4 + \beta^8$

Theorem 7.4 *Let β be a root in GF(p^m) of an irreducible polynomial $f(x)$ of degree m over GF(p). Then β^p, β^{p^2}, ..., $\beta^{p^{m-1}}$ are the other roots of $f(x)$.*

Since β, β^p, β^{p^2}, ..., $\beta^{p^{m-1}}$ are linearly independent over GF(p^m), they form a basis, and each element of GF(p^m) can be expressed uniquely in the form

$$b_0\beta + b_1\beta^p + b_2\beta^{p^2} + \cdots + b_{m-1}\beta^{p^{m-1}} \qquad b_i \in \text{GF(p)}$$

with the binary computer representation $(b_0 b_1 \cdots b_{m-1})$, $b_i \in \{0, 1\}$, for GF(2^m). The multiplicative identity element is represented by[6] $(111 \cdots 1)$, and the additive identity element is represented by $(00 \cdots 0)$. Table 7.2 gives examples of normal-basis representations for GF(2^4).

We are interested primarily in the field GF(2^m), for which β in a good basis[7] is generally determined in one of three ways [6]:

- If $m + 1$ is prime and 2 is a primitive root of $m + 1$, then β is taken to be a primitive $(m + 1)$st root of unity. The basis in such a case is known as a *Type I optimal normal basis*.
- If
  - $2m + 1$ is prime and 2 is a primitive root of $2m + 1$, or
  - $2m + 1$ is prime, $2m + 1 \equiv 3 \,(\text{mod } 4)$, and 2 generates the quadratic residues of $2m + 1$,

  then β is taken to be $\alpha + \alpha^{-1}$, where α is a primitive $(2m + 1)$st root of unity. This type of basis is known as a *Type II optimal normal basis*.
- When neither of the above is the case, there are other good bases, known as *Gaussian normal bases* (Chap. 11).[8]

---

[6]This seemingly unusual choice is explained in Sect. 11.1.

[7]What exactly constitutes a good basis depends on the complexity of arithmetic operations with the basis and is explained in Chap. 11.

[8]The Types I and II bases are in fact just special cases of this class.

References

1. J. B. Fraleigh. 2002. *A First Course in Abstract Algebra*. Addison-Wesley, Boston, USA.
2. R. Lidl and H. Niederreiter. 1994. *Introduction to Finite Fields and their Applications*. Cambridge University Press, Cambridge, UK.
3. R. E. Blahut. 1983. *Theory and Practice of Error Control Codes*. Addison-Wesley, Reading, Massachusetts, USA.
4. D. E. Knuth. 1998. *The Art of Computer Programming, Vol. 2*. Addison-Wesley, Reading, Massachusetts, USA.
5. National Institute of Standards and Technology. 1999. Recommended Elliptic Curves for Federal Government Use. Gaithersburg, Maryland, USA.
6. D.W. Ash, I.F. Blake, and S.A. Vanstone. 1989. Low complexity normal bases. *Discrete Applied Mathematics*, 25:191–210.

Chapter 8
Elliptic-Curve Basics

Abstract This chapter covers the essentials of elliptic curves as used in cryptography. The first section of the chapter gives the basics concepts of elliptic curves: the main defining equations for the curves of interest and an explanation of the arithmetic operations of "addition" and "multiplication" in the context of elliptic curves. We shall follow standard practice and first define elliptic curves over the field of real numbers, with geometric and algebraic interpretations of the arithmetic operations in relation to points on a curve.

Elliptic curves over the field of real numbers are not useful in cryptography, but the initial interpretations given are useful as a means of visualizing and understanding the arithmetic operations and the derivation of the relevant equations that are ultimately used in practice. In cryptography, the elliptic curves used are defined over finite fields, and the second section of the chapter covers that, with a focus on the two most commonly used fields: $GF(p)$, with p prime, and $GF(2^m)$, with m a positive integer (The main aspect of the first two sections is the definition of *point addition* and *point multiplication*, the latter being the primary operation in elliptic-curve cryptosystems.) The third section is on the implementation of point multiplication. And the last section is on *projective coordinates*, which simply inversion relative to the "normal" *affine coordinates*.

The following is a very basic introduction to elliptic curves. The reader will find much more in the published literature, e.g., [1–4].

8.1 Basic Curves

An *elliptic curve over a field F* is a cubic defined by the equation

$$y^2 + a_1xy + a_2y = x^3 + a_3x^2 + a_4x + a_5 \tag{8.1}$$

where $a_1, \ldots, a_5$ are constants in F that satisfy certain conditions.

© Springer Nature Switzerland AG 2020

A. R. Omondi, *Cryptography Arithmetic*, Advances in Information Security 77,
https://doi.org/10.1007/978-3-030-34142-8_8

The points of the curve consist of all (x, y) that satisfy the equation together with a designated point $\varnothing$—called *the point at infinity*—whose role is analogous to that of zero in ordinary addition.

In cryptography two "instances"[1] of Eqs. 8.1 are commonly used;

$$y^2 = x^3 + ax + b \qquad a, b \in F \tag{8.2}$$

and

$$y^2 + xy = x^3 + ax + b \qquad a, b \in F \tag{8.3}$$

subject to the conditions[2] $4a^3 + 27b^2 \neq 0$ for Eqs. 8.2 and $b \neq 0$ for Eqs. 8.3; the significance of these conditions will become apparent below. We shall assume only these curves in this and subsequent chapters, and past the preliminaries they will be associated with the fields GF(p), for Eqs. 8.2, and GF(2^m), for Eqs. 8.3.

The key arithmetic operation in elliptic-curve cryptography is *point multiplication*, whose role analogous to that of exponentiation in the modular-arithmetic cryptosystems of Part II of the book. As with ordinary arithmetic, "multiplication" here is defined in terms of "addition"—*point addition*, which at its core is defined in terms of ordinary arithmetic operations. We shall therefore first discuss the addition of two points and then proceed to point multiplication (which is of a scalar and a point).

8.1.1 Point Addition: Geometric Interpretation

In the geometric interpretation, with the field of real numbers, the definition of the addition of two points on an elliptic curve is based on the line through the two points and the point at which that line intersects the elliptic curve at hand. That gives the general case. It may, however, be the case that one of the points is $\varnothing$ or that the line through the two points does not intersect the curve; these may be considered special cases. For the curve $y^2 = x^3 + ax + b$, the details are as follows.

For $P \neq Q$ and neither point $\varnothing$, there are two subcases to consider:

- If $P \neq -Q$, where $-Q$ is the reflection of Q on the x-axis, then the line through P and Q will intersect the curve at exactly one point. That point is designated $-R$, and its reflection on the x-axis is taken as the result, R, of the addition: $P + Q = R$. (We may also imagine that the equation has been obtained from $P + Q - R = \varnothing$.)

---

[1] These are obtained from Eqs. 8.1 by a change of variables.
[2] This ensures that the polynomial has distinct roots and no singularities.

- If $P = -Q$, then, strictly, the line will not intersect the curve at any point. We may however take $\varnothing$ as the nominal point of intersection.[3] In this case, $R = P + (-P) = \varnothing$.

For $P = Q$ and neither point is $\varnothing$, the nominal line through the two points is taken to be the tangent[4] at P. There are two subcases to consider:

- If the tangent is not parallel to the y-axis, then it will intersect the curve at some point, $-R$, whose reflection on the x-axis is R, the result of the addition.
- Otherwise, there is no point intersection. Again, we may take $\varnothing$ as the nominal point of intersection.

The operation in the $P = Q$ case is known as *point doubling*, and for hardware implementation it has special significance that will become apparent later.

For the special case when one of the operands is $\varnothing$, we stipulate that $P + \varnothing = \varnothing + P = P$. We also stipulate that $-\varnothing = \varnothing$, which implies that $\varnothing + (-\varnothing) = \varnothing$. So $\varnothing$ is similar to 0 in ordinary arithmetic and to the origin in ordinary geometry.

For certain aspects of the implementation of point multiplication, it is useful to also have the operation of *point subtraction*. As in the case of ordinary integer arithmetic, subtraction may here too be defined as the addition of the negation of the subtrahend. That is, $P - Q$ is $P + (-Q)$.

With addition defined as above, we have an Abelian group:

- the elements of the group are $\varnothing$ and the points on the curve;
- the group operation is point addition;
- the identity element is $\varnothing$; and
- the inverse of a point P is its reflection on the x axis, denoted $-P$.

Much elliptic-curve cryptography is based on computations in very large cyclic subgroups of such groups.

The preceding geometric interpretation is useful in understanding the algebraic interpretation that follows—specifically, the form of the equations—and which is the basis of the equations used in implementation. One can formulate a geometric interpretation for the curve $y^2 + xy = x^3 + ax + b$ too, but we will not do so; what is given above is sufficiently exemplary.

---

[3]One may imagine that $\varnothing$ is the point where parallel lines meet.

[4]"Extrapolate" from the case where $P \neq Q$: if Q approaches P, then the line through the two points approaches the tangent and in the limit is that tangent.

8.1.2 Point Addition and Multiplication: Algebraic Interpretation

We now turn to the algebraic interpretation that corresponds to the preceding geometric interpretation and from which we obtain the basic derivations for the implementation of the point operations. The interpretation is in terms of ordinary arithmetic operations: $+$ will denote point addition or ordinary addition and $*$ (or $\times$ or $\cdot$ or just juxtaposition) will denote ordinary multiplication or point multiplication, depending on the context.

No additional details are necessary for the two special cases that involve $\varnothing$ as an operand or as a result. For any point P:

- $P + (-P) = \varnothing$
- $P + \varnothing = \varnothing + P = P$

Otherwise, the point addition is defined as follows.

The inverse of a point (x, y) will, by definition, be $(x, -y)$ for the curve $y^2 = x^3 + ax + b$ and $(x, x + y)$ for the curve $y^2 + xy = x^3 + ax + b$. Let $P = (x_P, y_P)$ and $Q = (x_Q, y_Q)$ be the two points to be added and $R = (x_R, y_R)$ be the result of the addition.

Addition $y^2 = x^3 + ax + b$

If $P \neq \pm Q$, then the slope of the line through P and Q is

$$\lambda = \frac{y_Q - y_P}{x_Q - x_P} \tag{8.4}$$

and that line intersects the curve at $(x_R, -y_R)$, where

$$x_R = \lambda^2 - x_P - x_Q \tag{8.5}$$

$$y_R = \lambda(x_P - x_R) - y_P \tag{8.6}$$

For point-doubling, taking derivatives:

$$2y\, dy = 3x^2\, dx + a\, dx$$

$$\frac{dy}{dx} = \frac{3x^2 + a}{2y}$$

and substituting $x = x_P$ and $y = y_P$:

$$\lambda = \frac{3x_P^2 + a}{2y_P} \tag{8.7}$$

The result of the addition is then given by

$$x_R = \lambda^2 - 2x_P \tag{8.8}$$

$$y_R = \lambda(x_P - x_R) - y_P \tag{8.9}$$

Addition $y^2 + xy = x^3 + ax + b$

If $P \neq \pm Q$, then the slope of the line through P and Q is

$$\lambda = \frac{y_Q + y_P}{x_Q + x_P} \tag{8.10}$$

and the result of the addition is given by

$$x_R = \lambda^2 + \lambda + x_P + x_Q + a \tag{8.11}$$

$$y_R = \lambda(x_P + x_R) + x_P + x_R \tag{8.12}$$

For point doubling the equations are

$$\lambda = x_P + \frac{y_P}{x_P} \tag{8.13}$$

$$x_R = \lambda^2 + s + a \tag{8.14}$$

$$y_R = x_P^2 + \lambda x_R + x_R \tag{8.15}$$

Multiplication: Both Curves

With the preceding definitions of point addition (+), we can now define *point multiplication*. Let k be a positive integer and P be a point on a given elliptic curve. Then the product of k and P is

$$kP = \overbrace{P + P + \cdots + P}^{k\ Ps}$$

The range of k may be extended to zero and negative numbers by defining

$$0P = \varnothing$$

$$(-k)P = k(-P)$$

The set of multiples of a point P under point addition form a cyclic subgroup of the group formed by the elliptic curve under point addition, with P being the generator for the subgroup. Elliptic curve cryptosystems are mostly based on computations in finite subgroups of this type.

The underlying field in all of the preceding is that of the real numbers, with addition and multiplication defined in the usual manner. Point addition and multiplication have been defined in terms of the usual real operations of addition, multiplication, division, and inversion. In general, the definition of these "basic operations" will depend on the field at hand.

8.2 Elliptic Curves Over Finite Fields

Arbitrary elliptic curves are not very useful for cryptography. Curves defined over real numbers, for example, would in a computer require floating-point arithmetic—with the usual difficulties, such as loss of speed in implementation and inaccuracy as a result of rounding errors.[5] For these reasons, elliptic curves in cryptography are defined over finite fields; so the core arithmetic is integer arithmetic. The precise nature of the basic arithmetic operations then depends on what the fields are and how their elements are represented.

The elliptic curves most often used in cryptography are

$$y^2 = x^3 + ax + b$$

with GF(p), p a large prime, and

$$y^2 + xy = x^3 + ax^2 + b$$

with GF(2^m), m a large positive integer. The latter may be generalized to GF(p^m), with p a prime and m a positive integer. ($p = 3$ has received relatively much attention.)

With finite fields, the geometric interpretations of Sect. 8.1.1 are no longer applicable, as we now have discrete points instead of continuous curves, but the algebraic interpretations carry through, with straightforward modifications.

The essence of security elliptic-curve cryptography is the *Elliptic-Curve Discrete-Logarithm Problem*[6]: For a point G—typically known as a *base point*—an elliptic curve over a finite field gives an additive cyclic group of some order n, with G as the generator. Given a point P the problem is to determine a k such that $0 \le k \le n - 1$ and $P = kG$. This is a very difficult problem if n is sufficiently large.

---

[5]It would be awkward to have, say, either encryption or decryption yield different results according to different errors from different sequences of basic arithmetic operations for the same point operation.

[6]The general Discrete-Logarithm Problem is stated in Sect. 7.1.

8.2.1 $y^2 = x^3 + ax + b$ Over GF(p)

The elements of GF(p) may be represented by the integers $0, 1, 2, \ldots, p-1$, which in a computer will be expressed in conventional binary. The additive operation for the field is addition modulo p, and the multiplicative operation is multiplication modulo p. Subtraction is interpreted as the addition of an additive inverse modulo p, and division is interpreted as multiplication by a multiplicative inverse modulo p. So, the basic arithmetic is just the modular arithmetic of Chaps. 2–6.

The curve consists of the point at infinity, $\varnothing$, together with all the points (x, y) that satisfy

$$y^2 = (x^3 + ax + b) \bmod p \qquad a, b \in \mathrm{GF}(p)$$

subject to the condition

$$4a^3 + 27b^2 \bmod p \neq 0$$

Example 8.1 Consider the curve $y^2 = x^3 + x + 6$ over GF(11). Example 2.5 shows that the quadratic residues of 11 are 1, 3, 4, 5, and 9. And Example 2.6 gives solutions y_1 and y_2 for the equation $y^2 \equiv a \pmod{11}$, where a is a quadratic residue. Summarizing:

x	y^2	y_1	y_2
0	6	–	–
1	8	–	–
2	5	4	7
3	3	5	6
4	8	–	–
5	4	2	9
6	8	–	–
7	4	2	9
8	9	3	8
9	7	–	–
10	4	2	9

So the points on the curve are $\varnothing$ and

$$(2, 4) \quad (3, 5) \quad (5, 2) \quad (7, 2) \quad (8, 3) \quad (10, 2)$$
$$(2, 7) \quad (3, 6) \quad (5, 9) \quad (7, 9) \quad (8, 8) \quad (10, 9)$$

$\square$

As examples of practical curves and parameters, the NIST standard includes the curve $y^2 = x^3 - 3x + b$, with the following values of p [5].

$$2^{192} - 2^{64} - 1$$

$$2^{224} - 2^{96} + 1$$

$$2^{256} - 2^{224} - 2^{192} + 2^{96} - 1$$

$$2^{384} - 2^{128} - 2^{96} + 2^{32} - 1$$

$$2^{521} - 1$$

With the basic arithmetic operations now defined as modular-arithmetic operations, for the point addition of $P = (x_P, y_P)$ and $Q = (x_Q, y_Q)$, with result $R = (x_R, y_R)$, Eqs. 8.4–8.6 are modified as follows.[7]

If $P \neq \pm Q$, then

$$\lambda = \left[(y_Q - y_P)(x_Q - x_P)^{-1} \right] \bmod p$$

$$x_R = \left(\lambda^2 - x_P - x_Q \right) \bmod p$$

$$y_R = [\lambda(x_P - x_R) - y_P] \bmod p$$

where $(x_Q - x_P)^{-1}$ is the modulo-p multiplicative inverse of $x_Q - x_P$, the subtraction of $-x_P$ is the addition of modulo-p additive inverse of x_P, and so forth.

Example 8.2 Take the curve $y^2 = x^x + x + 6$ over GF(11), as in Example 8.2. For the addition of $P(5, 9)$ and $Q(2, 7)$:

$$\lambda = \left[(7 - 9)(2 - 5)^{-1} \right] \bmod 11$$

$$= \left[(7 + 2)(2 + 6)^{-1} \right] \bmod 11$$

$$= \left(9 * 8^{-1} \right) \bmod 11$$

$$= (9 * 7) \bmod 11 = 8$$

$$x_R = \left(8^2 - 5 - 2 \right) \bmod 11 = 2$$

$$y_R = (8 \times (5 - 2) - 9) \bmod 11 = 4$$

So $P + Q = R = (2, 4)$.   □

---

[7]Except for GF(2) and GF(3), which need not concern us.

And for point-doubling the equations that correspond to Eqs. 8.7–8.9 are modified to

$$\lambda = \left[\left(3x_P^2 + a\right)(2y_P)^{-1}\right] \bmod p \tag{8.16}$$

$$x_R = \left(\lambda^2 - 2x_P\right) \bmod p \tag{8.17}$$

$$y_R = [\lambda(x_P - x_R) - y_P] \bmod p \tag{8.18}$$

where $(2y_P)^{-1}$ is the modulo-p multiplicative inverse of $2y_P$, the subtraction of $2x_P$ is the addition of the modulo-p additive inverse of $2x_P$, and so forth.

Example 8.3 For the same curve as in Example 8.2, the doubling of $P = (5, 2)$:

$$\lambda = \left[(73 * 5^2 + 1)(2 * 2)^{-1}\right] \bmod 11$$

$$= \left(10 * 4^{-1}\right) \bmod 11$$

$$= (10 * 3) \bmod 11 = 8$$

$$x_R = \left(8^2 - 2 * 5\right) \bmod 11 = 10$$

$$y_R = [8 * (5 - 10) - 2] \bmod 11 = 2$$

So $2P = (10, 2)$. □

8.2.2 $y^2 + xy = x^3 + ax + b$ Over GF(2^m)

The field GF(2^m), which is known as a *binary field*, permits representations that can greatly simplify the implementation of some aspects of point operations, by simplifying the implementation of the corresponding basic-arithmetic operations.

For the point addition of $P = (x_P, y_P)$ and $Q = (x_Q, y_Q)$, with result $R = (x_R, y_R)$, the relevant equations that correspond to Eqs. 8.10–8.12 are as follows. If $P \neq \pm Q$, then

$$\lambda = (y_P + y_Q)(x_P + x_Q)^{-1} \tag{8.19}$$

$$x_R = \lambda^2 + \lambda + x_P + x_Q + a \tag{8.20}$$

$$y_R = \lambda(x_R + x_P) + x_R + y_P \tag{8.21}$$

where $(x_P + x_Q)^{-1}$ is the multiplicative inverse of $x_P + x_Q$.

And for point doubling the equations that correspond to Eqs. 8.13–8.15 are

$$\lambda = x_P + y_P x_P^{-1} \tag{8.22}$$

$$x_R = \lambda^2 + \lambda + a \tag{8.23}$$

$$y_R = x_P^2 + \lambda x_R + x_R \tag{8.24}$$

where x_P^{-1} is the multiplicative inverse of x_P.

In contrast with the situation in Sect. 8.2.1, the definitions here of the arithmetic operations—addition, multiplication, and inversion—are not all straightforward. The definitions depend on the representations of the elements of the field, in ways that are not immediately apparent, and we will not give them in this introduction. Also, although squaring, which is required for Eqs. 8.20, 8.23, and 8.24, may be taken as just multiplication, treating it as a special case allows optimizations in its implementation as in ordinary arithmetic (Sect. 1.3).[8] We next briefly review the two most common representations of elements of $GF(2^m)$—a few more details are given in Sect. 7.4—and make a few remarks on the arithmetic operations.

The field $GF(2^m)$ may be viewed an m-dimensional vector space over $GF(2)$. That is, each element a of $GF(2^m)$ can be expressed in terms of some elements $\{\alpha_0, \alpha_1, \ldots, \alpha_{m-1}\}$—a *basis*—of $GF(2^m)$:

$$a = \sum_{i=0}^{m-1} a_i \alpha_i \qquad a_i \in GF(2)$$

That $GF(2^m)$ is defined in terms of $GF(2)$, which is a very simple field, greatly simplifies the implementation of some operations. We will see this in Chaps. 10 and 11.

In a computer, the 2^m elements of $GF(2^m)$ are easily represented by m-bit binary strings. There are, however, several different ways to interpret such strings; that is, different types of basis are possible. The two most common types of bases used in cryptography are *polynomial bases* and *normal bases*. (Each of the two has its advantages and disadvantages.)

Polynomial Basis A polynomial basis for $GF(2^m)$ is a set $\{x^{m-1}, x^{m-2}, \ldots, x^2, x, 1\}$ such that each element a of $GF(2^m)$ can be expressed as a polynomial

$$a = a_{m-1}x^{m-1} + a_{m-2}x^{m-2} + \cdots + a_2 x^2 + a_1 x + a_0 \qquad a_i \in GF(2)$$

with the binary representation $(a_{m-1}a_{m-2}\cdots a_0)$. The multiplicative identity element of the field is represented by $(00\cdots 1)$, and the additive identity element is represented by $(00\cdots 0)$.

Since the multiplication of two polynomials of degrees up to $m-1$ will yield a polynomial of degree up to $2m-2$, a reduction might be necessary to obtain

---

[8]And this has an effect on the choice of representation.

Table 8.1 Polynomial-basis
representations of GF(2^4)

(0000) 0	(1000) x^3
(0001) 1	(1001) $x^3 + 1$
(0010) x	(1010) $x^3 + x$
(0011) $x + 1$	(1011) $x^3 + x + 1$
(0100) x^2	(1100) $x^3 + x^2$
(0101) $x^2 + 1$	(1101) $x^3 + x^2 + 1$
(0110) $x^2 + x$	(1110) $x^3 + x^2 + x$
(0111) $x^2 + x + 1$	(1111) $x^3 + x^2 + x + 1$

a polynomial of degree at most $m - 1$. That will be done through a *reduction polynomial*, which is a monic irreducible polynomial of degree m over GF(2). A reduction polynomial therefore has a role that is similar to that of p in the GF(p) arithmetic of Part II of the book.

For a given value of m, there will be several irreducible polynomials that can serve as reduction polynomials. For example, for GF(2^4) there are three such polynomials—$x^4 + x + 1$, $x^3 + x^3 + 1$, and $x^4 + x^4 + x^2 + x + 1$—and any one of these may be used because the sets of results they produce are isomorphic to one another. Two types of reduction polynomial are most commonly used in cryptography: *trinomials*, which have the form $x^m + x^k + 1$ ($1 \le k \le m - 1$) and *pentanomials*, which have the form $x^m + x^k + x^j + x^i + 1$ ($\le k < j < i \le m - 1$. These choices, because they contain very few terms, facilitate the implementation of the corresponding arithmetic.

As examples of polynomial-basis representations, take the field GF(2^4). With the polynomial basis $\{x^3, x^2, x, 1\}$, the "binary" and polynomial representations of the 2^4 elements of the field are as shown in Table 8.1.

Normal Basis A *normal basis* for GF(2^m) is a set of m linearly independent elements $\{\beta, \beta^2, \ldots, \beta^{2^{m-1}}\}$ such that each element a of GF(2^m) can be expressed as

$$a = \sum_{i=0} a_i \beta^{2^i} \qquad a_i \in GF(2)$$

with the binary representation[9] $(a_0 a_1 \cdots a_{m-1})$. The multiplicative identity element is represented by[10] $(111 \cdots 1)$, and the additive identity element is represented by $(00 \cdots 0)$. Table 8.2 gives examples of normal-basis representations for GF(2^4).

---

[9]Note that a polynomial-basis element could just as easily be represented by the string $(a_0 a_1 \cdots a_{m-1})$ and a normal-basis element by $(a_{m-1} a_{m-2} \cdots a_0)$. The choices here are simply a matter of convention.

[10]This seemingly unusual choice is explained in Sect. 11.1.

Table 8.2 Normal-basis representations of GF(2^4)

(0000) 0	(1000) β
(0001) β^8	(1001) $\beta + \beta^8$
(0010) β^4	(1010) $\beta + \beta^4$
(0011) $\beta^4 + \beta^8$	(1011) $\beta + \beta^4 + \beta^8$
(0100) β^2	(1100) $\beta + \beta^2$
(0101) $\beta^2 + \beta^8$	(1101) $\beta + \beta^2 + \beta^8$
(0110) $\beta^2 + \beta^4$	(1110) $\beta + \beta^2 + \beta^4$
(0111) $\beta^2 + \beta^4 + \beta^8$	(1110) $\beta + \beta^2 + \beta^4 + \beta^8$

As examples of practical curves and parameters, the NIST standard includes the following specifications [5]:

- $m = 163, 233, 283, 409, 571$
- curves $y^2 + xy = x^3 + x^2 + b$ and $y^2 + xy = x^3 + ax^2 + 1$ ($a = 0$ or 1),

with both normal basis and polynomial basis. The second type of curve is known as a *Koblitz curve*, and its choice of parameters facilitates more efficient arithmetic than with the more general type of curve.

For polynomial basis, the NIST reduction polynomials are

$$m = 163 : x^{163} + x^7 + x^6 + x^3 + 1$$

$$m = 233 : x^{233} + x^{74} + 1$$

$$m = 283 : x^{283} + x^{12} + x^7 + x^5 + 1$$

$$m = 409 : x^{409} + x^{187} + 1$$

$$m = 571 : x^{571} + x^{10} + x^5 + x^2 + 1$$

8.3 Point-Multiplication Implementation

The algorithms given in Chap. 1, for ordinary binary multiplication, can be modified in a straightforward manner for point multiplication. The algorithms are essentially the same, but point addition now replaces ordinary addition, and point multiplication replaces ordinary multiplication. Of significance is that a left shift (i.e., multiplication by two) in ordinary multiplication is now replaced with point doubling, which is why that operation is of special importance.

For the computation of $Z = kP$, where k is a scalar with the binary representation $k_{n-1}k_{n-2} \cdots k_0$ of some positive number k, and P is a point, modifying Eqs. 1.22–1.24 gives us[11]

$$Z_{n-1} = \varnothing \qquad \text{point at infinity} \qquad (8.25)$$

$$Z_{i-1} = 2Z_i + k_i P \qquad \text{point doubling and point addition} \qquad (8.26)$$

$$Z = Z_{-1} \qquad i = n - 1, n - 2, \ldots, 0 \qquad (8.27)$$

An example multiplication is given in Table 8.3, and a corresponding architecture for a sequential implementation is shown in Fig. 8.1. The operational details for the latter are roughly similar to those of Fig. 1.16, except that the operand registers for P and Z_i are of a different sort (as they must hold pairs that define points); left shifting is replaced with point doubling; and the additions are point operations. The k register is an ordinary shift register that shifts one bit-position to the right in each cycle; so k_0 in cycle i is bit i of k. The Multiple Formation consists of a set of AND gates whose output is zero or P. Addition depends on the field at hand and is discussed in detail in Chaps. 10 and 11.

As with ordinary multiplication, the process here can be speeded up by using a computational radix larger than two, with and without recoding. Let r be the radix. And suppose n is divisible by r; if it is not, suitable value can be obtained by extending the representation of k with 0s at the most significant end. Then $k_{n-1}k_{n-2} \cdots k_0$ may be replaced with m radix-r digits: $K_{m-1}K_{m-2} \cdots K_0$, where $m = n/r$.

The modification of the algorithm of Eqs. 8.25–8.27 to a radix-2^r algorithm is

$$Z_{n-1} = \varnothing \qquad \text{point at infinity} \qquad (8.28)$$

$$Z_{i-1} = 2^r Z_i + K_i P \qquad r \text{ point doublings, a point addition} \qquad (8.29)$$

Table 8.3 Example of point multiplication

	$k = 27_{10} = 011011_2$	
i	k_i	Z_i
5	0	$\varnothing$
4	1	$\varnothing$
3	1	P
2	0	$3P$
1	1	$6P$
0	1	$13P$
-1	$-$	$27P$

[11] We leave it to the reader to ascertain that here scanning the multiplier, k, from right to left gives an algorithm that is "awkward" for high-radix computation.

$$Z = Z_{-1} \qquad\qquad i = m-1, m-2, \ldots, 0 \qquad\qquad (8.30)$$

An example multiplication is given in Table 8.4.

Recoding too is straightforward to effect here: the digit K_i in Eqs. 8.29 is replaced with a digit K_i' that is obtained according to the description for Table 1.8. Thus, for example, for radix-4 recoding, Table 8.5 corresponds to Table 1.8a, and an application of the former is given in Table 8.6. With recoding here, the cost of computing negations, an issue that does not arise in ordinary multiplication, should be taken into account.

8.4   Projective Coordinates

In what we have discussed so far, inversion is required to effect division—as multiplication by a multiplicative inverse—and it will be a costly operation however it is implemented. The need for division can, however, be eliminated if the form of the coordinates is changed—from *affine*, which is what is used above, to *projective*.

Fig. 8.1 Sequential point multiplier

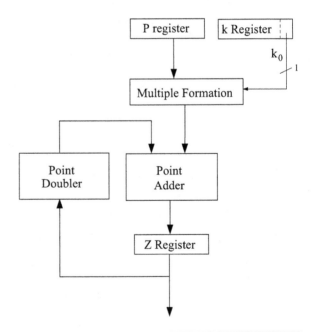

Table 8.4 Example of point multiplication

$k = 27_{10} = 011011_2$, $r = 2$, K_i Shown in binary equivalent		
i	K_i	Z_i
2	01	$\varnothing$
1	10	$\varnothing + P = P$
0	11	$4P + 2P = 6P$
-1	$-$	$24P + 3P = 27P$

Table 8.5 Radix-4 recoding

$k_{i+1}k_i$	k_{i-1}	K_i'
00	0	0
00	1	1
01	0	1
01	1	2
10	0	-2
10	1	-1
11	0	-1
11	1	0

Table 8.6 Example of radix-4 point multiplication

$k = 27_{10} = 011011_2\ \ r = 2,\ K_i$ Shown in decimal equivalent		
i	K_i'	Z_i
2	2	$\varnothing$
1	-1	$\varnothing + 2P = 2P$
0	-2	$8P - P = 7P$
-1	-	$28P - P = 27P$

The basic idea in the use of projective coordinates is that the replacement of a single number with a ratio allows the replacement of division with multiplications. Let us momentarily suppose that field in use is that of the real numbers. (The changes required from that are described in Sect. 8.2.) Suppose the affine coordinates x_P and x_Q are replaced with the ratios X_P/Z_P and X_Q/Z_Q, expressed as $(X_P : Z_P)$ and $(X_Q : Z_Q)$. Then the division of the two is

$$\frac{x_P}{x_Q} = \frac{X_P Z_Q}{X_Q Z_P} = (X_P Z_Q : X_Q Z_P)$$

That is, the division x_P/x_Q may be replaced with the two multiplications $X_P Z_Q$ and $X_Q Z_P$.

A point (x, y) in affine coordinates therefore gets replaced with a point $(X : Y : Z)$, with $Z \neq 0$, in projective coordinates. Computationally, the practical value of such replacement depends on the relative costs of inversion and the multiplications, and this will depend on the particular implementation.

Since $X/Z = (kX)/(kZ)$ and $Y/Z = (kY)/(kZ)$, for all $k \neq 0$, the point $(X : Y : Z)$ is equivalent to all points $(kX : kY : kZ)$. One of a given set of equivalent points may be taken as representative of the former set; that one is usually taken to be the point with $Z = 1$. The points $(X : Y : Z)$ with $Z = 0$ do not correspond to any affine point and constitute the *line at infinity*. Those of such points that lie on a given curve are the *points at infinity* on that curve.

In general, for affine points (x, y) the corresponding projective coordinates have the replacements $x = X/Z^i$ and $y = Y/Z^j$—a "representative" point is $(X/Z^i : Y/Z^j : 1)$—and different types of systems are obtained according to the values of i and j. For example:

- *Standard projective coordinates:* $i = 1, j = 1$
- *Jacobian projective coordinates:* $i = 2, j = 3$
- *Lopez-Dahab projective coordinates:* $i = 1, j = 2$

As an example, we consider the Jacobian system in slightly more detail.

In the Jacobian system $(X : Y : Z)$ represents the affine point $(X/Z^2, Y/Z^3)$. Thus, for example, the curve $y^2 = x^3 + ax + b$ becomes

$$\left(\frac{Y}{Z^3}\right)^3 = \left(\frac{X}{Z^2}\right)^2 + a\left(\frac{X}{Z^2}\right) + b$$

that is,

$$Y^2 = X^3 + aZ^4 + bZ^6$$

$\varnothing$, the point at infinity, corresponds to $(1 : 1 : 0)$, and the negation of $(X : Y : Z)$ is $(X : -Y : Z)$.

In the point addition of the points P and Q to obtain the point R, i.e.,

$$(X_P : Y_P : Q_P) + (X_Q : Y_P : Q_P) = (X_R : Y_R : Q_R)$$

X_R, Y_R, and Z_R are computed as follows.

If $P \neq Q$ and $P \neq -Q$:

$$r = X_P Z_Q^2$$

$$s = X_Q Z_P^2$$

$$u = Y_Q Z_P^3$$

$$v = s - r$$

$$w = u - t$$

$$X_R = -v^3 - 2rv^2 + w^2$$

$$Y_R = -tv^3 + (rv^2 - X_R)w$$

$$Z_R = vZ_P Z_Q$$

which requires 12 multiplications, 4 squarings, and some relatively easy additions.

If $P = Q$, i.e., point doubling:

$$v = 4X_P Y_P^2$$

$$w = 3X_P^2 + aZ_P^4$$

$$X_R = -2v + w^2$$

$$Y_R = -8Y_P^4 + (v - X_R)w$$

$$Z_R = 2Y_P Z_P$$

which requires 4 multiplications, 6 squarings, and some additions.

For $P = -Q$, we have $P + Q = \varnothing$, as usual.

When $a = -3$, point doubling can be made more efficient: 4 multiplications, 4 squarings, and some additions. In this case

$$w = 3\left(X_P^2 + Z_P^4\right)$$

$$= 3\left(X_P + Z_P^2\right)\left(X_P - Z_P^2\right)$$

It is for this reason that for curve $y^2 = x^3 - x + b$, $a = -3$ in the NIST standard [5].

In each of the cases above substantially more multiplications and squarings are required than when affine coordinates are used. Therefore, the extent to which the use of projective coordinates might be beneficial will depend on the particular implementation and the relative time required for inversion relative to multiplication and squaring.

References

1. D. Hankerson, A. Menezes, and S. Vanstone. 2004. *Guide to Elliptic Curve Cryptography*. Springer-Verlag, New York.
2. H. Cohen, G. Frey, et al. 2005. *Handbook of Elliptic and Hyperelliptic Curve Cryptography*. Chapman-Hall/CRC. Boca Raton, USA.
3. I. Blake, G. Seroussi, and N. Smart. 1999. *Elliptic Curves in Cryptography*. London Mathematical Society 265, Cambridge University Press.
4. L. Washington. 2003. *Elliptic Curves: Number Theory and Cryptography*. Chapman-Hall/CRC, Boca Raton, USA.
5. National Institute of Standards and Technology. 1999. Recommended Elliptic Curves for Federal Government Use. Gaithersburg, Maryland, USA.

Chapter 9
Elliptic-Curve Cryptosystems

Abstract This chapter consists of short descriptions of a few elliptic-curve cryptosystems. Examples of three types of cryptosystem are given: *message encryption*, *key agreement*, and *digital signatures*. The descriptions are intended to provide no more than a context for the arithmetic, and the reader who wishes to properly learn about the systems should consult the relevant literature.

Elliptic-curve cryptosystems are generally based on the Elliptic-Curve Discrete Logarithm Problem (Sect. 8.2), which may be phrased thus: If the various parameters have been chosen appropriately, then given a point P that has been computed as the scalar-point product kG, where G is a generator (also known as a *base point*) for the elliptic curve at hand, it is (without additional information) extremely difficult to determine the value of the integer k, even if G is known [2].

Assuming an elliptic curve E of the equation $y^2 + xy = x^3 + ax + b$ or the equation $y^2 = x^3 + ax^2 + b$, an elliptic-curve cryptosystem will typically be specified in several parameters that may be presumed to be known by all parties involved in the use of the system:

- The constants a and b.
- A base point, $G = (x_G, y_G)$.
- p, if the field is GF(p).
- m, if the field is GF(2^m).
- An irreducible polynomial r, if the field is GF(2^m).
- N, the order of E, i.e., the number of points on E.
- n, the order of G, i.e., the smallest number n such that $nG = \emptyset$.
- $h \overset{\triangle}{=} N/n$, a value that should be small and ideally equal to 1.

By way of concrete example, we next give some details from the standard issued by the National Institute of Standards and Technology (NIST) [1].

The NIST standard specifies five curves for prime fields GF (p) and the equation $y^2 = x^3 + ax + b$ and ten curves for the binary fields GF (2^m) and the equation $y^2 + xy = x^3 + ax^2 + b$. For the prime fields the bit-lengths of p are 80, 224, 256,

© Springer Nature Switzerland AG 2020

A. R. Omondi, *Cryptography Arithmetic*, Advances in Information Security 77,
https://doi.org/10.1007/978-3-030-34142-8_9

384, and 521, which values simplify modular reduction (Sect. 4.4); and $a = -3$, which simplifies addition in Jacobian coordinates (Sect. 8.3). For the binary fields, the values of m are 163, 233, 283, 409, and 571, each of which corresponds to the curves $y^2 + xy = x^3 + ax^2 + b$, with $a = 0$ or $a = 1$. That choice of a together with $b = 1$—the *Koblitz curves*—simplifies point multiplication.

As an example, for the curve over the prime field with p of bit-length 192:

- $p = 6277101735386680763835789423207666416083908700390324961279$
- $n = 6277101735386680763835789423176059013767194773182842284081$
- $b = 64210519E59C80E70FA7E9AB72243049FEB8DEECC146B9B1$
- $x_G = 188DA80EB03090F67CBF20EB43A18800F4FF0AFD82FF1012$
- $y_G = 07192B95FFC8DA78631011ED6B24CDD573F977A11E794811$

(The first two values are in decimal, and the other two are in hexadecimal)

As another example, for the binary field with $m = 163$, a curve of the first type, and polynomial basis representation:

- reduction polynomial is $r(\alpha) = \alpha^{163} + \alpha^7 + \alpha^6 + \alpha^3 + 1$
- $n = 5846006549323611672814742448276390689256843201587$
- $b = 20A601907B8C953CA1481EB10512F78744A3205FD$
- $x_G = 3F0EBA16286A2D57EA0991168D4994637E8343E36$
- $y_G = 0D51FBC6C71A0094FA2CDD545B11C5C0C797324F1$

A few more details on NIST parameters are given at the end of Sects. 8.2.1–8.2.2.

Note: In what follows we shall, as in Chap. 2, use $|x^{-1}|_m$ to denote the multiplicative inverse of x modulo m and simply write x^{-1} when the modulus is apparent from the context.

9.1 Message Encryption

In *message encryption*, the problem is the standard one of sending a message that is encrypted in such a way that decryption can be carried out only by the intended recipient. Any information that might be acquired by another, "unauthorized," party should be insufficient for decryption.

A message is assumed to be represented in binary and interpreted as an integer. Within a computer there will be a bound on the magnitude of representable integers; so an excessively long message will be split into several smaller pieces, each of which gets encrypted separately.

Some cryptosystems require a *message-embedding* function F that explicitly maps a messages to a point on an elliptic curve and a corresponding function F^{-1} that reverses the effect of F. An example of such a function is given in Sect. 9.4.

9.1.1 El-Gamal System

This system may be viewed as an elliptic-curve variant of the system described in Section 3.3.1 [3]. With an elliptic curve E and a base point G of order n, the encryption and decryption are as follows.

A receiver randomly selects a number k such that $1 \leq k \leq n - 1$ and then computes

$$P = kG$$

k is kept secret, and P is made public.

To encrypt a message, M, a sender:

(i) Computes $P_M = F(P)$, a point on E to represents M, where F is a message-embedding function.
(ii) Randomly selects a number d (kept secret) such that $1 \leq d \leq n - 1$ and then computes

$$Q = dG$$
$$R = P_M + dP$$

The encrypted message is sent as the pair (Q, R).

To decrypt the message, the receiver that computed the above k and P:

(i) Computes $R - kQ$.
(ii) Applies F^{-1} to the result from (i).

These yield the point that represents the original message:

$$R - kQ = (P_M + dP) - k(dG)$$
$$= P_M + dkG - kdG$$
$$= P_M$$

and $F^{-1}(P_M) = M$.

That the arrangement is secure follows from the fact that the only information that can be seen by a third party consists of kG, dG, and $P_M + dP$. But to extract P_M, it is necessary to also have k, and this is not easily obtained from the public information.

9.1.2 Massey-Omura System

With an elliptic curve E of order N, the encryption of a message M consists of four main steps:

(i) Using a message-embedding function F, the sender represents the message as the point $P_M = F(M)$ on E.

(ii) The sender chooses an integer s such that $\gcd(s, N) = 1$, computes $P = sP_M$, and sends that to the receiver. s is kept secret.

(iii) The receiver chooses an integer r such that $\gcd(r, N) = 1$, computes $Q = rP$, and sends that to the sender. r is kept secret.

(iv) The sender computes $|s^{-1}|_N$ and $R = s^{-1}Q$ and sends the latter to receiver as the encrypted message.

In decryption, the receiver computes $|r^{-1}|_N$ and $r^{-1}R$ (modulo N):

$$r^{-1}R = r^{-1}s^{-1}Q$$
$$= r^{-1}s^{-1}rP_M$$
$$= r^{-1}s^{-1}rsP_M$$
$$= P_M$$

from which the application of F^{-1} yields the original message.

The system is secure because the only information that a third party can acquire consists of P, M, and R, and it is not easy to obtain M from any of that.

9.1.3 Menezes-Vanstone System

This system [4], which is for an elliptic curve over a prime field $GF(p)$, is very similar to the El-Gamal system. The result of encrypting a message consists of a point on an elliptic curve E (with base point G of order n and two values in $GF(p)$).

A receiver first computes a public key and a private key. The private key is a randomly selected number d such that $1 \le d < n$, and the public key is the point $P = dG$.

To encrypt a message M, the sender first randomly selects a number k such that $1 \le k \le n - 1$, splits M into two parts—M_1 and M_2 in $GF(p)$—and then computes

- the point $Q = kG = (x_Q, y_Q)$
- the point $R = kG$
- $x = x_Q M_1 \mod p$, an element of $GF(p)$
- $y = y_Q M_2 \mod p$, an element of $GF(p)$

The encrypted message is the triplet (R, x, y).

For decryption, the receiver first uses d to compute

$$dR = d(kG) = k(dG) = Q$$
$$= (x_Q, y_Q)$$

and then recovers the original message by computing (modulo p):

$$\left|x_Q^{-1}\right|_p x = x_Q^{-1} x_Q M_1 = M_1$$

$$\left|y_Q^{-1}\right|_p y = y_Q^{-1} y_Q M_2 = M_2$$

where x_Q^{-1} and y_Q^{-1} are the multiplicative inverses of x_Q and y_Q modulo p.

The system is secure because the decryption requires d, which cannot be easily obtained from the information that is readily available—R, x, y, and, in particular, the public key.

9.2 Key Agreement

The basic problem in a *key agreement* system is as follows. Two entities wish to agree on a "secret" to be used as a key, or to generate a key, for secure communications. The two communicate through a channel that might be insecure, allowing a third party to eavesdrop. It is necessary to agree on the secret in such a way that any information acquired by the eavesdropper would be insufficient to determine the secret. We next give brief descriptions of three key-agreement systems.

9.2.1 Diffie-Hellman System

This system may be viewed as an elliptic-curve variant of the system described in Section 3.1.1 [5]. The two parties A and B agree on an elliptic curve and a base point G of order n on that curve and then establish a shared secret key as follows.

(i) A randomly selects a number d such that $1 \le d \le n - 1$, computes the point $P = dG$, and sends that to B. d remains known only to A.

(ii) B similarly randomly selects a number k such that $1 \le k \le n - 1$, computes $Q = kG$, and sends that to A. k remains known only to B

(iii) A computes dQ, and B computes kP.

The values computed in (iii) are equal and constitute the shared secret key:

$$dQ = d(kG) = k(dG) = kP$$

The only information that is communicated over the shared channel, and which a third party might have access to, consists of P and Q. In order to determine the secret key, C would have to determine d and k from those, and that is a difficult task.

9.2.2 Matsumoto-Takashima-Imai System

Let G be a base point on the elliptic curve used and n be the order of G. The two parties A and B randomly select two numbers a and b such that $1 \leq a, b \leq n - 1$, and each then establishes a pair of its own keys: a and $P = aG$ for A, b and $Q = bG$ for B. The key agreement is then as follows [6].

First:

(i) A randomly selects a number d such that $1 \leq d \leq n - 1$, computes the point $R = dQ$, and sends that to B.

(ii) B randomly selects a number k such that $1 \leq k \leq n - 1$, computes the point $S = dP$, and sends that to A.

Each party then computes (modulo n) the shared secret key, which is kdG:

(i) A computes $|a^{-1}|_n dS = a^{-1}dkP = a^{-1}dkaG = kdG$.

(ii) B computes $|b^{-1}|_n kR = b^{-1}kdQ = b^{-1}kdbG = kdG$.

where a^{-1} and b^{-1} are the inverses of a and b modulo n.

The system is secure because the only information that a third party can obtain consists of P, Q, R, and S; but a and a^{-1}, b and b^{-1}, k, and d are not easily obtained from that information.

9.2.3 Menezes-Qu-Vanstone System

Let G be a base point on the elliptic curve used and n be the order of G. Each of the two users, A and B, that seek to agree on a shared key first produces two pairs of keys: (a_1, A_1) and (a_2, A_2) for A, (b_1, B_1) and (b_2, B_2) for B, where $1 \leq a_1, a_2, b_1, b_2 \leq n - 1$, and

$$A_1 = a_1 G$$

$$A_2 = a_2 G \overset{\triangle}{=} (x_A, y_A)$$

$$B_1 = b_1 G$$

$$B_2 = b_2 G \overset{\triangle}{=} (x_B, y_B)$$

The keys a_1, a_2, b_1, and b_2 are secret; the other keys are exchanged by the users. The procedure also makes use of a function f [8]:

$$f(x, y) = \left(x \bmod 2^k \right) + 2^k \qquad \text{where } k = \left\lceil \frac{\log_2 n}{2} \right\rceil$$

(If P is a point on an elliptic curve, then $f(P)$ constitute the first k bits of the P's x coordinate.)

At the start, A computes

$$t_A = f(A_2) = f(x_A, y_A)$$
$$e_A = (t_A a_1 + a_2) \bmod n$$

and B computes

$$t_B = f(B_2) = f(x_B, y_B)$$
$$e_B = (t_B b_1 + b_2) \bmod n$$

The values t_A and t_B are exchanged.
    A then computes

$$R_A = e_A(t_B B_1 + B_2) = (x_A, y_A)$$

and B computes

$$R_B = e_B(t_A A_1 + A_2) = (x_B, y_B)$$

$R_A = R_B$, and the shared key is $x_A = x_B$:

$$
\begin{aligned}
R_A &= e_A(t_B B_1 + B_2) \\
&= [(t_A a_1 + a_2) \bmod n](t_B B_1 + B_2) \\
&= [(t_A a_1 + a_2) \bmod n](t_B b_1 G + b_2 G) \\
&= [(t_A a_1 + a_2) \bmod n][(t_B b_1 + b_2) \bmod n]G \quad \text{since } G \text{ is of order } n \\
&= [(t_B b_1 + b_2) \bmod n][(t_A a_1 G + a_2 G) \bmod n]G \\
&= [(t_B b_1 + b_2) \bmod n][(t_A a_1 + a_2) \bmod n] \\
&= [(t_B b_1 + b_2) \bmod n](t_A a_1 + a_2) \\
&= e_B(t_A A_1 + A_2) \\
&= R_B
\end{aligned}
$$

The system is described in more detail in [7].

9.3   Digital Signatures

The *digital-signature* problem is that of "signing" communications—i.e., appending some information to a message—in such a way that the "signature" can be verified as being from the true sender, cannot be forged, cannot be altered, cannot be associated with a message other than the one to which it is attached, and cannot be repudiated by the sender. Of these requirements, we will consider only verification; for the others, the reader is referred to the published literature. One algorithm for digital signatures is as follows [8].

Let G be a base point G of order n on the curve used. The sender then randomly selects a value d such that $1 \leq d \leq n - 1$ and computes

- the point $H = dG$ and
- the value $e = h(M)$

where h is a secure hash function and M is the message to be signed.

To produce the signature, the sender:

(i) Randomly selects a number k such that $1 \leq k \leq n - 1$.
(ii) Computes $P = kG = (x_P, y_P)$.
(iii) Computes $r = x_P \bmod n$.
(iv) If $r = 0$, repeats from (i); otherwise computes

$$s = \left[\left| k^{-1} \right|_n (e + rd) \right] \bmod n$$

(v) If $s = 0$, repeats from (i).

The pair (r, s) is the signature that is appended to the message.

To verify the signature, the receiver computes

(i) $e = h(M)$
(ii) $u = \left(\left| s^{-1} \right|_n e \right) \bmod n$
(iii) $v = \left(s^{-1} r \right) \bmod n$
(iv) the point $Q = uG + vH = (x_Q, y_Q)$

and concludes that the signature is valid if

$$r = x_Q \bmod n$$

This check suffices because $Q = P$, and therefore $r = x_Q \bmod n = x_P \bmod n$: since

$$s = \left[k^{-1}(e + rd) \right] \bmod n$$

we have

$$ks = (e + rd) \bmod n$$

$$k = \left[s^{-1}(e + rd) \right] \bmod n$$

and therefore

$$Q = uG + vH$$
$$= uG + vdG$$
$$= \left[\left(s^{-1}e\right) \bmod n\right]G + \left[\left(s^{-1}rd\right) \bmod n\right]G$$
$$= \left[s^{-1}(e + rd) \bmod n\right]G$$
$$= kG$$
$$= P$$

So, the Q that is computed in step (iv) of the verification phase is exactly the P computed in step (ii) of the signature phase. And a signature cannot be forged because "verifiable" values of r and s requires k, which is known only to the sender and cannot be obtained easily from other values that a third party might be able to acquire.

9.4   Message Embedding

The encryption of a message may require that it be mapped onto a point on an elliptic curve; *message embedding* is a process for that. As an example, the following is a probabilistic algorithm for the curve $y^2 = x^3 + ax + b$ over the field GF(p) [9].

A message M, represented as a binary string, may be interpreted as a positive number and an element of GF(p). The probability that a given positive integer less than p is a quadratic residue modulo p is 1/2. So, if j checks are made to determine whether or not a such number is a quadratic residue of p, the probability of failure is $1/2^j$. Now, suppose that K is large positive number such that for every possible message M it is the case that $(M + 1)K < p$ and that a failure probability of $1/2^K$ is acceptable. Let x_M be a value $MK + k$, $0 \le k < K$, such that $x_M^3 + ax_M + b$ is a quadratic residue of p. If y_M is a corresponding square root, then the message is represented by the point (x_M, y_M) on the elliptic curve. Since $X_M = MK + l$, with $0 \le l < K$, the original message can be recovered by computing $\lfloor x_M/K \rfloor$.

The search for a quadratic residue can be accomplished easily—by, say, running through the values MK, $MK+1$, $Mk+2$, ..., $(M+1)K-1$ and making appropriate checks until a suitable value is found.

References

1. National Institute of Standards and Technology. 1999. Recommended Elliptic Curves for Federal Government Use. Gaithersburg, Maryland, USA.
2. Institute of Electrical and Electronics Engineers. 2000. Standard Specifications For Public-Key Cryptography. Piscataway, New Jersey, USA.

3. T. ElGamal. 1985. A Public-Key Cryptosystem and a Signature Scheme Based on Discrete Logarithms. *IEEE Transactions on Information Theory*, 31(4):469–472.
4. A. J. Menezes and S. A. Vanstone. 1993. Elliptic curve cryptosystems and their implementation. *Journal of Cryptography*, 6(4):209–224.
5. W. Diffie and M. Hellman. 1976. New directions in cryptography. *IEEE Transactions on Information Theory*, 22(6):644–654.
6. T. Matsumoto, Y. Takashima, and H. Imai. 1986. On seeking smart public-key-distribution systems. *IEICE Transactions*, E69-E(2):99–106.
7. A.J. Menezes, M. Qu, M, and S. A. Vanstone. 2005. Some new key agreement protocols providing implicit authentication. In: *Proceedings, 2nd Workshop on Selected Areas in Cryptography* (Ottawa, Canada), pp. 22–32.
8. American National Standards Institute. 1999. ANSI X9.62: Public Key Cryptography for the Financial Services Industry: the Elliptic Curve Digital Signature Algorithm (ECDSA). Washington, D.C., USA.
9. N. Koblitz. 1987. Elliptic curve cryptosystems. *Mathematics of Computation*, 48(177):203–209.

Chapter 10
Polynomial-Basis Arithmetic

Abstract This chapter is on arithmetic and related operations over the field $GF(2^m)$ with polynomial-basis representations. The first section is on addition, subtraction, multiplication, and squaring; although subtraction is just an instance of addition, optimal squaring is not just multiplication with the same operand. The second section is on reduction. And the third section is on exponentiation, inversion, and division.

Polynomial-basis representation is discussed in more detail in Sect. 7.4. Briefly, a polynomial basis is a set $\{\alpha^{m-1}, \alpha^{m-2}, \ldots, \alpha^2, \alpha, 1\}$ such that each element a of $GF(2^m)$ can be expressed uniquely as a polynomial:

$$a = a_{m-1}\alpha^{m-1} + a_{m-2}\alpha^{m-2} + \cdots + a_2\alpha^2 + a_1\alpha + a_0 \qquad a_i \in GF(2)$$

with the binary representation $(a_{m-1}a_{m-2}\cdots a_0)$. As an example, Table 10.1 gives the representations of the sixteen elements of $GF(2^4)$.

The multiplicative identity element of the field is represented by $(00\cdots 1)$, and the additive identity element is represented by $(00\cdots 0)$. For operations other than addition and subtraction, reduction is required with respect to an irreducible polynomial of degree m—the polynomial used to define the field.

For visual clarity, we shall frequently drop the polynomial indeterminate and write $a, b, c, \ldots$ for $a(\alpha), b(\alpha), c(\alpha), \ldots$ the latter mostly to refer to the binary forms. We shall also sometimes "abuse" the language and—for brevity and where no confusion is possible—refer to a "polynomial" in some instances where the precise term ought to be "... representation of ... polynomial."

10.1 Addition and Subtraction

The operands will be

© Springer Nature Switzerland AG 2020

A. R. Omondi, *Cryptography Arithmetic*, Advances in Information Security 77,
https://doi.org/10.1007/978-3-030-34142-8_10

Table 10.1 Representation
of GF(2^4)

Polynomial	Binary
0	(0000)
1	(0001)
α	(0010)
$\alpha + 1$	(0011)
α^2	(0100)
$\alpha^2 + 1$	(0101)
$\alpha^2 + \alpha$	(0110)
$\alpha^2 + \alpha + 1$	(0111)
α^3	(1000)
$\alpha^3 + 1$	(1001)
$\alpha^3 + \alpha$	(1010)
$\alpha^3 + \alpha + 1$	(1011)
$\alpha^3 + \alpha^2$	(1100)
$\alpha^3 + \alpha^2 + 1$	(1101)
$\alpha^3 + \alpha^2 + \alpha$	(1110)
$\alpha^3 + \alpha^2 + \alpha + 1$	(1111)

$$a(\alpha) = \sum_{i=0}^{m-1} a_i \alpha^i = (a_{m-1} a_{m-2} \cdots a_0)$$

$$b(\alpha) = \sum_{i=0}^{m-1} b_i \alpha^i = (b_{m-1} b_{m-2} \cdots b_0)$$

and the result will be

$$c(\alpha) = \sum_{i=0}^{m-1} c_i x_i = (c_{m-1} c_{m-2} \cdots c_0) \qquad i = 0, 1, 2, \ldots, m-1$$

Addition here is quite straightforward: the coefficients of the polynomial
operands are added modulo 2. That is, the result $c = a + b$ is given by

$$c_i = (a_i + b_i) \bmod 2$$

which is just the exclusive-OR operation on a_i and b_i.

Subtraction is the addition of the additive inverse of the subtrahend and is exactly
the same as addition, because a polynomial with coefficients in GF(2) is its own
inverse.

Fig. 10.1 Polynomial-basis
adder

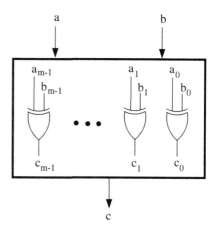

Example 10.1 In GF(2^4), whose elements are given in Table 10.1:

$$(0101) + (1111) = (\alpha^2 + 1) + (\alpha^3 + \alpha^2 + \alpha + 1)$$
$$= \alpha^3 + 2\alpha^2 + \alpha + 2$$
$$= \alpha^3 + \alpha \qquad \text{coefficients modulo 2}$$
$$= (1010)$$
$$(1010) - (0111) = (\alpha^3 + \alpha) - (\alpha^2 + \alpha + 1)$$
$$= (\alpha^3 + \alpha) + (\alpha^2 + \alpha + 1)$$
$$= \alpha^3 + \alpha^2 + 2\alpha + 1$$
$$= \alpha^3 + \alpha^2 + 1 \qquad \text{coefficients mod 2}$$
$$= (1101)$$

□

An architecture for addition will consist of just a set of gates, as shown in
Fig. 10.1.

10.2 Multiplication and Squaring

We consider two types of multiplication. The first is direct multiplication and
squaring, in which the task is to compute $a(\alpha)b(\alpha) \bmod r(\alpha)$ for operands $a(\alpha)$ and
$b(\alpha)$ and irreducible polynomial $r(\alpha)$. The second is Montgomery multiplication,
which here is the polynomial version of the integer Montgomery multiplication of
Sect. 5.2.3 and is the computation of $a(\alpha)b(\alpha)r^{-1}(\alpha)$, where $r(\alpha)$ is an element

of the field and the inversion is with respect to some irreducible polynomial. Montgomery multiplication is especially useful in exponentiation. It is also possible to devise a multiplication algorithm that corresponds to the Barrett-reduction algorithm of Sect. 5.2.2.

10.2.1 Direct Multiplication

Multiplication is more complicated than addition, because the immediate product of two polynomials can require a reduction. That is, the product $c = a * b$ is given by

$$c(\alpha) = \left(\sum_{i=0}^{m-1} a_i \alpha^i \sum_{i=0}^{m-1} b_i \alpha^i \right) \bmod r(\alpha) \tag{10.1}$$

where $r(\alpha)$ is an irreducible polynomial of degree m.

Equation 10.1 may be effected in one of two main ways. The first is to multiply the polynomials and then reduce the result with respect to $r(\alpha)$. And the second is to reduce the partial products in the multiplication as they are formed. We start with a discussion of the former.

Example 10.2 In GF(2^4), as given in Table 10.1, with the reduction polynomial $r(\alpha) = \alpha^4 + \alpha + 1$:

$$
\begin{aligned}
(1001) * (1101) &= (\alpha^3 + 1)(\alpha^3 + \alpha^2 + 1) \\
&= \alpha^6 + \alpha^5 + 2\alpha^3 + \alpha^2 + 1 \\
&= \alpha^6 + \alpha^5 + \alpha^2 + 1 && \text{coefficients mod 2} \\
&= \alpha^3 + \alpha^2 + \alpha + 1 && \text{reduction modulo } r(\alpha) \\
&= (1111)
\end{aligned}
$$

□

The pre-reduction multiplication can be implemented in a manner similar to ordinary binary multiplication—i.e., as a sequence of shifts and additions—but with more simplicity, because the addition here is free of carries. As an example, the pre-reduction multiplication of Example 10.2 is shown in Fig. 10.2, which corresponds to a right-to-left scan of the multiplier (as in paper-and-pencil multiplication).

As indicated in Sect. 7.2, in some respects α plays a role that is similar to the radix in ordinary arithmetic. To the extent that is so, the algorithms for the pre-reduction multiplication here will have the same general forms as those in Chap. 1 for ordinary integer multiplication. But there is one major difference: addition over GF(2) is a very simple operation, which means that the algorithms here, as well as their implementations, will be much simpler.

Fig. 10.2 Pre-reduction
multiplication array

$$\alpha^6\ \alpha^5\ \alpha^4\ \alpha^3\ \alpha^2\ \alpha\ 1$$

```
                    1  0  0  1
                    1  1  0  1
                    1  0  0  1
                 0  0  0  0
              1  1  0  1
           1  1  0  1
           1  1  0  0  1  0  1    column sums mod 2
```

Converting the algorithm of Eqs. 1.25–1.29 for the ordinary, radix-2 integer computation of $z = xy$ from n-bit operands, i.e.,

$$Z_0 = 0 \tag{10.2}$$

$$X_0 = x \tag{10.3}$$

$$Z_{i+1} = Z_i + y_i X_i \qquad i = 0, 1, 2, \ldots, n-1 \tag{10.4}$$

$$X_{i+1} = 2X_i \tag{10.5}$$

$$z = Z_n \tag{10.6}$$

into an algorithm for the polynomial computation of $c = ab$, with reduction by r, yields

$$Z_0 = 0 \tag{10.7}$$

$$A_0 = a \tag{10.8}$$

$$Z_{i+1} = Z_i + b_i A_i \qquad i = 0, 1, 2, \ldots, m-1 \tag{10.9}$$

$$A_{i+1} = \alpha A_i \tag{10.10}$$

$$\tilde{c} = Z_m \tag{10.11}$$

$$c = \tilde{c} \bmod r \tag{10.12}$$

An example of the application of this algorithm is given in Table 10.2.

The algorithm of Eqs. 10.2–10.6 may also be converted into one in which the partial products are reduced along the way:

$$Z_0 = 0 \tag{10.13}$$

$$A_0 = a \tag{10.14}$$

$$Z_{i+1} = (Z_i + b_i A_i) \bmod r \qquad i = 0, 1, 2, \ldots, m-1 \tag{10.15}$$

$$A_{i+1} = \alpha A_i \tag{10.16}$$

$$c = Z_m \tag{10.17}$$

Table 10.2 Example of multiplication followed by reduction

$a(\alpha) = \alpha^3 + 1$, $a = (1001)$; $b(\alpha) = (\alpha^3 + \alpha^2 + 1)$, $b = (1101)$				
$r(\alpha) = \alpha^4 + \alpha + 1$, $m = 4$, $j = m - i - 1$				
i	b_i	A_i	$b_i A_i$	Z_i
0	1	1001	1001	0000
1	0	10010	00000	1001
2	1	100100	100100	1001
3	1	1001000	1001000	101101
4	–	–	1100101	

$\tilde{c}(\alpha) = \alpha^6 + \alpha^5 + \alpha^2 + 1 \overset{\triangle}{=} (1100101)$

$c(\alpha) = \tilde{c}(\alpha) \bmod r(\alpha) \overset{\triangle}{=} \alpha^3 + \alpha^2 + \alpha + 1 \overset{\triangle}{=} (1111)$

Table 10.3 Example of multiplication with interleaved reductions

$a(\alpha) = (\alpha^3 + 1)$ $a = (1001)$; $b(\alpha) = (\alpha^3 + \alpha^2 + 1)$ $b = (1101)$				
$r(\alpha) = \alpha^4 + \alpha + 1$ $m = 4$, $j = m - i - 1$				
i	b_i	A_i	$b_i A_i$	Z_i
0	1	001	1001	0000
1	0	10010	00000	1001
2	1	100100	100100	1011
3	1	1001000	1001000	1111

$c(\alpha) = (1111) = \alpha^3 + \alpha^2 + \alpha + 1$

The application of this algorithm to the operands in Table 10.2 is given in Table 10.3. The main difference between the two corresponding algorithms and tables is in the magnitudes of the Z_i values.

And the algorithm of Eqs. 1.22–1.24, i.e.,

$$Z_0 = 0 \tag{10.18}$$

$$Z_{i+1} = 2Z_i + b_{n-i-1}a \qquad i = 0, 1, 2, \ldots, n-1 \tag{10.19}$$

$$\tilde{c} = Z_n \tag{10.20}$$

may be converted into the algorithm

$$Z_0 = 0 \tag{10.21}$$

$$Z_{i+1} = \alpha Z_i + b_{m-i-1}a \qquad i = 0, 1, 2, \ldots, m-1 \tag{10.22}$$

$$\tilde{c} = Z_m \tag{10.23}$$

$$c = \tilde{c} \bmod r \tag{10.24}$$

An application of this algorithm will largely be similar to that of the algorithm of Eqs. 10.7–10.12, except that the multiplicand multiples and partial products will now be produced in "reverse" order. In particular, the magnitudes of the values produced over the computations will be similar. The reduction that is required for both algorithms is discussed in Sect. 10.2. We next consider the algorithm for which an architecture will be given.

In ordinary, sequential multiplication a right-to-left scan of the multiplier is better than a left-to-right scan, because the algorithm of Eqs. 10.2–10.6 can be optimized in implementation: at step i, the low-order i bits of the partial product are bits of the final product and so need not be included in the addition at that step. Such an optimization is not possible with the algorithm of Eqs. 10.18–10.20 because all bits of the partial product must be involved in the addition. In modular multiplication (Sect. 5.2) the corresponding version of the basic algorithm works quite well with a left-to-right scan, in so far as and the resulting intermediate values are smaller than in the original algorithm, but the reductions are not easy to carry out. The situation is different here: the reductions required with the interleaving of additions and reductions are easy.

$$Z_0 = 0 \tag{10.25}$$

$$Z_{i+1} = (\alpha Z_i + b_{m-i-1}a) \bmod r \qquad i = 0, 1, 2, \ldots, m-1 \tag{10.26}$$

$$c = Z_m \tag{10.27}$$

The equation that is effected is

$$c(\alpha) = (\alpha(\cdots(\alpha(\alpha b_{m-1}a + b_{m-2}a) \bmod r(\alpha) + b_{m-3}a) \bmod r(\alpha) \cdots$$

$$b_1 a) + b_0 a) \bmod r(\alpha) \tag{10.28}$$

As an example, Table 10.4 shows the application of the algorithm to the operands of Example 10.2 and Table 10.1.

Recall that the role of the reduction polynomial in polynomial-basis arithmetic is similar to that of the modulus in modular arithmetic (Chap. 5). But in the modular version of the algorithm of Eqs. 10.25–10.27 the reduction is not easy. It requires determining if the modulus has been exceeded and then subtracting

Table 10.4 Example multiplication with interleaved reductions

$a(\alpha) = \alpha^3 + 1$, $a = (1001)$; $b(\alpha) = \alpha^3 + \alpha^2 + 1$, $b = (1101)$			
$r(\alpha) = \alpha^4 + \alpha + 1$, $m = 4$, $j = m - i - 1$			
i	b_j	$b_j a$	Z_i
0	1	1001	0000
1	1	1001	1001
2	0	0000	1000
3	1	1001	0011
4	–	–	1111

that, which means the full use of carry-propagate adders (very slow) or a complex approximation if carry-save adders are used. The situation is very different here.

In Eq. 10.26, a is a polynomial of degree $m - 1$, as is $b_{m-i-1}a$, because b_{m-i-1} is 0 or 1. And αZ_i is a polynomial of degree m. Therefore, $u \triangleq \alpha Z_i + b_{m-i-1}a$ is a polynomial of degree m. So, to determine whether a reduction is necessary, it suffices to check if bit m in the representation of Z_i is 1. If a reduction is necessary, then it is carried out by subtracting the reduction polynomial, i.e., adding it, since in $GF(2^m)$ the negation of the subtrahend is the subtrahend itself.

Figure 10.3 shows an architecture for the implementation of the algorithm. The b register is a shift register whose contents are shifted left by one bit-position in each cycle; the Z register is initialized to zero; αZ_i is formed by a wired left shift of one bit-position. Each multiple $b_{m-i-1}a$ is formed in a set of AND gates whose outputs are either zero or a. The rest of the diagram is self-explanatory.

Beyond the use of carry-save adders, ordinary multiplication is speeded up in two main ways: using an implementation radix larger than two—i.e., taking several bits of the multiplier operand at one go—and having parallel additions of multiplicand multiples (Sects. 1.2.2 and 1.2.3). These techniques can be used here too—and with greater simplicity because addition is simpler. On the basis that α plays a similar role

Fig. 10.3 Sequential
polynomial-basis multiplier

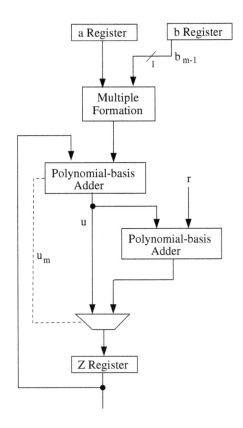

to the radix in ordinary multiplication, all of the algorithms above can be modified easily for "high-radix" multiplication. Moreover, here there are no hard-to-generate multiples of the multiplicand. The following discussion is for the last algorithm above.

Suppose m is divisible by k, so that the multiplier b can be split into $l = m/k$ "pieces." (If that is not the case, then it can be made so by extending b with 0s at the most significant end.) Then the radix-α^k version of the algorithm of Eqs. 10.25–10.27 is

$$Z_0 = 0 \tag{10.29}$$

$$Z_{i+1} = (\alpha^k Z_i + d_{l-i-1}a) \bmod r \qquad i = 0, 1, 2, \ldots, l-1 \tag{10.30}$$

$$\tilde{c} = Z_l \tag{10.31}$$

where the nominal multiplication by α^k denotes a left shift of k bit-positions, and d_{l-i-1} is a k-bit digit of the multiplier operand: $d_j = b_{(j+1)k-1} \cdots b_{jk+1} b_{jk}$.

In ordinary multiplication the use of a radix larger than two poses the problem of "difficult" multiples of the multiplicand—i.e., those that are not powers of two—because of the carry-propagate additions required to form them. The (partial) solution is an on-the-fly recoding of the multiplier into a redundant signed-digit representation. There is no such problem here: addition is a simple operation, without carry propagation, and the multiples can be formed directly.

The "radix-α^2" version of the algorithm of Eqs. 10.25–10.27, which version corresponds to a radix-4 algorithm in ordinary multiplication, is

$$Z_0 = 0 \tag{10.32}$$

$$Z_{i+1} = \left[\alpha^2 Z_i + (b_{jk+1} b_{jk})a \right] \bmod r \qquad i = 0, 1, 2, \ldots, l-1 \tag{10.33}$$

$$c = Z_l \qquad\qquad j = l - i - 1 \tag{10.34}$$

The possible multiplier bit-pairs are 00, 01, 10, 11. The corresponding multiplicand multiples are

$$
\begin{aligned}
00:&\quad 0\\
01:&\quad a\\
10:&\quad \alpha a\\
11:&\quad \alpha a + a
\end{aligned}
$$

(We shall use m_0, m_1, m_2, and m_3 to denote these multiples.)

An example computation is shown in Table 10.5.

In implementing Eqs. 10.32–10.34 there are several possible arrangements for forming the multiplicand multiples formation, adding them, and reducing the intermediate results. We shall describe a straightforward version of the architecture of Fig. 10.3; the reader will find other designs in the literature, e.g., [2, 3].

Table 10.5 Example of high-radix multiplication

\multicolumn{5}{l}{$a(\alpha) = \alpha^5 + \alpha^3 + \alpha^2 + 1 \ a = (101101)$}
\multicolumn{5}{l}{$b(\alpha) = \alpha^4 + \alpha^3 + \alpha^2 + \alpha, \ b = (011110)$}
\multicolumn{5}{l}{$r(\alpha) = \alpha^6 + \alpha + 1$}
\multicolumn{5}{l}{$m = 6, \ k = 2, \ l = 3, \ j = l - i - 1$}
i	$(b_{jk+1}b_j)a$	$(b_{jk+1}b_j)a$	$\alpha^2 Z_i$	Z_i
0	01	101101	000000	000000
1	11	1110011	10110100	101101
2	10	1011010	00011000	000110
3	–	–	–	000001
\multicolumn{5}{l}{$a(\alpha)b(\alpha) \bmod r(\alpha) = 1$}

In modifying the design of Fig. 10.3, two more multiplicand multiples are now required of Multiple Formation unit: m_2 and m_3 above. The former is easily obtained through a wired left shift. The latter requires an addition, which is cheap here; so it can be computed and held in a register before the iterative process starts.

In Fig. 10.3 any required reduction is carried out by the single adder to the right. Here, an intermediate partial product can now be a polynomial of degree up to $m+1$; so a slightly more complex unit is required for reduction. The reduction can be done in two levels of polynomial-basis addition: up to two subtractions (i.e., additions here), one of $\alpha r(\alpha)$ and one of $r(\alpha)$, may be required to divide a polynomial of degree $m+1$ by one of degree m. The possible actions on $w \overset{\triangle}{=} \alpha^2 Z_i + (b_{jk+1}b_{jk})a$ are

- do nothing
- subtract $r(\alpha)$
- subtract $\alpha r(\alpha)$
- subtract $\alpha r(\alpha)$, then subtract $r(\alpha)$

So, here a Reduction Unit computes four values, and the "small" multiplication, $(b_{jk+1}b_{jk})a$, is just a selection of one of four possibilities.

The design of the unit is shown in Fig. 10.4. Let $w_{m+1}w_m$ be the two most significant bits of z, and u_m be the most significant bit the result of subtracting $\alpha r(\alpha)$ from z. Then the multiplexer controls, $c_1 c_0$, are obtained as

$$
\begin{array}{ll}
00: & \overline{w}_{m+1}\overline{w}_m \\
01: & \overline{w}_{m+1}w_m \\
10: & w_{m+1}\overline{u}_m \\
00: & w_{m+1}u_m
\end{array}
$$

Putting together all of the above, we have the architecture of Fig. 10.5. The b Register is a shift register that shifts by two bit-positions in each iteration. The Z register is initialized to zero. $\alpha^2 Z_i$ is computed as a wired left shift, of two bit-positions, of Z_i.

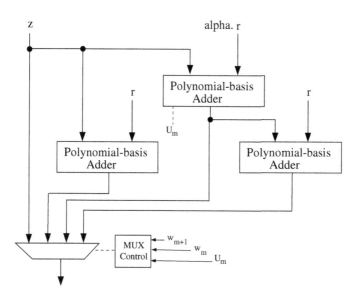

Fig. 10.4 "Radix-4" reduction unit

Fig. 10.5 "Radix-4"
sequential multiplier

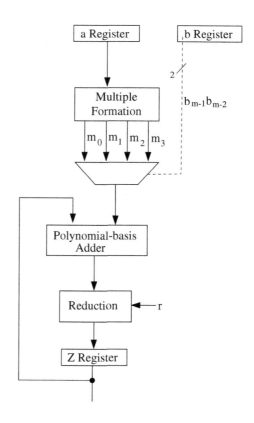

Relative to the design of Fig. 10.3, the delay per iteration has increased: the delay from the primary adder to the Z register is now four gate delays, one up from three. But the factor of increase is smaller than the factor in halving the number of iterations. Also—and this is more significant—the cumulative delay through the Z register—a delay that is substantial relative to that through an adder—is now halved.

The other way in which basic multiplication can be speeded up is by, essentially, unrolling the loop in a sequential algorithm, e.g., Eq. 10.22:

$$Z_{i+1} = (\alpha Z_i + b_{m-i-1}a) \bmod r$$

A straightforward arrangement for loop-unrolling is to have one adder for each multiplicand multiple. Another arrangement is one in which as many multiples as possible are added in parallel. The benefit is that the register delay (Fig. 10.3) is then eliminated. The considerations are exactly the same as in an ordinary multiplier, for which the corresponding architectures are those of Figs. 1.19 and 1.20 (Sect. 1.2.3).

As an example, for six multiples—M_0, M_1, ..., M_5—to be added, with Eq. 10.22, we have the structure of Fig. 10.6, which corresponds to Fig. 1.20. The multiplications by α are taken care of in wired shifts, and reduction is similar to that in Figs. 10.3 and 10.4. And one can similarly devise a design that corresponds to Fig. 1.19.

Figure 10.6 is not practical for very high precision. But the basic idea can be used in a more practical form in a sequential-parallel multiplier that is between the design of Fig. 10.3 and the present one; that is, in a design that corresponds to that of Fig. 1.18. A further enhancement would be to have a large implementation radix (as in Fig. 10.5).

High-performance implementations of multipliers can also be obtained by concurrently employing multipliers of lower precision than the target precision. This method is useful if "pre-built" low-precision multipliers are to be used or if high-precision multiplication is to be effected at reasonable (but not the best of either) performance and cost. The techniques described in Sect. 1.2.4 for ordinary multiplication are equally applicable here. The only notable differences are that the operands are now (representations of) polynomials, addition is a much simpler operation, and powers of the radix are replaced powers of α. The following is for pre-reduction multiplication followed by reduction, but it can be readily applied with interleaved reductions.

If the multiplication is to be of m-bit operands and the multipliers available are $m/2$-bit-by-$m/2$-bit ones, then two operands a and b are each split into two equal parts: $\mathbf{a}_h$ and $\mathbf{a}_l$, and $\mathbf{b}_h$ and $\mathbf{b}_l$:

$$a(\alpha) = \alpha^{m/2}\mathbf{a}_h(\alpha) + \mathbf{a}_l(\alpha) \tag{10.35}$$

$$b(\alpha) = \alpha^{m/2}\mathbf{b}_h(\alpha) + \mathbf{b}_l(\alpha) \tag{10.36}$$

and the product is

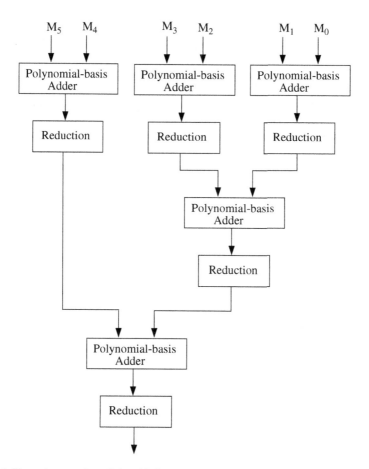

Fig. 10.6 Example core of parallel multiplier

$$\alpha^m \mathbf{a}(\alpha)_h(\alpha)\mathbf{b}(\alpha)_h + \alpha^{m/2}\,[\mathbf{a}_h(\alpha)\mathbf{b}_l(\alpha) + \mathbf{a}_l(\alpha)\mathbf{b}_h(\alpha)] + \mathbf{a}_l(\alpha)\mathbf{b}_l(\alpha) \quad (10.37)$$

$$\overset{\triangle}{=} \mathbf{z}_h(\alpha)\alpha^m + \mathbf{z}_m(\alpha)\alpha^{m/2} + \mathbf{z}_l(\alpha)$$

where the powers of α represent relative shifts, and the additions and multiplications are polynomial operations. (Other operands splittings may be applied similarly, according to the precisions of the multipliers to be used.)

The *Karatsuba–Ofman Algorithm* computes ab from a and b decomposed as in Eqs. 10.35 and 10.36, but with fewer multiplications and more additions than in Eq. 10.39: $\mathbf{z}_m$ is computed as

$$[\mathbf{a}_h(\alpha) + \mathbf{a}_l(\alpha)][\mathbf{b}_h(\alpha) + \mathbf{b}_l(\alpha)] - \mathbf{z}_h(\alpha) - \mathbf{z}_l(\alpha) \quad (10.38)$$

Both types of splittings may be applied "recursively" to produce smaller pieces.

Unlike the corresponding case in ordinary multiplication, where the computation of z_m gives rise to some complications (in implementations with carry-save adders), the simplicity of addition here makes the Karatsuba–Ofman Algorithm quite attractive.

Squaring may be included in multiplication; that is, it may be taken as just another polynomial multiplication followed by a polynomial modular reduction or multiplication with interleaved reductions, as described above. But because the operand is multiplied by itself, in the former case the pre-reduction multiplication can be made more efficient than arbitrary polynomial multiplication, just as is possible with ordinary squaring (Sect. 1.3).

For the multiplication part here:

$$\left(\sum_{i=0}^{m-1} a_i \alpha^i\right)^2 = \sum_{i=0}^{m-1} a_i \alpha^{2i} \tag{10.39}$$

This can be shown by induction. If it is so for $\sum_{i=0}^{n} a_i \alpha^i$, then it is for $\sum_{i=0}^{n+1} a_i \alpha^i$:

$$\left(\sum_{i=0}^{n+1} a_i \alpha^i\right)^2 = \left(a_{n+1}\alpha^{n+1} + \sum_{i=0}^{n} a_i \alpha^i\right)^2$$

$$= a_{n+1}^2 \alpha^{2(n+1)} + 2a_{n+1}\alpha^{n+1} \sum_{i=0}^{n} a_i \alpha^i + \left(\sum_{i=0}^{n} a_i \alpha^i\right)^2$$

$$= a_{n+1}^2 \alpha^{2(n+1)} + \sum_{i=0}^{n} a_i \alpha^{2i} \qquad \text{coefficients mod 2 and by hypothesis}$$

$$= a_{n+1}\alpha^{2(n+1)} + \sum_{i=0}^{n} a_i \alpha^{2i} \qquad a_{n+1}^2 = a_{n+1} \text{ since } a_{n+1} \text{ is 0 or 1}$$

$$= \sum_{i=0}^{n+1} a_i \alpha^{2i}$$

The representation of the result is $(a_{m-1}0a_{m-2}0\cdots 0a_10a_0)$, which is obtained by simply inserting 0s between the bits of the representation of the operand.

As a "confirmatory" an example, the full multiplication array[1] for the squaring $(a_4\alpha^4 + a_3\alpha^3 + a_2x_2 + a_1\alpha + a_0)^2 = a_4\alpha^8 + a_3\alpha^6 + a_2\alpha^4 + a_1\alpha^2 + a_0$ is shown in Fig. 10.7. The result of the full-array addition of multiplicand multiples is as expected.

---

[1] This corresponds Fig. 1.26 for ordinary squaring.

α^8	α^7	α^6	α^5	α^4	α^3	α^2	α	1
				a_4	a_3	a_2	a_1	a_0
				a_4	a_3	a_2	a_1	a_0
				a_4a_0	a_3a_0	a_2a_0	a_1a_0	a_0^2
			a_4a_1	a_3a_1	a_2a_1	a_1^2	a_0a_1	
		a_4a_2	a_3a_2	a_2^2	a_1a_2	a_0a_2		
	a_4a_3	a_3^2	a_2a_3	a_1a_3	a_0a_3			
a_4^2	a_3a_4	a_2a_4	a_1a_4	a_0a_4				
a_4^2	$2a_4a_3+$ a_3^2	$2a_4a_2$ $2a_3a_2$	$2a_4a_1$ $+2a_3a_1$	$2a_4a_0 + a_2^2$ $2a_2a_1$	$2a_3a_0+$	a_1^2	$2a_1a_0$	a_0^2
a_4^2	0	a_3^2	0	a_2^2	0	a_1^2	0	a_0^2
a_4	0	a_3	0	a_2	0	a_1	0	a_0

Fig. 10.7 Pre-reduction multiplication

In sum, the modular squaring of an element $a = (a_{n-1}a_{n-1}\cdots a_1a_0)$ is the computation

$$a(\alpha)^2 = \left(\sum_{i=0}^{n-1} a_i\alpha^i\right)^2 \mod r(\alpha)$$

$$= \left(\sum_{i=0}^{n-1} a_i\alpha^{2i}\right) \mod r(\alpha)$$

There is no real computation in the "multiplication" part. The reduction part is discussed in the next section.

10.2.2 Montgomery Multiplication and Squaring

The integer Montgomery multiplication algorithm computes $z = xyR^{-1} \mod m$, where m^{-1} is the multiplicative inverse of m with respect to R, $-m^{-1}$ is the additive inverse of that, and R^{-1} is the multiplicative inverse of R with respect to m. The algorithm is

$$\tilde{m} = -m^{-1} \tag{10.40}$$

$$u = xy \tag{10.41}$$

$$\tilde{q} = u\tilde{m} \bmod R \tag{10.42}$$

$$\tilde{y} = \frac{u + \tilde{q}m}{R} \tag{10.43}$$

$$y = \begin{cases} \tilde{y} & \text{if } \tilde{y} < m \\ \tilde{y} - m & \text{otherwise} \end{cases} \tag{10.44}$$

With a few changes the preceding algorithm and corresponding multiplication algorithms can be adapted to polynomial computation [1, 6]. (As per the note at the end of the chapter's introduction, we will frequently drop the polynomial indeterminate, writing, for example, a for $a(\alpha)$, etc.)

Let $n(\alpha)$ be an irreducible polynomial in the field, $r(\alpha)$ be an element of the field, $r^{-1}(\alpha)$ be the multiplicative inverse of $r(\alpha)$ with respect to $n(\alpha)$, and $n^{-1}(\alpha)$ be the multiplicative inverse of $n(\alpha)$ with respect to $r(\alpha)$. Specifically, such that:

$$r(\alpha)r^{-1}(\alpha) + n(\alpha)n^{-1}(\alpha) = 1 \tag{10.45}$$

Since $r(\alpha)$ is irreducible, $n(\alpha)$ and $r(\alpha)$ are relatively prime; so the inverses exist.

For the computation of $c = a(\alpha)b(\alpha)r^{-1}(\alpha) \bmod n(\alpha)$ the algorithm obtained from that of Eqs. 10.40–10.44 is

$$\tilde{n} = n^{-1} \tag{10.46}$$

$$u = ab \tag{10.47}$$

$$q = u\tilde{n} \bmod r \tag{10.48}$$

$$c = \frac{u + qn}{r} \tag{10.49}$$

A correction that corresponds to that of Eq. 10.44 is not necessary, as y is a polynomial of degree less than m. This is shown in the following correctness proof.

Example 10.3 Take the fields GF(2^4) and $n(\alpha) = \alpha^4 + \alpha + 1 \stackrel{\triangle}{=} (10011)$. Then $r(\alpha) = \alpha^4, r^{-1}(\alpha) = \alpha^3 + \alpha^2 + \alpha$, and $\tilde{n}(\alpha) = n^{-1}(\alpha) = \alpha^3 + \alpha^2 + \alpha + 1$.

If $a(\alpha) = \alpha^3 + 1 \stackrel{\triangle}{=} (1001)$ and $b(\alpha) = \alpha^3 + \alpha^2 + 1 \stackrel{\triangle}{=} (1101)$, then

$$u(\alpha) = a(\alpha)b(\alpha) = \left(\alpha^3 + \alpha^2 + 1\right)\left(\alpha^3 + 1\right)$$

$$= \alpha^6 + \alpha^5 + \alpha^2 + 1$$

$$q(\alpha) = u(\alpha)\tilde{n}(\alpha)$$

$$= \left(\alpha^6 + \alpha^5 + \alpha^2 + 1\right)\left(\alpha^3 + \alpha^2 + \alpha + 1\right) \bmod \alpha^4$$

$$= \alpha + 1$$

$$c(\alpha) = \frac{\left(\alpha^6 + \alpha^5 + \alpha^2 + 1\right) + (\alpha + 1)\left(\alpha^4 + \alpha + 1\right)}{\alpha^4}$$

$$= \alpha^2 + 1$$

So

$$a(\alpha)b(\alpha)r^{-1}(\alpha) \bmod n(\alpha) = \alpha^2 + 1 \overset{\triangle}{=} (0101)$$

$\square$

Since

$$q = u\tilde{n} \bmod r$$

there is a polynomial $K(\alpha)$ over GF(2) such that

$$q = u\tilde{n} + Kr$$

So, from Eq. 10.49:

$$c = \frac{u + nu\tilde{n} + nKr}{r}$$

From Eq. 10.45:

$$nn^{-1} = 1 + rr^{-1}$$

So, noting that the arithmetic is over GF(2):

$$c = \frac{u + n(1 + rr^{-1}) + nKr}{r}$$

$$= \frac{urr^{-1} + nKr}{r}$$

$$= ur^{-1} + nK$$

$$\equiv ur^{-1} \bmod n$$

$$\equiv abr^{-1} \bmod n$$

Since a and b are each of degree at most $m - 1$, u is of degree at most $2m - 2$. And r and n are each of degree m. Therefore the degree of q is of degree less than m, and the degree of c is

$$\deg(c) \leq \max[\deg(u), \deg(q) + \deg(n)] - \deg(r)$$
$$\leq \max[2m - 2, m - 1 + m] - m$$
$$\leq m - 1$$

In the integer Montgomery algorithm, actual division by R is avoided by picking an R that is a power of the implementation radix: a quotient is obtained simply by discarding some low-order digits of the operand, and a remainder is obtained by discarding some high-order digits. Since α plays a role similar to such a radix, $r(\alpha)$ is taken to be α^m. Then, a quotient (from the/operation) with respect to $r(\alpha)$ is obtained by discarding the terms of the dividend in which the powers of α are lower than m and the corresponding remainder (from the *mod* operation) by discarding the terms of lowers at least m.

That α plays a role similar to that of the radix in ordinary integer multiplication means that algorithms for the latter can be adapted in a straightforward manner for polynomial multiplication. One essentially just takes the former algorithms but interprets the arithmetic operations as polynomial ones. The serial-sequential Montgomery multiplication algorithm for the integer computation of $z = xy2^{-n} \bmod m$ ($R = 2^n$ in Eqs. 10.42–10.43) is

$$Z_0 = 0 \tag{10.50}$$

$$U_i = Z_i + y_i x \qquad i = 0, 1, 2, \ldots, n - 1 \tag{10.51}$$

$$\widetilde{q}_i = u_0 \tag{10.52}$$

$$Z_{i+1} = \frac{U_i + \widetilde{q}_i m}{2} \tag{10.53}$$

$$z = \begin{cases} Z_n & \text{if } Z_n < m \\ Z_n - m & \text{otherwise} \end{cases} \tag{10.54}$$

where y_i is bit i of u, and u_0 is bit 0 of U_i. From this we directly obtain the algorithm for the polynomial computation of $c = a(\alpha)b(\alpha)\alpha^{-m}(\alpha) \bmod n(\alpha)$ ($r = \alpha^m$ in Eqs. 10.47–10.48.)

$$C_0 = 0 \tag{10.55}$$

$$U_i = C_i + ab_i \tag{10.56}$$

$$q_i = u_0 \tag{10.57}$$

$$C_{i+1} = \frac{U_i + q_i n}{\alpha} \qquad i = 0, 1, 2, \ldots, m - 1 \tag{10.58}$$

$$c = C_m \tag{10.59}$$

where u_0 is the least significant bit of U_i. For a larger radix, we would have $q_i = u_0 \widetilde{n}_0 \bmod \alpha$.

Example 10.4 With the same operands, as in Example 10.3—i.e., $a = (1001)$, $b = (1101)$, and $n = (1001)$—the application of the algorithm, which gives the result $c = C_4 = (0101) \overset{\triangle}{=} \alpha^2 + 1$, is as follows.

i	b_i	U_i	q_i	$U_i + q_i n$	C_i
0	1	(1001)	1	(11010)	(0000)
1	0	(1101)	1	(11110)	(1101)
2	1	(0110)	0	(00110)	(1111)
3	1	(1010)	0	(01010)	(0011)
4	–	–	–	–	(0101)

□

An architecture for the algorithm of Eqs. 10.55–10.59 is shown in Fig. 10.8. The y register is a shift register that shifts by one bit-position in each cycle; so y_0 in cycle i is bit i of the operand b. The C register is initialized to zero. And the nominal division by α is effected through a wired shift that drops the least significant bit. As

Fig. 10.8 Montgomery polynomial multiplier

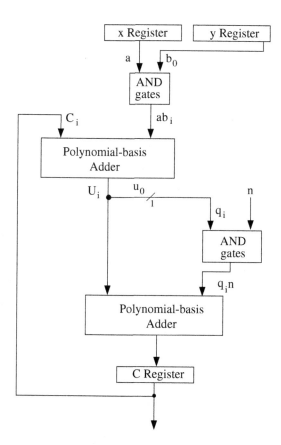

with the Montgomery multiplication algorithm on integers, it is straightforward to devise a "high-radix" version of the algorithm and architecture.

In squaring, the computation is that of $c(\alpha) = [a(\alpha)]^2 \alpha^{-m} \bmod n(\alpha)$. There is, however, no actual multiplication—see Eq. 10.39—so the algorithm of Eqs. 10.55–10.59 may be modified to directly compute the Montgomery reduction of the polynomial of degree $2m - 2$:

$$C_0 = \sum_{i=0}^{m-1} a_i \alpha^{2i}$$

$$C_{i+1} = \frac{C_i + c_0^* n}{\alpha} \qquad\qquad i = 0, 1, 2, \ldots, m - 1$$

$$c = C_m$$

where c_0^* is the least significant bit of C_i.

10.3   Reduction

Two main methods for multiplication are described in Sect. 10.1, and this section is on reduction for both methods. The first method is that in which the partial products are reduced as they are formed. The other is that in which two polynomials of degree up to $m - 1$ each are multiplied to produce an intermediate result of degree at up to $2m - 2$, which polynomial is then reduced to a final result of degree at most $m - 1$. The latter case includes squaring, for which there is no actual multiplication, but the result is a polynomial of degree $2m - 2$. Other cases in which reduction might be required are covered in Sect. 10.3.

Because subtraction (i.e., addition here) is a cheap operation with a polynomial basis, division is far less complex than ordinary division or modular division. Therefore, it is reasonable to consider reduction by direct division. The fundamental techniques used in modular reduction—e.g., in Barrett reduction (Sect. 4.1)—can also be applied here [4].

Sequential multiplication with interleaved reduction may be generalized into a radix-2^k algorithm—i.e., k bits at a time in the multiplier operand—for which the core equation is (Eq. 10.30)

$$z = (\alpha^k Z_i + d_{l-i-1} a) \bmod r \qquad\qquad (10.60)$$

where d_j is the next k-bit digit of the multiplier operand b. z will be of $m + k$ bits, and reduction modulo r can be accomplished by carrying out a reduction in a k-"level" version of the architecture of Fig. 10.4. That will, essentially, be a division of a polynomial of degree $m + k$ by one of degree k, with a very small value of k. A trade-off is to be made that involves the number of levels, the delay per cycle, and reduction in the number of cycles in the multiplier.

In the second type of multiplication algorithm, an intermediate result polynomial of degree up to $2m - 2$ must be reduced to one of degree at most $m - 1$. We next consider two methods for the reduction: direct division and table lookup.

Paper-and-pencil integer division starts with the alignment of the divisor with the most significant of the dividend followed by a sequence of steps, each of which step involves determining a digit of the quotient, subtracting a multiple of the divisor from the partial dividend (initially the dividend and finally the remainder), and shifting the divisor one digit down relative to the partial dividend. For a computer algorithm it is convenient to hold the divisor in place and instead shift the partial dividend, whence the algorithm of Eqs. 1.39–1.45, which is the starting point for the development of algorithms for integer division. By considering that division algorithm and the paper-and-pencil examples of Sect. 7.3, we have the following algorithm for the division of a polynomial $z(\alpha)$ of degree $2m - 2$ by a polynomial $r(\alpha)$ of degree m, with a remainder $R(\alpha)$ of degree less than m.

$$X_1 = z \tag{10.61}$$

$$r^* = \alpha^{m-2} r \tag{10.62}$$

$$q_i = \begin{cases} 1 & \text{if } X_{i,2m-2} = 1 \\ 0 & \text{otherwise} \end{cases} \qquad i = 1, 2, \ldots, m - 1 \tag{10.63}$$

$$X^*_{i+1} = X_i + q_i r^* \tag{10.64}$$

$$X_{i+1} = \alpha X^*_i \tag{10.65}$$

$$R = \alpha^{-(m-2)} X^*_m \tag{10.66}$$

X_i is the i-th partial dividend, and q_i is the i-th quotient bit. r^* is the result of aligning r with the most significant part of the partial dividend; this is a shift by the difference between $2m - 2$ and m. To determine if a partial dividend should be reduced, it is sufficient to examine its most significant bit, denoted $X_{i,2m-2}$. If that bit is a 1, then a subtraction (i.e., an addition here) is to be carried out and an intermediate partial dividend is computed accordingly. The relative shift required between the divisor and the next partial dividend is obtained by shifting the intermediate partial dividend. To compensate for the initial scaling of the divisor, the final partial dividend (which is the remainder from the division) is scaled, by a matching right shift.

An example computation is given in Table 10.6, and an architecture is shown in Fig. 10.9. In implementation the shifts in the preceding description are effected through wiring.

For table-lookup, we shall treat the reduction as a special case of the reduction of a polynomial of degree $n = k + m$, with $k \geq 1$:

$$z(\alpha) = z_n \alpha^n + z_{n-1} \alpha^{n-1} + \cdots + z_m \alpha^m + z_{m-1} \alpha^{m-1} + \cdots + z_1 \alpha + z_0$$

Table 10.6 Example of reduction by division

$z(\alpha) = \alpha^{10} + \alpha^6 + \alpha^5 + \alpha^3 + \alpha + 1$			
$r(\alpha) = \alpha^6 + \alpha^5 + \alpha + 1 \quad m = 2$			
$r^*(\alpha) = \alpha^{10} + \alpha^9 + \alpha^5 + \alpha^4$			
i	q_i	X_i^*	X_i
1	1	$\alpha^{10} + \alpha^6 + \alpha^5 + \alpha^3 + \alpha + 1$ (10001101011)	$-$
2	1	$\alpha^9 + \alpha^6 + \alpha^4 + \alpha^3 + \alpha + 1$ (01001011011)	$\alpha^{10} + \alpha^7 + \alpha^5 + \alpha^4 + \alpha^2 + \alpha$ (10010110110)
3	1	$\alpha^9 + \alpha^7 + \alpha^2 + \alpha$ (01010000110)	$\alpha^{10} + \alpha^8 + \alpha^3 + \alpha^2$ (10100001100)
4	1	$\alpha^9 + \alpha^8 + \alpha^5 + \alpha^4 + \alpha^3 + \alpha^2$ (01100111100)	$\alpha^{10} + \alpha^9 + \alpha^6 + \alpha^5 + \alpha^4 + \alpha^3$ (11001111000)
5	0	$\alpha^6 + \alpha^3$ (00001001000)	$\alpha^7 + \alpha^4$ (00010010000)
6	$-$	$\alpha^7 + \alpha^4$ (00010010000)	$-$
$R(\alpha) = \alpha^{-4}(\alpha^7 + \alpha^4) = \alpha^3 + 1$			

The reduction of $z(\alpha)$ may be done as for Eq. 10.60. But for multiplication and squaring k will be large, and the reduction might as well just be done through direct division. Since that is covered in Sect. 10.3, here we consider a different method. Because of the simplicity of addition here—specifically, the absence of carry propagation—methods and implementations that are problematic with the modular arithmetic of Part II of the text may now be reasonably considered. We next consider the particular case of the use of lookup tables.

The reduction polynomial will be

$$r(\alpha) = \alpha^m + r_{m-1}\alpha^{m-1} + r_{m-2}\alpha^{m-2} + \cdots + r_1\alpha + r_0$$

$$\stackrel{\triangle}{=} \alpha^m + \tilde{r}(\alpha) \tag{10.67}$$

We will make use of the fact that the division of α^m by $r(\alpha)$ produces the quotient 1 and remainder $\tilde{r}(\alpha)$:

$$\alpha^m \bmod r(\alpha) = \tilde{r}(\alpha)$$

Now, with $n = k + m$:

$$z(\alpha) \bmod r(\alpha) = z_n\alpha^n \bmod r(\alpha) + z_{n-1}\alpha^{n-1} \bmod r(\alpha) + \cdots + z_m\alpha^m \bmod r(\alpha)$$

$$+z_{m-1}\alpha^{m-1} + \cdots + z_1\alpha + z_0 \tag{10.68}$$

Fig. 10.9
Reduction-by-division unit

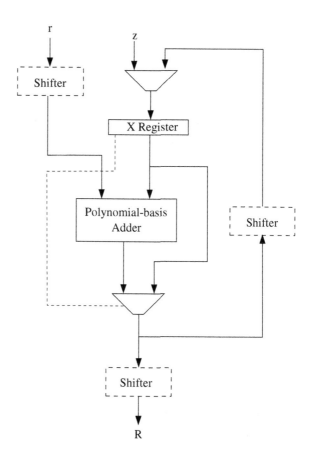

which is a summation of several m-bit values [5]. The number of values might be large, but the additions are very simple.

Example 10.5 Let $z(\alpha) = \alpha^5 + \alpha^4 + \alpha^2 + \alpha + 1$ and $r(\alpha) = \alpha^4 + \alpha^3 + 1$. Then

$$z(\alpha) \bmod r(\alpha) = \alpha^5 \bmod r(\alpha) + \alpha^4 \bmod r(\alpha) + \alpha^2 + \alpha + 1$$

$$= \left(\alpha^3 + \alpha + 1\right) + \left(\alpha^3 + 1\right) + \left(\alpha^2 + \alpha + 1\right)$$

$$\text{by Eq. 10.68}$$

$$= (1011) + (1001) + (0111)$$

$$= (0101)$$

$$= \alpha^2 + 1$$

□

The reduction polynomial typically will be fixed for a chosen field. So, it is efficient to "pre-compute" the (representations of the) polynomials $z_j \alpha^j \bmod r(\alpha)$, store them in a lookup table (LUT), and add them—one at a time or with some parallelism—according to the values of z_j for a given operand. Implementing Eq. 10.68 directly is not very practical; the additions are simple, but there is also the time to read from the LUT. The number of iterations can be reduced by changing from "radix-2"/"radix-α" (one bit at a time in the multiplier operand) to a larger "radix" (several bits at a time). Suppose the operand z is of n bits and the reduction polynomial r is of $m + 1$ bits. And suppose that $n - m$ is divisible by k. (If that is not so, then an appropriate value can be obtained by appending 0s at the most significant end of z.) The essence of the "radix-α^k" algorithm consists of splitting z into $j = (n - m)/k + 1$ "blocks," $\mathbf{z}_i$, and storing all the possible reduced values for each block:

$$z = \mathbf{z}_j \cdots \mathbf{z}_1 \mathbf{z}_0$$

where $\mathbf{z}_0 = z_{m-1} z_{m-1} \cdots z_0$ and $\mathbf{z}_i = z_{m+ik-1} z_{m+ik-2} \cdots z_{m+(i-1)k}$ $(i = 1, 2, \cdots l)$.

That is

$$z(\alpha) = \mathbf{z}_j(\alpha)\alpha^{m+(j-1)k} + \cdots \mathbf{z}_2(\alpha)\alpha^{m+k} + \mathbf{z}_1(\alpha)\alpha^m + \mathbf{z}_0(\alpha)$$

and so

$$z(\alpha) \bmod r(\alpha) = \left[\sum_{i=1}^{j} \left(\mathbf{z}_i(\alpha)\alpha^{m+(i-i)k} \right) \bmod r(\alpha) \right] + \mathbf{z}_0 \qquad (10.69)$$

where $\mathbf{z}_0 = z_{m-1}\alpha^{m-1} + \cdots + z_2\alpha^2 + z_1\alpha + z_0$.

A direct way to implement Eq. 10.69 is to "pre-compute" the terms $\left[\mathbf{z}_i(\alpha)\alpha^{m+(i-1)k} \right] \bmod r(\alpha)$ for the possible different values of $\mathbf{z}_i$, store them (i.e., the corresponding binary values), and then for a given operand make selections from the stored values and add up those.

Example 10.7 Consider Eq. 10.69 with $n = 10$, $m = 4$ and $k = 2$ (so $j = 3$) with $r(\alpha) = \alpha^4 + \alpha + 1$. The lookup table is constructed as shown below. The first two columns give the indices used to address the table, and the values stored are those in the last column.

Now, suppose the operand is $z(\alpha) = \alpha^9 + \alpha^8 + \alpha^7 + \alpha^4 + \alpha^3 + \alpha + 1$ $(n = 10)$. That is, $z = (z_9 z_8 z_7 z_6 z_5 z_4 z_3 z_2 z_1 z_0) = (1110011011)$, which gives $\mathbf{z}_3 = z_9 z_8 = 11$, $\mathbf{z}_2 = z_7 z_6 = 10$, and $\mathbf{z}_1 = z_5 z_4 01$. Then

$$z(\alpha) = \mathbf{z}_3(\alpha)\alpha^8 + \mathbf{z}_2(\alpha)\alpha^6 + \mathbf{z}_1(\alpha)\alpha^4 + \mathbf{z}_0$$

i	z_i $(z_{l+1}z_l)$	$z_i\alpha^l$ $(z_{l+1}\alpha^{l+1}z_l\alpha^l)$	$z_i\alpha^l \bmod r(\alpha)$	
3	00	0	0	(0000)
	01	α^8	$\alpha^2 + 1$	(0100)
	10	α^9	$\alpha^3 + \alpha$	(1001)
	11	$\alpha^9 + \alpha^8$	$\alpha^3 + \alpha^2 + \alpha + 1$	(1111)
2	00	0	0	(0000)
	01	α^6	$\alpha^3 + \alpha^2$	(1100)
	10	α^7	$\alpha^3 + \alpha + 1$	(1101)
	11	$\alpha^7 + \alpha^6$	$\alpha^2 + \alpha + 1$	(0111)
1	00	0	0	(0000)
	01	α^4	$\alpha + 1$	(0011)
	10	α^5	$\alpha^2 + \alpha$	(0110)
	11	$\alpha^5 + \alpha^4$	$\alpha^2 + 1$	(0101)
			$l = (m+)i - 1$	

Using the table and including z_0:

$$z \bmod r = (1111) + (1011) + (0011) + 1011$$

$$= 1100$$

So

$$z(\alpha) \bmod r(\alpha) = \alpha^3 + \alpha^2$$

□

An architecture for the implementation of radix-α^k LUT-based reduction is shown in Fig. 10.10. In each of j cycles, after the first one to add z_0 to a running value that is initially zero, the k least significant bits of the operand register are used as an address to select an entry from the LUT; the value from the LUT is added to the value in the result register; and the contents of the operand register are shifted k places to the right.

10.4 Exponentiation, Inversion, and Division

The task in exponentiation is the computation of $b(\alpha) = a(\alpha)^e \bmod f(\alpha)$ for an n-bit integer, $e \triangleq e_{n-1}e_{n-2}\cdots e_0$ and an irreducible polynomial $f(\alpha)$. Just as the polynomial multiplication algorithms above have been derived from the basic algorithms for integer multiplication, so too can polynomial exponentiation algorithms be derived from those for ordinary integer exponentiation.

Fig. 10.10 LUT-based
reduction unit

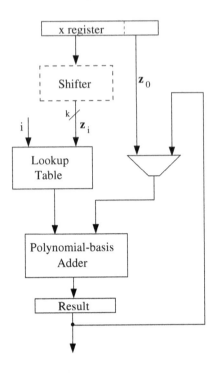

The two standard integer exponentiation algorithms are the two *square-and-multiply* algorithms of Sect. 6.1. One algorithm takes the exponent from the most significant bit to the least significant bit, and the other does the reverse. With a change in variables from those of Sect. 6.1, the algorithms for the integer computation of $b = a^e$ are

$$B_0 = 1 \tag{10.70}$$

$$A_0 = a \tag{10.71}$$

$$B_{i+1} = \begin{cases} B_i A_i & \text{if } e_i = 1 \quad i = 0, 1, 2, \ldots n-1 \\ B_i & \text{otherwise} \end{cases} \tag{10.72}$$

$$A_{i+1} = A_i^2 \tag{10.73}$$

$$b = B_n \tag{10.74}$$

and

$$A_n = 1 \tag{10.75}$$

$$B_{i-1} = \begin{cases} a A_i & \text{if } e_i = 1 \quad i = n, n-1, \ldots, 1 \\ A_i & \text{otherwise} \end{cases} \tag{10.76}$$

$$A_{i-1} = B_i^2 \tag{10.77}$$

$$b = B_0 \tag{10.78}$$

The main practical difference between the two algorithms is that in the two multiplications in each case can be carried out in parallel (at a slightly higher cost in hardware).

The two algorithms can be adapted directly for polynomial exponentiation, simply by interpreting the operands as polynomials and the arithmetic operations as polynomial ones. Then for modular exponentiation either the multiplications (Eqs. 10.72–10.73 and 10.76–10.77) are replaced with modular multiplications or the results B_n and B_0 are reduced. Both will be costly to implement: the former requires reductions in every iterations, and the latter involves increasingly very large intermediate operands followed by a reduction on a very large value. The solution in integer modular exponentiation is to replace the multiplications with Montgomery ones, and such a solution is equally applicable here.

The Montgomery integer multiplication, which will be denoted $\otimes$, of x and y, with respect to R and m is (Sect. 5.2.3)

$$x \otimes y = xyR^{-1} \bmod m$$

where R^{-1} is the multiplicative inverse of R with respect to m.

With the use of Montgomery multiplications, the reductions are interleaved with the multiplications, but the reductions are less costly than with the direct modifications of the two algorithms above. The key in this is to keep the intermediate results in a particular form—the *Montgomery residues* in Chap. 5. (We will use the same terminology here.) Consider the Montgomery multiplication of the Montgomery residues $\widetilde{a(\alpha)} \overset{\triangle}{=} a(\alpha)r(\alpha) \bmod f(\alpha)$ and $\widetilde{b(\alpha)} \overset{\triangle}{=} a(\alpha)r(\alpha) \bmod f(\alpha)$ with respect to the polynomial $f(\alpha)$:

$$\widetilde{a(\alpha)} \otimes \widetilde{a(\alpha)} = a(\alpha)r(\alpha) \bmod f(\alpha) \otimes b(\alpha)r(\alpha) \bmod f(\alpha)$$

$$= [a(\alpha)r(\alpha) \bmod f(\alpha)b(\alpha)r(\alpha) \bmod f(\alpha)]r^{-1} \bmod f(\alpha)$$

$$= a(\alpha)b(\alpha)r(\alpha) \bmod f(\alpha)$$

$$= \widetilde{a(\alpha)b(\alpha)}$$

Thus the product of two Montgomery residues is a Montgomery residue, and if the initial operands are in that form, then the intermediate and final results too will be. The latter must be converted to "ordinary" form, and this can be done through another Montgomery multiplication (with 1 as the other operand):

$$u(\alpha)r(\alpha) \bmod f(\alpha) \otimes 1 = u(\alpha)r(\alpha) * 1 * r^{-1}(\alpha) \bmod f(\alpha)$$

$$= u(\alpha) \bmod f(\alpha)$$

So, for the computation of $b(\alpha) = a(\alpha)^e \bmod r(\alpha)$, we obtain the following algorithms from Eqs. 10.70–10.74 and 10.75–10.78. (For notational clarity we drop the "(α)"s.)

$$B_0 = 1 * r \bmod f \tag{10.79}$$

$$A_0 = a * r \bmod f \tag{10.80}$$

$$B_{i+1} = \begin{cases} B_i \otimes A_i & \text{if } e_i = 1 \quad\quad i = 0, 1, 2, \ldots n-1 \\ B_i & \text{otherwise} \end{cases} \tag{10.81}$$

$$A_{i+1} = A_i \otimes A_i \tag{10.82}$$

$$b = B_n \otimes 1 \tag{10.83}$$

and

$$A_n = 1 * r \bmod f \tag{10.84}$$

$$B_{i-1} = \begin{cases} a \otimes A_i & \text{if } e_i = 1 \quad\quad i = n, n-1, \ldots, 1 \\ A_i & \text{otherwise} \end{cases} \tag{10.85}$$

$$A_{i-1} = B_i \otimes B_i \tag{10.86}$$

$$b = B_0 \tag{10.87}$$

(The squaring is shown as a multiplication, but it should be noted that squaring may be implemented more efficiently, as shown above.)

An example computation with the algorithm of Eqs. 10.79–10.83 is given in Table 10.7. An architecture for the same algorithm is shown in Fig. 10.11. This is a straightforward architecture, and the reader can easily devise variations. For example, A_0 and B_0, which are assumed to be "pre-computed," can also be computed in the primary datapath (with some changes), and it is possible to have a B-path with only one multiplexer. The contents of the e register are shifted right, once in each cycle; so in each cycle e_0 corresponds to e_i in the operand. The rest of the architecture is self-explanatory.

Table 10.7 Example of Montgomery polynomial computation

$e = 11010_2 = 26$			
i	e_i	B_i	A_i
0	0	$r \bmod f$	$ar \bmod f$
1	1	$r \bmod f$	$a^2 r \bmod f$
2	0	$a^2 r \bmod f$	$a^4 r \bmod f$
3	1	$a^2 r \bmod f$	$a^8 r \bmod f$
4	1	$a^{10} r \bmod f$	$a^{16} r \bmod f$
5	–	$a^{26} r \bmod f$	$a^{32} r \bmod f$
$b = [(a^{26} r \bmod f)(1 \bmod f)]r^{-1} \bmod f = a^{26} \bmod f$			

Fig. 10.11 Polynomial
exponentiation unit

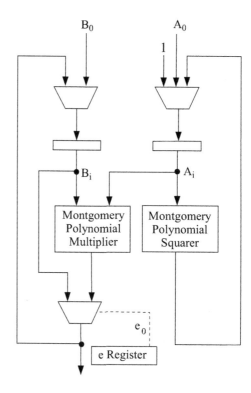

For multiplicative inversion we wish to compute $b(\alpha) = a^{-1}(\alpha) \bmod f(\alpha)$, where $f(\alpha)$ is an irreducible polynomial. And division is just multiplication by an inverse: $a(\alpha)b(\alpha) \bmod f(\alpha) = a(\alpha)b^{-1}(\alpha) \bmod f(\alpha)$.

One way to do this is through exponentiation, on the basis of the following result.[2]

Theorem 10.1 *For every a in GF(q), $a^q = a$.*

With $q = 2^m$, and multiplying both sides of the Eq. 10.88 by a^{-2}, we have

$$a^{-1} = a^{2^m - 2}$$

That is, for polynomial computations:

$$a^{-1}(\alpha) \bmod f(\alpha) = a^{2^m - 2}(\alpha) \bmod f(\alpha)$$

The other standard method for computing integer inverses is the Extended Euclidean Algorithm (Sect. 6.2), which can be readily adapted for polynomial computations. The basis of the integer algorithm is the following result.

---

[2]This may be viewed as a generalization of Fermat's Little Theorem (Theorem 2.1, Section 2.2.2).

Theorem 10.2 (Bezout's Lemma) *If a and b are nonzero integers, then there exist integers x and y such that*

$$\gcd(a, b) = ax + by \tag{10.88}$$

If $\gcd(m, u) = 1$, then

$$mx + uy = 1$$

$$(mx + uy) \bmod m = 1$$

$$uy \bmod m = 1$$

$$y = u^{-1} \bmod m$$

So an algorithm that computes the x and y in Equation 10.88 may be used to compute inverses.

Theorem 10.2 holds if the integers are replaced with polynomials and the arithmetic operations are interpreted as polynomial arithmetic operations. Therefore, with a similar interpretation, the integer algorithms may be directly used for polynomial computations. The integer Extended Euclidean Algorithm is

$$R_0 = a$$
$$R_1 = b$$
$$X_0 = 1$$
$$X_1 = 0$$
$$Y_0 = 0$$
$$Y_1 = 1$$
repeat
$$\qquad Q_i = R_{i-1} \div R_i$$
$$\qquad R_{i+1} = R_{i-1} - Q_i R_i \qquad\qquad i = 1, 2, 3 \ldots,$$
$$\qquad X_{i+1} = X_{i-1} - Q_i X_i$$
$$\qquad Y_{i+1} = Y_{i-1} - Q_i Y_i$$
until $R_{i+1} = 0$

The values computed by the algorithm satisfy the condition (see also Theorem 10.2)

$$\gcd(a, b) = aX_i + bY_i$$

If on termination $R_{n+1} = 0$, then

$$aX_n + bY_n = R_n = \gcd(a, b)$$

For inversion, $\gcd(a, b) = 1$, so:

$$aX_n + bY_n = 1$$

$$bY_n \bmod a = 1$$

$$Y_n = b^{-1} \bmod a$$

To use the preceding algorithm for polynomial computation, we simply suppose that the "(α)" has been dropped from the variable names. That is, R_i is actually $R_i(\alpha)$, a is actually $a(\alpha)$, "$-$" is polynomial subtraction, and so forth. a is the irreducible "modulus" polynomial, and b is the polynomial to be inverted. With that interpretation, an example computation is given in Table 10.8. Note that for the computation of just inverses, the values X_i are not necessary.

The following version of the Extended Euclidean Algorithm does not involve direct division and is therefore more suitable for hardware implementation [7]. "$\deg(\ldots)$" is the degree of $\ldots$, $f(\alpha)$ is an irreducible polynomial, and result is $U_n = a^{-1}(\alpha) \bmod f(\alpha)$ for some n.

An example application of the algorithm is given in Table 10.9. For hardware implementation, the algorithm is better than the preceding one—and the published literature has some hardware designs for such an algorithm—it is far from ideal. The nominal multiplications by α^{δ_i+1} are effected as shifts, but they require actual shifters (or slow shift registers). The degree of a polynomial is determined by shifting the binary representation to locate the leading 1 and then encoding its

Table 10.8 Example of polynomial inversion

$a(\alpha) = \alpha^7 + \alpha^6 + \alpha^3 + \alpha + 1, \ \ b(\alpha) = \alpha^6 + \alpha^4$

i	Q_i	R_i	X_i	Y_i
0	–	$\alpha^7 + \alpha^6 + \alpha^3 + \alpha + 1$	1	0
1	$\alpha + 1$	$\alpha^6 + \alpha^4$	0	1
2	$\alpha + 1$	$\alpha^5 + \alpha^4 + \alpha^3 + \alpha + 1$	1	$\alpha + 1$
3	α	$\alpha^4 + \alpha^3 + \alpha^2 + 1$	$\alpha + 1$	α^2
4	$\alpha^4 + \alpha^3 + \alpha^2 + 1$	1	$\alpha^2 + \alpha + 1$	$\alpha^3 + \alpha + 1$
5	–	0	$\alpha^6 + \alpha^4$	$\alpha^7 + \alpha^6 + \alpha^3 + \alpha + 1$

$\gcd(a(\alpha), b(\alpha)) = R_4 = 1, \ b^{-1}(\alpha) \bmod a(\alpha) = Y_4 = \alpha^3 + \alpha + 1$

Table 10.9 Example of polynomial inversion

$f(\alpha) = \alpha^7 + \alpha^6 + \alpha^3 + \alpha + 1, \ \ a(\alpha) = \alpha^6 + \alpha^4$

i	δ_i^*	δ_i	R_i	S_i	U_i	V_i
0	–	–	$\alpha^6 + \alpha^4$	$\alpha^7 + \alpha^6 + \alpha^3 + \alpha + 1$	1	0
1	1	1	$\alpha^6 + \alpha^4$	$\alpha^6 + \alpha^5 + \alpha^3 + \alpha + 1$	1	α
2	0	0	$\alpha^6 + \alpha^4$	$\alpha^5 + \alpha^4 + \alpha^3 + \alpha + 1$	1	$\alpha + 1$
3	−1	1	$\alpha^5 + \alpha^4 + \alpha^3 + \alpha + 1$	$\alpha^5 + \alpha^2 + \alpha$	$\alpha + 1$	$\alpha^2 + \alpha + 1$
4	0	0	$\alpha^5 + \alpha^4 + \alpha^3 + \alpha + 1$	$\alpha^4 + +\alpha^3 + \alpha^2 + 1$	$\alpha + 1$	α^2
5	−1	1	$\alpha^4 + \alpha^3 + \alpha^2 + 1$	1	α^2	$\alpha^3 + \alpha + 1$
6	−4	4	1	$\alpha^3 + \alpha^2 + 1$	$\alpha^3 + \alpha + 1$	$\alpha^6 + \alpha^3 + \alpha + 1$

$a^{-1}(\alpha) \bmod f(\alpha) = U_6 = \alpha^3 + \alpha + 1$

$$R_0 = a$$
$$S_0 = f$$
$$V_0 = 0$$
$$U_0 = 1$$
while $\deg(R_i) \neq 0$ **do** $i = 0, 1, 2, \ldots$
$\qquad \delta_{i+1}^* = \deg(S_i) - \deg(R_i)$

$\qquad$ **if** $\delta_{i+1}^* < 0$ **then**
$\qquad\qquad (R_{i+1}, S_i) = (S_i, R_i)$
$\qquad\qquad (U_{i+1}, V_i) = (V_i, U_i)$
$\qquad\qquad \delta_{i+1} = -\delta_{i+1}^*$
$\qquad$ **else**
$\qquad\qquad \delta_{i+1} = \delta_{i+1}^*$
$\qquad\qquad R_{i+1} = R_i$
$\qquad\qquad U_{i+1} = U_i$
$\qquad$ **end if**
$\qquad S_{i+1} = S_i + \alpha^{\delta_{i+1}} R_{i+1}$
$\qquad V_{i+1} = V_i + \alpha^{\delta_{i+1}} U_{i+1}$
end while

position; the logic required is costly. Lastly, computing δ_{i+1}^* requires a full length carry-propagate addition, and the computation of its negation is also not without cost.

We next give an algorithm for which we will provide a hardware architecture. Unlike the preceding algorithm, the number of iterations in this third one is fixed but potentially very large. So, in comparing the two algorithms, with respect to hardware implementation, the higher hardware costs and cycle time of the first should be weighed against the number of cycles in the second.

The third algorithm [8, 9]:

$$R_0 = a, \; S_0 = f$$
$$U_0 = 0, \; V_0 = 1$$
$$\delta_0 = 0$$
if $r_m = 0$ **then** $i = 0, 1, 2, \ldots, 2m - 1$
$\qquad R_{i+1} = \alpha R_i$
$\qquad U_{i+1} = \alpha U_i$
$\qquad \delta_{i+1} = \delta_i + 1$
$\qquad S_{i+1} = S_i$
$\qquad V_{i+1} = V_i$
else
$\qquad$ **if** $s_m = 1$ **then**
$\qquad\qquad S_{i+1}^* = S_i - R_i$
$\qquad\qquad V_{i+1}^* = V_i - U_i$
$\qquad$ **else**
$\qquad\qquad S_{i+1}^* = S_i$
$\qquad\qquad V_{i+1}^* = V_i$

end if

$$S_{i+1}^{**} = \alpha S_{i+1}^*$$

if $\delta_i = 0$ **then**

$$R_{i+1} = S_{i+1}^{**}$$
$$S_{i+1} = R_i$$
$$U_{i+1} = \alpha V_{i+1}^*$$
$$V_{i+1} = U_i$$
$$\delta_{i+1} = 1$$

else

$$R_{i+1} = R_i$$
$$S_{i+1} = S_{i+1}^{**}$$
$$V_{i+1} = V_{i+1}^*$$
$$U_{i+1} = U_i/\alpha$$
$$\delta_{i+1} = \delta_i - 1$$

end if

end if

Table 10.10 Example of polynomial inversion

$f(\alpha) = \alpha^7 + \alpha^6 + \alpha^3 + \alpha + 1, \ a(\alpha) = \alpha^6 + \alpha^4$					
i	R_i	S_i	U_i	V_i	δ_i
0	$\alpha^6 + \alpha^4$	$\alpha^7 + \alpha^6 + \alpha^3 + \alpha + 1$	1	0	0
1	$\alpha^7 + \alpha^5$	$\alpha^7 + \alpha^6 + \alpha^3 + \alpha + 1$	α	0	1
2	$\alpha^7 + \alpha^5$	$\alpha^7 + \alpha^6 + \alpha^4 + \alpha^2 + \alpha$	1	α	0
3	$\alpha^7 + \alpha^6 + \alpha^5 + \alpha^3 + \alpha^2$	$\alpha^7 + \alpha^5$	$\alpha^2 + \alpha$	1	1
4	$\alpha^7 + \alpha^6 + \alpha^5 + \alpha^3 + \alpha^2$	$\alpha^7 + \alpha^4 + \alpha^3$	$\alpha + 1$	$\alpha^2 + \alpha + 1$	0
5	$\alpha^7 + \alpha^6 + \alpha^5 + \alpha^3$	$\alpha^7 + \alpha^6 + \alpha^5 + \alpha^3 + \alpha^2$	α^3	$\alpha + 1$	1
6	$\alpha^7 + \alpha^6 + \alpha^5 + \alpha^3$	α^3	α^2	$\alpha^3 + \alpha + 1$	0
7	α^4	$\alpha^7 + \alpha^6 + \alpha^5 + \alpha^3$	$\alpha^4 + \alpha^2 + \alpha$	α^2	1
8	α^5	$\alpha^7 + \alpha^6 + \alpha^5 + \alpha^3$	$\alpha^5 + \alpha^3 + \alpha^2$	α^2	2
9	α^6	$\alpha^7 + \alpha^6 + \alpha^5 + \alpha^3$	$\alpha^6 + \alpha^4 + \alpha^3$	α^2	3
10	α^7	$\alpha^7 + \alpha^6 + \alpha^5 + \alpha^3$	$\alpha^7 + \alpha^5 + \alpha^4$	α^2	4
11	α^7	$\alpha^7 + \alpha^6 + \alpha^4$	$\alpha^6 + \alpha^4 + \alpha^3$	$\alpha^7 + \alpha^5 + \alpha^4 + \alpha^1$	3
12	α^7	$\alpha^7 + \alpha^5$	$\alpha^5 + \alpha^3 + \alpha^2$	$\alpha^7 + \alpha^6 + \alpha^5 + \alpha^3 + \alpha^2$	2
13	α^7	α^6	$\alpha^4 + \alpha^2 + \alpha$	$\alpha^7 + \alpha^6$	1
14	α^7	α^7	$\alpha^3 + \alpha + 1$	$\alpha^7 + \alpha^6$	0
$a^{-1}(\alpha) \bmod f(\alpha) = U_{14} = \alpha^3 + \alpha + 1$					

where r_m denotes bit m in the representation of R_i, and s_m denotes bit m in the representation of S_i. At the end of the iterations, $U_{2m}(\alpha) = a^{-1}(\alpha) \bmod f(\alpha)$. An example computation is given in Table 10.10.

Fig. 10.12 Polynomial
inversion unit

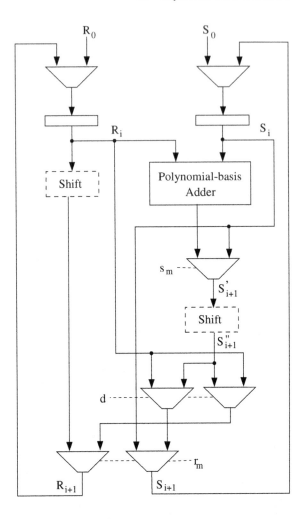

One part of a straightforward architecture for the last algorithm is shown in
Fig. 10.12—for R_i and S_i. The part for U_i and V_i is mostly similar, and its design
is left the reader, as is the part for δ_i. The nominal multiplications and divisions by
α are effected as wired left and right shifts, of a single bit-position each. d is the
signal for $\delta_i = 0$.

The logic to compute δ_i is, in principle, straightforward. But it is problematic, in
that adding 1, subtracting 1, and checking for a zero value will be quite costly (in
both logic and time) relative to the logic for the other parts. The three computations
can be carried out concurrently, but the delay will still be the dominant part of a
cycle.

References

1. S. S. Erdem, T. Yanık, and Çetin K. Koç. 2006. Polynomial basis multiplication over GF(2^m). *Acta Applicandae Mathematica*, 93(1–3):33-55.
2. L. Song and K. K. Parhi. 1998. Low energy digit-serial parallel finite field multipliers. *Journal of VLSI Signal Processing*, 19(2):149–166.
3. J. Guajardo, T. Güneysu, S. S. Kumar, C. Paar, and Pelzl. 2006. Efficient hardware implementation of finite fields with applications to cryptography. *Acta Applicandae Mathematica*, 93(1–3):75–118.
4. J.-F. Dhem 2003. Efficient Modular Reduction Algorithm in $\mathbb{F}[x]$ and its application to "left to r" modular multiplication in $\mathbb{F}_2[x]$. *Proceedings, International Workshop on Cryptographic Hardware and Embedded Systems*, pp. 203–213.
5. D. Hankerson, A. Menezes, and S. Vanstone. 2004. *Guide to Elliptic Curve Cryptography*. Springer-Verlag, New York.
6. C. K. Koc and T. Acar. 1998. Montgomery Multiplication in GF(2^k). *Designs, Codes and Cryptography*, 14(1):57–69.
7. D. Hankerson, J. Hernandez, and A. Menezes. 2000. Software implementation of elliptic curve cryptography over binary fields. *Proceedings, Conference on Cryptographic Hardware and Embedded Systems*, pp. 1–24.
8. H. Brunner, A. Curiger, and M. Hofstetter. 1993. On computing multiplicative inverses in GF(2^m). *IEEE Transactions on Computers*, 42(8):1010–1015.
9. K. Kobayashi, N. Takagi, and K. Takagi. 2007. An algorithm for inversion in GF(2^m) suitable for implementation using a polynomial multiply instruction on GF(2). *Proceedings, 18th IEEE Symposium on Computer Arithmetic*, pp. 105–112.

Chapter 11
Normal-Basis Arithmetic

Abstract This chapter consists of three sections on arithmetic operations in the field $GF(2^m)$ with normal-basis representations and the implementation of those operations. The first section—a short one—is on addition and squaring; both are very simple operations with a normal basis. The second section is on multiplication, a more complicated operation than the preceding two. And the last section is on exponentiation, inversion, and division.

A normal basis here is a set $\{\beta, \beta^2, \beta^4, \ldots, \beta^{2^{m-1}}\}$ of linearly independent elements of $GF(2^m)$. Each element a of $GF(2^m)$ can be expressed uniquely as

$$a = a_0\beta + a_1\beta^2 + a_2\beta^4 + \cdots + a_{m-1}\beta^{2^{m-1}} \qquad a_i \in GF(2)$$

with the binary representation $(a_0a_1a_2\cdots a_{m-1})$. The multiplicative identity element is represented by $(111\cdots 1)$, and the additive identity element is represented by $(00\cdots 0)$. (See Sect. 7.4.)

As an example, Table 11.1 gives the normal-basis representations for the elements of $GF(2^4)$.

For what follows, the operands for the arithmetic operations will be

$$a = (a_0a_1\cdots a_{m-1}) = \sum_{i=0}^{m-1} a_i\beta^{2^i}$$

$$b = (b_0b_1\cdots b_{m-1}) = \sum_{i=0}^{m-1} b_i\beta^{2^i}$$

and the result will be

$$c = (c_0c_1\cdots c_{m-1}) = \sum_{i=0}^{m-1} c_i\beta^{2^i}$$

© Springer Nature Switzerland AG 2020

A. R. Omondi, *Cryptography Arithmetic*, Advances in Information Security 77,
https://doi.org/10.1007/978-3-030-34142-8_11

Table 11.1 Normal-basis
representations of the field
GF(2^4)

(0000) 0	(1000) β
(0001) β^8	(1001) $\beta + \beta^8$
(0010) β^4	(1010) $\beta + \beta^4$
(0011) $\beta^4 + \beta^8$	(1011) $\beta + \beta^4 + \beta^8$
(0100) β^2	(1100) $\beta + \beta^2$
(0101) $\beta^2 + \beta^8$	(1101) $\beta + \beta^2 + \beta^8$
(0110) $\beta^2 + \beta^4$	(1110) $\beta + \beta^2 + \beta^4$
(0111) $\beta^2 + \beta^4 + \beta^8$	(1110) $\beta + \beta^2 + \beta^4 + \beta^8$

Fig. 11.1 Normal-basis
adder

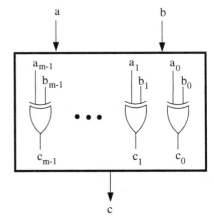

11.1 Addition, Subtraction, and Squaring

Addition is straightforward and very simple: the result $c = a + b$ is given by

$$c_i = (a_i + b_i) \bmod 2 \qquad i = 0, 1, 2, \ldots, m - 1$$

which is just the exclusive-OR operation on a_i and b_i. Thus the implementation is
as shown in Fig. 11.1.

Subtraction is as the addition of the additive inverse of the subtrahend. Here, the
additive inverse of an element is just that element, so subtraction is just the addition
of the subtrahend.

Example 11.1 In GF(2^4), as shown in Table 11.1:

$$\begin{aligned}
(0101) + (1111) &= (\beta^2 + \beta^8) + (\beta + \beta^2 + \beta^4 + \beta^8) \\
&= \beta + 2\beta^2 + \beta^4 + 2\beta^8 \\
&= \beta + \beta^4 \qquad \text{coefficients mod 2} \\
&= (1010)
\end{aligned}$$

$$(1100) - (0101) = (1100) + (0101)$$
$$= (\beta + \beta^2) + (\beta^2 + \beta^8)$$
$$= \beta + \beta^8$$
$$= (1001)$$

□

Squaring may be taken as just an instance of multiplication, i.e., one in which the multiplicand and multiplier operands happen to be the same. In this case, however, squaring can be effected in a way that is much simpler than multiplication, and this has significant implications—e.g., in exponentiation. Two important facts are used in the explanations.

The first is that for any a and b in GF(2^m)

$$(a + b)^2 = a^2 + 2ab + b^2$$
$$= a^2 + b^2 \qquad \text{since } 2ab \bmod 2 = 0. \qquad (11.1)$$

The second is the following theorem.

Theorem 11.1 *If a is an element of GF(q), then*

$$a^q = a \qquad (11.2)$$

Now, each $a_i \beta^{2^i}$ is an element of GF(2^m), and $a_i^2 = a_i$, since a_i is 0 or 1. Therefore:

$$a^2 = (a_0 a_1 \cdots a_{m-1})^2$$

$$\stackrel{\triangle}{=} \left(\sum_{i=0}^{m-1} a_i \beta^{2^i} \right)^2$$

$$= \left(a_0 \beta + a_1 \beta^2 + \cdots a_2 \beta^4 + \cdots + a_{m-1} \beta^{2^{m-1}} \right)^2$$

$$= a_0^2 \beta^2 + a_1^2 \beta^4 + \cdots a_{m-2}^2 \beta^{2^{m-1}} + a_{m-1}^2 \beta^{2^m} \qquad \text{by Eq. 11.1}$$

$$= a_0 \beta^2 + a_1 \beta^4 + \cdots a_{m-2} \beta^{2^{m-1}} + a_{m-1} \beta^{2^m} \qquad \text{since } a_i^2 = a_i$$

$$= a_{m-1} \beta + a_0 \beta^2 + a_1 \beta^4 + \cdots a_{m-2} \beta^{2^{m-1}} \qquad \text{by Eq. 11.2}$$

$$\stackrel{\triangle}{=} (a_{m-1} a_0 a_1 \cdots a_{m-2})$$

Fig. 11.2 Normal-basis
squarer

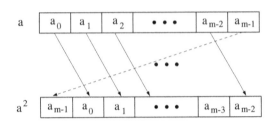

Thus squaring is effected with just a one-place cyclic right rotation of the operand's representation.[1] In implementation, this can be realized easily through appropriate wiring, as shown in Fig. 11.2.

Generalizing, the computation of a^{2^k} is

$$a^{2^k} = \sum_{i=0}^{m-1} a_i \beta^{2^{i+k}}$$

$$= (a_k a_{k+1} \cdots a_{k-1}) \tag{11.3}$$

which is just a cyclic right shift of k places. That this is such a simple computation is very significant in both exponentiation and squaring.

11.2   Multiplication

Multiplication requires a special type of basis—a *Gaussian normal basis*—if it is to be implemented efficiently. We shall first derive a basic and straightforward multiplication algorithm and then explain what such a basis is.

A direct expression for the product $c = a * b$ is

$$c = a * b = \sum_{i=0}^{m-1} a_i \beta^{2^i} \sum_{j=0}^{m-1} b_j \beta^{2^j}$$

$$= \sum_{i=0}^{m-1} \sum_{j=0}^{m-1} a_i b_j \beta^{2^i} \beta^{2^j} \tag{11.4}$$

---

[1]Note that this is another reason why the representation of the multiplicative identity element must be $(11 \cdots 1)$, given that $(00 \cdots 0)$ is already taken for the additive identity element.

$$= \sum_{k=0}^{m-1} c_k \beta^{2^k} \quad \text{for some } c_k \in \text{GF(2)} \tag{11.5}$$

$$= (c_0 c_1 c_2 \cdots c_{m-1}) \tag{11.6}$$

Since $\beta^{2^i} \beta^{2^j} = \beta^{2^{i+j}}$ is an element $\text{GF}(2^m)$, it may be expressed as

$$\beta^{2^i} \beta^{2^j} = \sum_{k=0}^{m-1} \lambda_{i,j}^{(k)} \beta^{2^k} \tag{11.7}$$

for some values $\lambda_{i,j}^{(k)} \in \text{GF(2)}$.

Also, for any integer n:

$$\beta^{2^i} \beta^{2^j} = \left(\beta^{2^{i-n}} \beta^{2^{j-n}} \right)^{2^n}$$

$$= \sum_{k=0}^{m-1} \left(\lambda_{i-n,j-n}^{(k)} \beta^{2^k} \right)^{2^n} \quad \text{by Eq. 11.7}$$

$$= \sum_{k=0}^{m-1} \lambda_{i-n,j-n}^{(k)} \beta^{2^{k+n}}$$

$$= \sum_{k=0}^{m-1} \lambda_{i-n,j-n}^{(k-n)} \beta^{2^k} \tag{11.8}$$

with the indices taken modulo m, since $\beta^{2^m} = \beta$ (by Eq. 11.2). All the indices in what follows are similarly taken modulo m.

Equating coefficients in Eqs. 11.7 and 11.8, we get

$$\lambda_{i,j}^{(k)} = \lambda_{i-n,j-n}^{(k-n)}$$

and in particular, with $k = n$:

$$\lambda_{i,j}^{(k)} = \lambda_{i-k,j-k}^{(0)}$$

Substituting from Eq. 11.8 into Eqs. 11.4 and 11.5:

$$c = \sum_{i=0}^{m-1} \sum_{j=0}^{m-1} a_i b_j \sum_{k=0}^{m-1} \lambda_{i-n,j-n}^{(k-n)} \beta^{2^k}$$

$$= \sum_{k=0}^{m-1} \left(\sum_{i=0}^{m-1} \sum_{j=0}^{m-1} \lambda_{i-n,j-n}^{(k-n)} a_i b_j \right) \beta^{2^k}$$

That is:

$$c_k = \sum_{i=0}^{m-1} \sum_{j=0}^{m-1} \lambda_{i-n,j-n}^{(k-n)} a_i b_j \qquad k = 0, 1, 2, \ldots, m-1 \qquad (11.9)$$

$$= \sum_{i=0}^{m-1} \sum_{j=0}^{m-1} \lambda_{i-k,j-k}^{(0)} a_i b_j \qquad \text{with } n = k \qquad (11.10)$$

$$= \sum_{i=0}^{m-1} \sum_{j=0}^{m-1} \lambda_{i,j}^{(0)} a_{i+k} b_{j+k} \qquad (11.11)$$

with the indices taken modulo m.

We will use $\mathbf{M}$ to denote the matrix in which row i and column j is the $\lambda_{i,j}^{(0)}$ of Eq. 11.11 and refer to this matrix as the *multiplication matrix*. With this notation, the computation of c_k is the vector-matrix-vector computation:

$$c_k = (a_k a_{k+1} \cdots a_{k-1}) \mathbf{M} (b_k b_{k+1} \cdots b_{k-1})^{\mathrm{T}} \qquad k = 0, 1, 2, \ldots, m-1 \qquad (11.12)$$

where $(\cdots)$ are the row vectors associated with the two operands a and b, and "T" denotes transposition.

We next explain a significant practical implication of Eq. 11.12, after first introducing some notation that we shall use again elsewhere. Let $\mathbf{x}$ denote the row vector $(x_0 x_1 \cdots x_{m-1})$ and $S_k(\mathbf{x})$ denote the k-place cyclic right shift of $\mathbf{x}$; so $S_0(\mathbf{x}) = \mathbf{x}$. The computation of c_k may then be expressed as

$$c_k = S_k(\mathbf{a}) \mathbf{M} S_k(\mathbf{b})^{\mathrm{T}} \qquad k = 0, 1, 2, \ldots, m-1 \qquad (11.13)$$

We may think of this as the application of some function F to $S_k(\mathbf{a})$ and $S_k(\mathbf{b})$:

$$c_k = F(S_k(\mathbf{a}), S_k(\mathbf{b})) \qquad k = 0, 1, 2, \ldots, m-1 \qquad (11.14)$$

with

$$F(\mathbf{x}, \mathbf{y}) = \mathbf{x} \mathbf{M} \mathbf{y}^{\mathrm{T}} \qquad (11.15)$$

A complete expression for F is easily obtained from the expression for c_0:

$$c_0 = F(S_0(\mathbf{a}), S_0(\mathbf{b})) = F(\mathbf{a}, \mathbf{b}) \qquad (11.16)$$

(Equation 11.14 constitutes a direct multiplication algorithm that is commonly referred to as the *Massey-Omura Algorithm*, first described in [1].)

The implication of Eqs. 11.13–11.16 for hardware implementation is that a single logic circuit devised for the computation of c_0 (i.e., for F) can, with the inclusion of appropriate logic—including a shifter or shift registers—be used to compute all of $c_0, c_1, c_2, \ldots c_{m-1}$.

The key aspect in all of the preceding is the multiplication matrix **M**, and we now turn to its construction. The construction of $GF(2^m)$ is based on an irreducible polynomial (Sect. 7.4), known in this context as a *normal polynomial*. Let $r(x) = x^m + r_{m-1}x^{m-1} + \cdots + r_2x^2 + r_1x + r_0$ be such a polynomial. The algorithm for **M** is as follows, with all the basic arithmetic done modulo 2 [2].

1. Compute the values $u_{i,j}$ from

$$x \equiv \left(u_{0,0} + u_{0,1}x + u_{0,2}x^2 + \cdots + u_{0,m-1}x^{m-1}\right) \ (\mathrm{mod} \ r(x))$$

$$x^2 \equiv \left(u_{1,0} + u_{1,1}x + u_{1,2}x^2 + \cdots + u_{1,m-1}x^{m-1}\right) \ (\mathrm{mod} \ r(x))$$

$$x^4 \equiv \left(u_{2,0} + u_{2,1}x + u_{2,2}x^2 + \cdots + u_{2,m-1}x^{m-1}\right) \ (\mathrm{mod} \ r(x))$$

$$\cdots$$

$$x^{2^{m-1}} \equiv \left(u_{m-1,0} + u_{m-1,1}x + u_{m-1,2}x^2 + \cdots + u_{m-1,m-1}x^{m-1}\right) \ (\mathrm{mod} \ r(x))$$

and set

$$\mathbf{U} = \begin{pmatrix} u_{0,0} & u_{0,1} & \cdots & u_{0,m-1} \\ u_{1,0} & u_{1,1} & \cdots & u_{1,m-1} \\ \vdots & \vdots & \ddots & \vdots \\ u_{m-1,0} & u_{m-1,1} & \cdots & u_{m-1,m-1} \end{pmatrix}$$

2. Compute the matrix $\mathbf{V} = \mathbf{U}^{-1}$. (If **U** has no inverse, then no normal basis exists with the present polynomial, and the procedure should be restarted with a different polynomial.[2])
3. Set

$$\mathbf{W} = \begin{pmatrix} 0 & 1 & 0 & \cdots & 0 \\ 0 & 0 & 1 & \cdots & 0 \\ \vdots & \vdots & \vdots & \ddots & \vdots \\ 0 & 0 & 0 & \cdots & 1 \\ r_0 & r_1 & r_2 & \cdots & r_{m-1} \end{pmatrix}$$

---

[2]If $m \leq 2000$, then it is not hard to find a suitable polynomial [2].

4. Compute $\mathbf{Z} = \mathbf{U}W V$. (Hereafter $z_{i,j}$ will denote the element in row i and column j of Z.)
5. Set $\mu_{i,j} = z_{j-i,-i}$, where $0 \le i, j \le m - 1$ and with indices taken modulo m and set

$$
\mathbf{M} = \begin{pmatrix}
\mu_{0,0} & \mu_{0,1} & \cdots & \mu_{0,m-1} \\
\mu_{1,0} & \mu_{1,1} & \cdots & \mu_{1,m-1} \\
\vdots & \vdots & \ddots & \vdots \\
\mu_{m-1,0} & \mu_{m-1,1} & \cdots & \mu_{m-1,m-1}
\end{pmatrix}
$$

Example 11.2 With $GF(2^4)$ and the normal polynomial $r(x) = x^4 + x^3 + x^2 + x + 1$, the computation of the multiplication matrix is as follows.

1. The computation of $\mathbf{U}$:

$$x \bmod r(x) = x = (0100)$$

$$x^2 \bmod r(x) = x^2 = (0010)$$

$$x^4 \bmod r(x) = x^3 + x^2 + x + 1 = (1111)$$

$$x^8 \bmod r(x) = x^3 = (0001)$$

So

$$
\mathbf{U} = \begin{pmatrix}
0 & 1 & 0 & 0 \\
0 & 0 & 1 & 0 \\
1 & 1 & 1 & 1 \\
0 & 0 & 0 & 1
\end{pmatrix}
$$

2. Inversion:

$$
\mathbf{V} = \mathbf{U}^{-1} = \begin{pmatrix}
1 & 1 & 1 & 1 \\
1 & 0 & 0 & 0 \\
0 & 1 & 0 & 0 \\
0 & 0 & 0 & 1
\end{pmatrix}
$$

3. The matrix $\mathbf{W}$:

$$
\mathbf{W} = \begin{pmatrix}
0 & 1 & 0 & 0 \\
0 & 0 & 1 & 0 \\
0 & 0 & 0 & 1 \\
1 & 1 & 1 & 1
\end{pmatrix}
$$

4. The matrix $\mathbf{Z}$:

$$\mathbf{Z} = \mathbf{U}W V = \begin{pmatrix} 0 & 1 & 0 & 0 \\ 0 & 0 & 0 & 1 \\ 1 & 1 & 1 & 1 \\ 0 & 0 & 1 & 0 \end{pmatrix}$$

5. The matrix $\mathbf{M}$:

$$\mu_{0,0} = z_{0,0} = 0 \quad \mu_{0,1} = z_{1,0} = 0 \quad \mu_{0,2} = z_{2,0} = 1 \quad \mu_{0,3} = z_{3,0} = 0$$
$$\mu_{1,0} = z_{3,3} = 0 \quad \mu_{1,1} = z_{0,3} = 0 \quad \mu_{1,2} = z_{1,3} = 1 \quad \mu_{1,3} = z_{2,3} = 1$$
$$\mu_{2,0} = z_{2,2} = 1 \quad \mu_{2,1} = z_{3,2} = 1 \quad \mu_{2,2} = z_{0,2} = 0 \quad \mu_{2,3} = z_{1,2} = 0$$
$$\mu_{3,0} = z_{1,1} = 0 \quad \mu_{3,1} = z_{2,1} = 1 \quad \mu_{3,2} = z_{3,1} = 0 \quad \mu_{3,3} = z_{0,1} = 1$$

That is

$$\mathbf{M} = \begin{pmatrix} 0 & 0 & 1 & 0 \\ 0 & 0 & 1 & 1 \\ 1 & 1 & 0 & 0 \\ 0 & 1 & 0 & 1 \end{pmatrix}$$

Now, suppose we wish to multiply $a = (a_0 a_1 a_2 a_3)$ and $b = (b_0 b_1 b_2 b_3)$. Then

$$c_0 = (a_0 \ a_1 \ a_2 \ a_3) \begin{pmatrix} 0 & 0 & 1 & 0 \\ 0 & 0 & 1 & 1 \\ 1 & 1 & 0 & 0 \\ 0 & 1 & 0 & 1 \end{pmatrix} \begin{pmatrix} b_0 \\ b_1 \\ b_2 \\ b_3 \end{pmatrix}$$

$$= a_0 b_2 + a_1(b_2 + b_3) + a_2(b_0 + b_1) + a_3(b_1 + b_3)$$

$$c_1 = (a_1 \ a_2 \ a_3 \ a_0) \begin{pmatrix} 0 & 0 & 1 & 0 \\ 0 & 0 & 1 & 1 \\ 1 & 1 & 0 & 0 \\ 0 & 1 & 0 & 1 \end{pmatrix} \begin{pmatrix} b_1 \\ b_2 \\ b_3 \\ b_0 \end{pmatrix}$$

$$= a_1 b_3 + a_2(b_3 + b_0) + a_3(b_1 + b_2) + a_0(b_2 + b_0)$$

$$c_2 = (a_2 \ a_3 \ a_0 \ a_1) \begin{pmatrix} 0 & 0 & 1 & 0 \\ 0 & 0 & 1 & 1 \\ 1 & 1 & 0 & 0 \\ 0 & 1 & 0 & 1 \end{pmatrix} \begin{pmatrix} b_2 \\ b_3 \\ b_0 \\ b_1 \end{pmatrix}$$

$$= a_2b_0 + a_3(b_0 + b_1) + a_0(b_2 + b_3) + a_1(b_3 + b_1)$$

$$c_3 = (a_3\ a_0\ a_1\ a_2) \begin{pmatrix} 0\ 0\ 1\ 0 \\ 0\ 0\ 1\ 1 \\ 1\ 1\ 0\ 0 \\ 0\ 1\ 0\ 1 \end{pmatrix} \begin{pmatrix} b_3 \\ b_0 \\ b_1 \\ b_2 \end{pmatrix}$$

$$= a_3b_2 + a_0(b_1 + b_2) + a_1(b_3 + b_0) + a_2(b_0 + b_2)$$

Note that we can here obtain the expressions for c_1, c_2, and c_3 in a more direct manner, by applying Eq. 11.16: the equation for c_0 gives

$$F(\mathbf{x}, \mathbf{y}) = x_0y_2 + x_1(y_2 + y_3) + x_2(y_0 + y_1) + x_3(y_1 + y_3)$$

□

Example 11.3 The multiplication of $a = (0100)$ and $b = (1101)$ in the field of Example 11.2:

$$c_0 = 0(0) + 1(0 + 1) + 0(1 + 1) + 0(1 + 1) = 1$$
$$c_1 = 1(1) + 0(1 + 1) + 0(1 + 0) + 0(0 + 1) = 1$$
$$c_2 = 0(1) + 0(1 + 1) + 0(0 + 1) + 1(1 + 1) = 1$$
$$c_3 = 0(1) + 0(1 + 0) + 1(1 + 1) + 0(0 + 1) = 0$$

That is, $c = a * b = (1110) = \beta + \beta^2 + \beta^4$.
In terms of the function F:

$$c_0 = F((0\ 1\ 0\ 0), (1\ 1\ 0\ 1)) = 1$$
$$c_1 = (F(1\ 0\ 0\ 0), (1\ 0\ 1\ 1)) = 1$$
$$c_2 = F((0\ 0\ 0\ 1), (0\ 1\ 1\ 1)) = 1$$
$$c_3 = F((0\ 0\ 1\ 0), (1\ 1\ 1\ 0)) = 0$$

□

From the above, it is apparent that multiplication will be simple to the extent that the elements of the multiplication matrix are 0s. The number of such values is always at least $2m - 1$, and that number may be used as a measure of the complexity of multiplication with respect to a given basis. A basis for which the complexity is exactly $2m - 1$ is known as an *optimal normal basis* [3]. There are two types of optimal normal bases for GF(2^m), designated as *Type I* and *Type II*. A Type I basis exists when

- $m + 1$ is prime
- 2 is a primitive root of $m + 1$

And a Type II basis exists when

- $2m + 1$ is prime
- 2 is a primitive root of $2m + 1$

or

- $2m + 1$ is prime,
- $2m + 1 \equiv 3 \pmod 4$, and
- 2 generates the quadratic residues of $2m + 1$

The multiplication matrices—i.e., the values of $\lambda_{i,j}^{(0)}$ (Eq. 11.11)—for these types of basis are easily produced [3]. For a Type I basis, $\lambda_{i,j}^{(0)} = 1$ if and only if i and j satisfy one of the two conditions:

$$2^i + 2^j \equiv 1 \pmod{m + 1}$$
$$2^i + 2^j \equiv 0 \pmod{m + 1}$$

And for a Type II basis, $\lambda_{i,j}^{(0)} = 1$ if and only if i and j satisfy one of the four conditions:

$$2^i + 2^j \equiv 1 \pmod{2m + 1}$$
$$2^i + 2^j \equiv -1 \pmod{2m + 1}$$
$$2^i - 2^j \equiv 1 \pmod{2m + 1}$$
$$2^i - 2^j \equiv -1 \pmod{2m + 1}$$

The normal polynomial for a Type I basis is $r(x) = x^m + x^{m-1} + \cdots + x + 1$. That for a Type II basis is produced from the following sequence, with modulo-2 arithmetic.

$$r_0(x) = 1$$
$$r_1(x) = x + 1$$
$$r_{i+1}(x) = x r_i(x) + r_{i-1}(x) \qquad i = 1, 2, \ldots, m - 1$$
$$r(x) = r_m(x)$$

These conditions given above imply that there will be values of m for which optimal normal bases do not exist, which indeed is the case: for example, they exist for only 23% of the values of $m < 2000$ [3]. For other values of m, there is a broader class of low-complexity bases—the *Gaussian normal bases*—of which the optimal normal bases are just special cases [4].

A Gaussian normal basis is characterized by an integer T—the *type*—that is a measure of the complexity of multiplication with that basis: the smaller the value of T, the simpler the multiplication. For given m and T, either $GF(2^m)$ has no Gaussian normal basis or it has exactly one of type T. The latter is the case when

- m is not divisible by 8,
- $mT + 1$ is a prime, and
- $\gcd(m, mT/k) = 1$, where k is the multiplicative order of 2 modulo $mT + 1$

The Gaussian normal bases with $T = 1$ and $T = 2$ are exactly the Type I and Type II optimal normal bases.

For examples, Table 11.2 gives the values of m and T in the NIST standards on elliptic-curve cryptography [5].

For a Gaussian normal basis the multiplication matrix $\mathbf{M}$ can be obtained in a more direct and simpler way than through the procedure given above, by directly constructing the function F and then "reading off" $\mathbf{M}$ from that. The details are as follows.

Given m and T such that a Gaussian normal basis exists, let p be $mT + 1$ and u be an element of multiplicative order T modulo p. And let the operands be $(a_0 a_1 \cdots a_{m-1})$ and $(b_0 b_1 \cdots b_{m-1})$ and the result be $(c_0 c_1 \cdots c_{m-1})$. The multiplication function F, which gives the expression for the computation of c_0 (Eq. 11.16), is constructed in two steps. The first step consists of computing the values $f(1), f(2), \ldots, f(p-1)$ as

$$f(2^i u^j \bmod p) = i \qquad i = 0, 1, \ldots, m-1, \quad j = 0, 1, \ldots, T-1$$

The second step consists of formulating the expression for c_0, which gives an expression for F:

$$c_0 = \sum_{k=1}^{p-2} a_{f(k+1)} b_{f(p-k)} \qquad \text{if } T \text{ is even}$$

$$c_0 = \sum_{k=1}^{p-2} a_{f(k+1)} b_{f(p-k)} + \sum_{k=1}^{m/2} a_{k-1} b_{m/2+k-1} + a_{m/2+k-1} b_{k-1} \qquad \text{if } T \text{ is odd}$$

where all indices are taken modulo m. The multiplication matrix $\mathbf{M}$ can be then "read off" directly from the expression for c_0.

Table 11.2 NIST elliptic-curve parameters	m	Type (T)
	163	4
	233	2
	283	6
	409	4
	571	10

Example 11.4 For the Type 3 basis for $GF(2^4)$, $p = 13$, and $u = 2$ has order 12 modulo 13. So

$$f(1) = 0 \quad f(2) = 1 \quad f(3) = 0 \quad f(4) = 2$$
$$f(5) = 1 \quad f(6) = 1 \quad f(7) = 3 \quad f(8) = 3$$
$$f(9) = 0 \quad f(10) = 2 \quad f(11) = 3 \quad f(12) = 2$$

which gives

$$c_0 = a_0(b_1 + b_2 + b_3) + a_1(b_0 + b_2) + a_2(b_0 + b_1) + a_3(b_0 + b_3)$$

and

$$F(\mathbf{x}, \mathbf{y}) = x_0(y_1 + y_2 + y_3) + x_1(y_0 + y_2) + x_2(y_0 + y_1) + x_3(y_0 + y_3)$$

Therefore:

$$c_1 = a_1(b_2 + b_3 + b_0) + a_2(b_1 + b_3) + a_3(b_1 + b_2) + a_0(b_1 + b_0)$$
$$c_2 = a_2(b_3 + b_0 + b_1) + a_3(b_2 + b_0) + a_0(b_2 + b_3) + a_1(b_2 + b_1)$$
$$c_3 = a_3(b_0 + b_1 + b_2) + a_0(b_3 + b_1) + a_1(b_3 + b_0) + a_2(b_3 + b_2)$$

Thus, for example, the product of (1000) and (10101) is (0010):

$$c_0 = F((1, 0, 0, 0), (1, 1, 0, 1)) = 0$$
$$c_1 = (F(0, 0, 0, 1), (1, 0, 1, 1)) = 0$$
$$c_2 = F((0, 0, 1, 0), (0, 1, 1, 1)) = 1$$
$$c_3 = F((0, 1, 0, 0), (1, 1, 1, 0)) = 0$$

□

Implementation

The expression for c_k (Eqs. 11.13–11.14) consists of terms for the multiplication of bit pairs and the addition (modulo 2) of the results of the multiplications. The logic for the multiplication of a bit pair is just an AND gate, and the logic for the modulo-2 addition of two bits is just an XOR gate. Thus the logic to compute c_k consists of just a set of AND gates feeding a tree of XOR gates. Figure 11.3 shows this for the c_0 of Example 11.2.

Fig. 11.3 Example of
F-logic

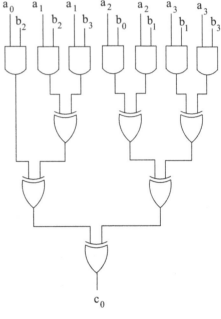

Fig. 11.4 Sequential
normal-basis multiplier

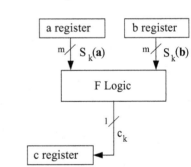

As described above, the logic for c_0, which implements the function F of
Eq. 11.14, is the same logic for all c_k, but with different inputs (the results of the
shift functions S_k) for each c_k. So a sequential multiplier is obtained with a single
F-logic block and shift registers for the operands, as shown in Fig. 11.4. The a and
b registers are shift registers that shift one bit-position to the right in each cycle; the
c register too is a shift register that similarly shifts one bit-position to the left in each
cycle.

A parallel multiplier is obtained by replicating the F-logic, once for each c-
value to be computed. The registers are now "ordinary" registers, and the shifts
are effected through wiring. The architecture is shown in Fig. 11.5. Relative to the
design of Fig. 11.4, the operational time of the registers is no longer a factor in
performance.

Fig. 11.5 Parallel
normal-basis multiplier

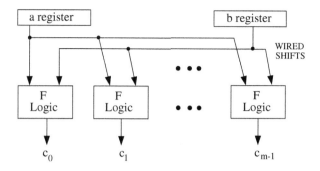

11.3 Exponentiation, Inversion, and Division

The ordinary square-and-multiply algorithm (Sect. 6.1) is equally applicable here
for exponentiation, but with the arithmetic operations now normal-basis ones over
$GF(2^m)$. Indeed, the algorithms in Chap. 6 (Eqs. 6.6–6.10 and Eqs. 6.11–6.14) may
be adopted as they are, with the only change being the interpretation of the
arithmetic operations: addition, multiplication, and squaring of integers now become
addition, multiplication, and squaring of elements of $GF(2^m)$. What is especially
significant here, though, is that normal-basis squaring is practically "for free," as it
involves just a simple one bit-place shift.

Suppose the binary representation of e is $e_{n-1} \cdots e_3 e_2 e_1 e_0$, where $e_i = 0$ or
$e_i = 1, i = 0, 1, 2, \ldots, n-1$); that is, $e = \sum_{i=0}^{n-1} e_i 2^i$. Then

$$a^e = ((\cdots (a^{e_{n-1}})^2 a^{e_{n-2}})^2 \cdots a^{e_3})^2 a^{e_2})^2 a^{e_1})^2 a^{e_0}$$

The corresponding algorithm to compute $b = a^e$ is (see [8])

$$Z_n = 1$$

$$Y_{i-1} = \begin{cases} a Z_i & \text{if } e_i = 1 \quad i = n, n-1, \ldots, 2, 1 \\ Z_i & \text{otherwise} \end{cases}$$

$$Z_{i-1} = Y_i^2$$

$$b = Y_0$$

An example computation is given in Table 11.3, for $GF(2^4)$ as shown in
Table 11.3, and the architecture for an implementation is shown in Fig. 11.6. Squaring is as discussed in Sect. 11.1, from which it should be noted that implementing
the operation is essentially cost-free.

Table 11.3 Example
computation of x^e

$a^e = a^{25} = a^{11001_2}$			
i	e_i	Y_i	Z_i
5	–	–	1
4	1	a	a^2
3	1	$a * a^2 = a^3$	a^6
2	0	a^6	a^{12}
1	0	a^{12}	a^{24}
0	1	$a * a^{24} = a^{25}$	–

Table 11.4 Example
computation of x^e

$a^e = a^{25} = a^{11001_2}$			
i	e_i	Y_i	Z_i
0	1	1	a
1	1	$a * 1 = a$	a^2
2	0	a	a^4
3	0	a	a^8
4	1	$a * a^8 = a^9$	a^{16}
5	–	$a^9 * a^{16} = a^{25}$	–

The algorithm for a right-to-left scan of the exponent bits is

$$Y_0 = 1$$

$$Z_0 = a$$

$$Y_{i+1} = \begin{cases} Y_i Z_i & \text{if } e_i = 1 \\ Y_i & \text{otherwise} \end{cases} \qquad i = 0, 1, 2, \ldots n - 1$$

$$Z_i = Z_i^2$$

$$y = Y_n$$

with an example given in Table 11.4.

Inversion is easily done as exponentiation, and division is just the multiplication of the dividend and the multiplicative inverse of the divisor. The basis for the former is the Theorem 11.1 (Eq. 11.2). For GF(2^m) that gives

$$a^{-1} = a^{2^m - 2}$$

This can be computed using the general exponentiation, but a slightly more efficient method exists for this particular exponent [11].

Since

$$2^m - 2 = 2^1 + 2^2 + 2^3 + \cdots 2^{m-1}$$

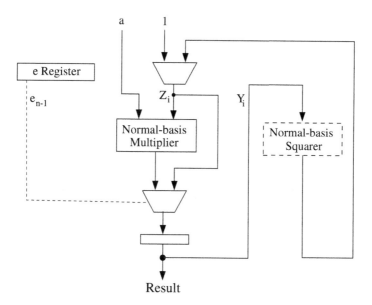

Fig. 11.6 Normal-basis exponentiation unit

we have

$$I \overset{\triangle}{=} a^{2^m-2} = \left(a^{2^1}\right)\left(a^{2^2}\right)\left(a^{2^3}\right)\cdots\left(a^{2^{m-1}}\right)$$

This may be expressed as

$$I = \prod_{i=1}^{m-1} a^{2^i}$$

$$= \left(a * \prod_{i=1}^{m-2} a^{2^i}\right)^2$$

$$= (a * I_{m-1})^2$$

with $I_1 = a^{2^1} = a^2$.

The corresponding algorithm:

$$I_1 = a^2 \tag{11.17}$$

$$I_{i+1} = (a * I_i)^2 \qquad i = 2, \cdots, m-1 \tag{11.18}$$

$$I = I_{m-1} \tag{11.19}$$

Table 11.5 Example
square-and-multiply
computation of a^{2^m-2}

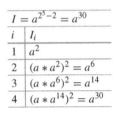

$I = a^{2^5-2} = a^{30}$	
i	I_i
1	a^2
2	$(a * a^2)^2 = a^6$
3	$(a * a^6)^2 = a^{14}$
4	$(a * a^{14})^2 = a^{30}$

Fig. 11.7
Square-and-multiply inverter

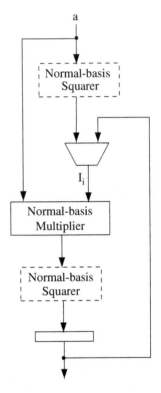

An example computation is given in Table 11.5, and an architecture for implementation is shown in Fig. 11.7. The details of such an implementation are straightforward. The nominal squarings are, of course, just a wiring arrangement.

The *Itoh-Tsujii Algorithm* is a variant of the square-and-multiply algorithm that is generally more efficient than the basic version given above [6]. Noting that

$$a^{2^m-2} = \left(a^{2^{m-1}-1}\right)^2 \tag{11.20}$$

the key in the algorithm is the following decomposition of $2^{m-1} - 1$ [10].

Suppose $m - 1$ is represented in $n \triangleq \lfloor \log_2(m - 1) \rfloor$ bits, $m_{n-1}m_{n-2}m_{n-3}\cdots m_1m_0$, where $m_{n-1} = 1$.

That is,

$$m - 1 = 2^{n-1} + m_{n-2}2^{n-2} + m_{n-3}2^{n-3} + \cdots + m_1 2^1 + m_0 2^0$$

Then

$$2^{m-1} - 1 = 2^{2^{n-1}+m_{n-2}2^{n-2}+m_{n-3}2^{n-3}+\cdots+m_1 2^1+m_0 2^0} - 1$$

$$= 2^{2^{n-1}} 2^{m_{n-2}2^{n-2}+\cdots+m_1 2^1+m_0 2^0}$$

$$-2^{m_{n-2}2^{n-2}+\cdots+m_1 2^1+m_0 2^0}$$

$$+2^{m_{n-2}2^{n-2}+\cdots+m_1 2^1+m_0 2^0}$$

$$-1$$

$$= \left(2^{2^{n-1}} - 1\right) 2^{m_{n-2}2^{n-2}+\cdots+m_1 2^1+m_0 2^0}$$

$$+2^{m_{n-2}2^{n-2}+\cdots+m_1 2^1+m_0 2^0} - 1$$

And since

$$2^{2^{n-1}} - 1 == \left(1 + 2^{2^0}\right)\left(1 + 2^{2^1}\right)\cdots\left(1 + 2^{2^{n-2}}\right)$$

we have

$$2^{m-1} - 1 = \left(1 + 2^{2^{n-2}}\right)\left(1 + 2^{2^{n-3}}\right)\cdots\left(1 + 2^{2^0}\right) 2^{m_{n-2}2^{n-2}+\cdots+m_1 2^1+m_0 2^0}$$

$$+2^{m_{n-2}2^{n-2}+\cdots+m_1 2^1+m_0 2^0} - 1 \tag{11.21}$$

Proceeding as above, the last two terms of this equation are

$$2^{m_{n-2}2^{n-2}+\cdots+m_1 2^1+m_0 2^0} - 1$$

$$= m_{n-2}\left(2^{2^{n-2}-1}\right) 2^{m_{n-3}2^{n-3}+\cdots+m_1 2^1+m_0 2^0}$$

$$+ 2^{m_{n-3}2^{n-3}+\cdots+m_1 2^1+m_0 2^0} - 1$$

Applying a similar reduction recursively in Eq. 11.21:

$$2^{m-1} - 1 = \left(\left(\left(\cdots\left(\left(\left(1 + 2^{2^{n-2}}\right)2^{m_{n-2}2^{n-2}} + m_{n-2}\right)\left(1 + 2^{2^{n-3}}\right)\right.\right.\right.\right.$$

$$2^{m_{n-3}2^{n-3}} + m_{n-3}\right)\cdots\right)\left(1 + 2^{2^2}2^{m_2 2^2}\right) + m_2\right)$$

$$\left(1 + 2^{2^1}\right)2^{m_1 2^1} + m_1\right)\left(1 + 2^{2^0}\right)2^{m_0 2^0} + m_0$$

Therefore:

$$a^{2^{m-1}-1}$$

$$= \left(\left(\left(\cdots\left(\left(a^{(1+2^{2^{n-2})})2^{2^{n-2}}m_{n-2}} \times a^{m_{n-2}}\right)^{(1+2^{2^{n-3}})2^{2^{n-3}}m_{n-3}} \times a^{m_{n-3}}\right)\right.\right.\right.$$

$$\cdots\right)^{(1+2^{2^2})2^{2^2}m_2} \times a^{m_2}\right)^{(1+2^{2^1})2^{2^1}m_1} * a^{m_1}\right)^{(1+2^{2^0})2^{2^0}m_0} \times a^{m_0}$$

from which a squaring yields a^{-1} (Eq. 11.20). The computation requires $\lfloor \log_2(m - 1)\rfloor + H(m - 1) - 1$ multiplications, where $H(m - 1)$ is the number[3] of 1s in the binary representation of $m - 1$.

The computation above is essentially based on a recursive decomposition of $2^{m-1} - 1$:

$$1+2^1+2^2+\cdots 2^{m-1} = \begin{cases} (1 + 2)\left(1 + 2^2 + 2^4 + \cdots + 2^{m-3}\right) & \text{if } m - 1 \text{ is even} \\ 1 + 2(1 + 2)\left(1 + 2^2 + 2^4 + \cdots + 2^{m-4}\right) & \text{if } m - 1 \text{ is odd} \end{cases}$$

As examples, for GF(2^{31}) and GF(2^{244}):

$$1 + 2^1 + \cdots + 2^{29}$$

$$= (1 + 2)\left(1 + 2^2\left(1 + 2^2\right)\left(1 + 2^4\left(1 + 2^4\right)\left(1 + 2^8\left(1 + 2^8\right)\right)\right)\right)$$

$$1 + 2^1 + \cdots + 2^{242}$$

$$= \left(1 + 2(1 + 2)\left(1 + 2^2\left(1 + 2^2\right)\left(1 + 2^4\right)\left(1 + 2^8\right)\right.\right.$$

$$\left(1 + 2^{16}\left(1 + 2^{16}\right)\left(1 + 2^{32}\left(1 + 2^{32}\right)\left(1 + 2^{64}\left(1 + 2^{64}\right)\right)\right)\right)\right)$$

---

[3] The *Hamming weight*.

The algorithm:

$$k_{n-2} = 1 \tag{11.22}$$

$$I_{n-2} = a \tag{11.23}$$

$$\widetilde{I}_{i-1} = I_i * I_i^{2^{k_i}} \qquad i = n-2, n-3, \ldots, 1, 0 \tag{11.24}$$

$$\widetilde{k}_{i-1} = 2k_i \tag{11.25}$$

$$I_{i-1} = \begin{cases} a * \left(\widetilde{I}_{i-1}\right)^2 & \text{if } m_i = 1 \\ \widetilde{I}_{i-1} & \text{otherwise} \end{cases} \tag{11.26}$$

$$k_{i-1} = \begin{cases} \widetilde{k}_{i-1} + 1 & \text{if } m_i = 1 \\ \widetilde{k}_{i-1} & \text{otherwise} \end{cases} \tag{11.27}$$

$$I = I_{-1}^2 \tag{11.28}$$

An example is given in Table 11.6, and an architecture is shown in Fig. 11.8. The squaring is just a wiring arrangement, and the multiplication by two is a left shirt of one bit-position. The 2^k-Power unit is a cyclic shifter that computes $I_i^{2^{k}}$ according to Eq. 11.3.

The basic Itoh-Tsujii algorithm has been generalized through the use of *addition chains* and also somewhat improved in other ways; see, for example, [7, 9, 10]. We next briefly describe the use of addition chains.

Define

$$I_j = a^{2^j - 1}$$

$$I_k = a^{2^k - 1}$$

Table 11.6 Example of Itoh-Tsujii computation of $a^{2^{m-1}-1}$

GF(2^{12}), $m - 1 = 1011_2$, $n - 2 = 2$, $a^{2^{11}-1} = a^{2047}$

i	m_i	$\widetilde{k}_i$	k_i	$\widetilde{I}_i$	I_i
2	0	–	1	–	a
1	1	2	2	$a * a^{2^1} = a^3$	a^3
0	1	4	5	$a^3 * \left(a^3\right)^{2^2} = a^{15}$	$a * \left(a^{15}\right)^2 = a^{31}$
−1	1	4	5	$a^{31} * \left(a^{31}\right)^{2^5} = a^{1023}$	$a * \left(a^{1023}\right)^2 = a^{2047}$

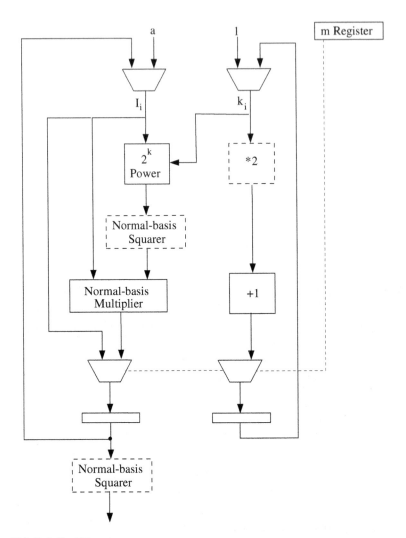

Fig. 11.8 Itoh-Tsujii inverter

then

$$I_{j+k} = a^{2^{j+k}-1}$$

$$= \left(a^{2^{j+k}}a^{2^{-k}}\right)a^{2^k-1}$$

$$= \left(a^{2^j-1}\right)^{2^k}a^{2^k-1}$$

$$= I_j^{2^k} * I_k \qquad\qquad (11.29)$$

So

$$a^{-1} = a^{2^m - 2}$$

$$= \left(a^{2^{m-1}-1}\right)^2$$

$$= I_{m-1}^2$$

Therefore, Eq. 11.28 may be used to compute $I = a^{-1}$ for a given m, by finding sequences of j and k and iterating to end with I_{m-1} whose square yields the sought value:

$$I_1 = a \tag{11.30}$$

$$I_{j+k} = I_j^{2^k} * I_k \tag{11.31}$$

$$I = I_{m-1}^2 \tag{11.32}$$

An example computation, for inversion in $GF(2^{233})$, is given in Table 11.7.

The question, then, is how to obtain the indices j and k in Eq. 11.28. That is where addition chains come in. An *addition chain for an integer n* is a sequence of integers

$$u_0, \ u_1, \ u_2, \ \ldots, \ u_t$$

such that $u_0 = 1$, $u_t = n$, and

Table 11.7 Example of generalized Itoh-Tsujii computation of $a^{2^{m-1}-1}$

j	k	I_{j+k}
–	–	$I_1 = a$
1	1	$I_2 = I_1^{2^1} * I_1 = a^{2^1 - 1}$
2	1	$I_3 = I_2^{2^1} * I_1 = a^{2^3 - 1}$
3	3	$I_6 = I_3^{2^3} * I_3 = a^{2^6 - 1}$
6	1	$I_2 = I_6^{2^1} * I_1 = a^{2^7 - 1}$
7	7	$I_{14} = I_7^{2^7} * I_7 = a^{2^{14} - 1}$
14	14	$I_{28} = I_{14}^{2^7} * I_{14} = a^{2^{28} - 1}$
28	1	$I_{29} = I_{28}^{2^1} * I_1 = a^{2^{29} - 1}$
29	29	$I_{58} = I_{29}^{2^{29}} * I_{29} = a^{2^{58} - 1}$
58	58	$I_{116} = I_{58}^{2^{58}} * I_{59} = a^{2^{116} - 1}$
116	116	$I_{232} = I_{116}^{2^{116}} * I_{116} = a^{2^{232} - 1}$

Table 11.8 Example of an
addition chain

i	u_i
0	$u_0 = 1$
1	$u_1 = u_0 + u_0 = 2$
2	$u_2 = u_1 + u_0 = 3$
3	$u_3 = u_2 + u_2 = 6$
4	$u_4 = u_3 + u_0 = 7$
5	$u_5 = u_4 + u_4 = 14$
6	$u_6 = u_5 + u_5 = 28$
7	$u_7 = u_6 + u_0 = 29$
8	$u_8 = u_7 + u_7 = 58$
9	$u_9 = u_8 + u_8 = 116$
10	$u_{10} = u_9 + u_9 = 232$

$$u_i = u_j + u_k \qquad k \le j < i, \quad i = 1, 2, 3, \ldots, t$$

We further add a constraint that is necessary for most useful addition chains—that

$$u_0 < u_1 < u_2 < \cdots < u_t$$

As an example, Table 11.8 gives addition chain for $n = 232$ and the computation of that chain. (See also Table 11.7.)

Let us suppose that each u_i (except the first) is represented as a pair (u_j, u_k) such that $u_i = u_j + u_k$. Then from Eqs. 11.29–11.31, we obtain the following algorithm for $I = a^{-1}$.

$$I_{u_0} = 1$$

$$I_{u_i} = I_{u_j}^{2^{u_k}} * I_{u_k} \qquad i = 1, 2, 3, \ldots, t$$

$$I = I_{u_t}^2$$

This requires t multiplications; so we want the addition chain to be as short as possible. The general problem of finding the shortest addition chain is difficult (NP-complete) [12]. Nevertheless, good heuristics exist. Moreover, most practical applications involve only a few, known, values of m—for example those in the NIST standard [5]—and for optimal addition chains can be determined for these.

References

1. J. K. Omura and J. L. Massey. 1986. Computational method and apparatus for finite field arithmetic. U. S. Patent 4,587,627. United States Patent Office, Alexandria, Virginia, USA.

2. Institute of Electrical and Electronics Engineers. 2000. P1363: Standard Specifications For Public-Key Cryptography (Annex A). New York, New York, USA.

3. R. C. Mullin, S. A. Vanstone, I. M. Onyszchuk, and R.M. Wilson. 1988–1989. Optimal normal bases in GF(p^n). *Discrete Applied Mathematics*, 22:149–161.

4. D. W. Ash, I. F. Blake, and S. A. Vanstone. 1989. Low complexity normal bases. *Discrete Applied Mathematics*, 25:191–210.

5. National Institute of Standards and Technology. 1999. Recommended Elliptic Curves for Federal Government Use. Gaithersburg, Maryland, USA.

6. T. Itoh and S. Tsujii. 1988. A fast algorithm for computing multiplicative inverses in GF(2^m) using normal basis. *Information and Computation*, 78:171–177.

7. F. Rodríguez-Henriquez, G. Morales-Luna, N. A. Saqib, and N. Cruz-Cortes. 2007. Parallel Itoh–Tsujii multiplicative inversion algorithm for a special class of trinomials. *Designs, Codes and Cryptography*, 45(1):19–37.

8. D. E. Knuth. 1998. *The Art of Computer Programming, Vol. 2*. Addison-Wesley, Reading, Massachusetts, USA.

9. J. Hu, W. Guo, J. Wei, and R. C. C. Cheung. 2015. Fast and generic inversion architectures over GF(2^m) using modified Itoh-Tsujii algorithms. *IEEE Transactions on Circuits and Systems–II: Express Briefs*, 62(4):367–371.

10. N. Takagi, J.-I. Yoshiki, and K. Takagi. 2001. A fast algorithm for multiplicative inversion in GF(2^m) using normal basis. *IEEE Transactions on Computers*, 50(6):394–398.

11. C. C. Wang, T. K. Truong, H. M. Shao, I. J. Deutsch, J. K. Omura, and I. S. Reed. 1985. VLSI architecture for computing multiplications and inverses in GF(2^m). *IEEE Transactions on Computers*, 34(8):709–716.

12. P. Downey, B. Leong, and R. Sethi. 1982. Computing sequences with addition chains. *SIAM Journal of Computing*, 10:638–646.

Appendix A
Mathematical Proofs

This appendix consists of proofs of some of the main mathematical results used in the main text. These proofs will be found, in one form or another, in standard texts, e.g., [1–6]. The main results are numbered as in the main text: "1.1," "1.2," and so forth. Those results that appear only here are numbered "A.1," "A.2," and so forth.

A.1 Part II of Main Text

All variables—$a, b, c, \ldots, x, y, z$—are of integers. A number that is a modulus will be assumed to be greater than one.

The proof of the Theorem 2.1 makes use of Lemma A.1.

Lemma A.1 *If* $kx \equiv ky \pmod{m}$ *and* $\gcd(k, m) = 1$, *then* $x \equiv y \pmod{m}$.

Proof By the definition of congruence, if $kx \equiv ky \pmod{m}$, then $kx - ky = k(x - y)$ is divisible by m. So, either k or $x - y$ is divisible by m. Since $\gcd(k, m) = 1$, it must be $x - y$ that is divisible by m. Therefore $x \equiv y \pmod{m}$.

$\Diamond$

Theorem 2.1 (Fermat's Little Theorem) *If* p *is prime and* $\gcd(a, p) = 1$, *then* $a^{p-1} \equiv 1 \pmod{p}$.

Proof Consider the following $p - 1$ multiples of a.

$$a, 2a, 3a, \ldots, (p - 1)a$$

Since $\gcd(a, p) = 1$, if for some k and j such that $1 \le k < j \le p - 1$, it was the case that

$$ka \equiv ja \pmod{p}$$

© Springer Nature Switzerland AG 2020
A. R. Omondi, *Cryptography Arithmetic*, Advances in Information Security 77,
https://doi.org/10.1007/978-3-030-34142-8

then, by Lemma A.1, we would have

$$k \equiv j \pmod{p}$$

which is impossible, given that $k < j$. Therefore, the multiples above of a are all distinct, and, as none is congruent (modulo p) to zero, they must be congruent (modulo p) to $1, 2, 3, \ldots, p - 1$ in some order. Their product is

$$a * 2a * 3a * \cdots * (p - 1)a \equiv 1 * 2 * 3 * \cdots * (p - 1) \pmod{p}$$

That is,

$$a^{p-1}(p - 1)! \equiv (p - 1)! \pmod{p}$$

Since $\gcd((p - 1)!, p) = 1$, by Lemma A.1

$$a^{p-1} \equiv 1 \pmod{p}$$

$\Diamond$

Corollary 2.1 *If p is prime, then* $a^p \equiv a \pmod{p}$.

Proof If $\gcd(a, p) = 1$, then we have the theorem. And if $\gcd(a, p) \neq 1$, then $a^p \equiv 0 \pmod{p}$ and $a \equiv 0 \pmod{p}$.

$\Diamond$

Theorem 2.2 *If a is of order k modulo m, then n is divisible by k if and only if* $a^n \equiv 1 \pmod{m}$.

Proof Suppose n is divisible by k; i.e., $n = jk$ for some integer k. Then

$$a^n = \left(a^k\right)^j$$
$$\equiv 1^j \pmod{m}$$
$$\equiv 1 \pmod{m}$$

Conversely, suppose $a^n \equiv 1 \pmod{m}$. There are integers q and r such that $n = qk + r$, with $0 \leq r < k$, so:

$$a^n = a^{qk+r}$$
$$= \left(a^k\right)^q a^r$$
$$\equiv a^r \pmod{m}$$
$$\equiv 1 \pmod{m} \qquad \text{by assumption}$$

Since $0 \leq r < k$, and k is the smallest integer such that $a^k \equiv 1 \pmod{m}$, it must be that $r = 0$. Therefore, $n = qk$; i.e., n is divisible by k.

◊

Theorem 2.3 (Euler's Criterion) *If p is an odd prime and $\gcd(a, p) = 1$, then a is a quadratic residue of p if and only if*

$$a^{(p-1)/2} \equiv 1 \pmod{p}$$

Proof Suppose a is a quadratic residue of p. Then there is a b such that $b^2 \equiv a \pmod{p}$. Since $\gcd(a, p) = 1$, we have $\gcd(b, p) = 1$. Therefore:

$$a^{(p-1)/2} \equiv \left(b^2\right)^{(p-1)/2} \pmod{p}$$

$$\equiv b^{p-1} \pmod{p}$$

$$\equiv 1 \pmod{p} \qquad \text{by Fermat's Little Theorem}$$

Conversely, suppose that $a^{(p-1)/2} \equiv 1 \pmod{p}$. Let g be a primitive root of p. Since $a \equiv g^j \pmod{p}$ for some positive integer j:

$$a^{(p-1)/2} \equiv g^{j(p-1)/2} \pmod{p}$$

$$\equiv 1 \pmod{p} \qquad \text{by assumption}$$

Since the order of g is $p - 1$, by Theorem 2.2, $j(p - 1)/2$ is divisible by $p - 1$. Therefore, it must be that $j = 2l$ for some integer l. Now, if $b \equiv g^l \pmod{p}$, then

$$b^2 \equiv g^{2l} \pmod{p}$$

$$\equiv g^j \pmod{p}$$

$$\equiv a$$

which confirms that a is a quadratic residue of p.

◊

Corollary 2.3 *If p is an odd prime and $\gcd(a, p) = 1$, then a is a quadratic nonresidue of p if and only if*

$$a^{(p-1)/2} \equiv -1 \pmod{p}$$

Proof By Fermat's Little Theorem, $a^{p-1} \equiv 1 \pmod{p}$. Therefore $a^{(p-1)/2} \equiv 1 \pmod{p}$ or $a^{(p-1)/2} \equiv -1 \pmod{p}$. The former is the case when a is a quadratic residue, so the latter must be the case when a is a quadratic nonresidue.

◊

In terms of the Legendre symbol, Euler's Criterion and its corollary may be expressed as

$$\left(\frac{a}{p}\right) \equiv a^{(p-1)/2} \pmod{p} \tag{A.1}$$

The proof of Theorem 2.4 makes use of the Lemma A.2.

Lemma A.2 *If* $\gcd(k, m) = 1$, *then* $a \equiv b \pmod{km}$ *if and only if* $a \equiv b \pmod{k}$ *and* $a \equiv b \pmod{m}$.

Proof Suppose $a \equiv b \pmod{km}$. Then, by the definition of congruence, $a - b$ is divisible by km and therefore divisible by k and by m. So, $a \equiv b \pmod{k}$ and $a \equiv b \pmod{m}$.

Conversely, suppose $a \equiv b \pmod{k}$ and $a \equiv b \pmod{m}$. Then, by the definition of congruence, there are i and j such that $a - b = ik$ and $a - b = jm$, which implies that $ik = jm$ and, therefore, that jm is divisible by k. Since $\gcd(k, m) = 1$, it must be the case that j is divisible by k; i.e., there is some l such that $j = lk$. So. $a - b = jm = l(km)$, from which we conclude that $a \equiv b \pmod{km}$.

$\Diamond$

Theorem 2.4 *If* p *and* q *are distinct primes, then* a *is a quadratic residue of* pq *if and only if* a *is a quadratic residue of* p *and* a *is a quadratic residue of* q.

Proof Suppose a is a quadratic residue of pq; i.e., $x^2 \equiv a \pmod{pq}$ has solutions. Then $x^2 - a$ is divisible by pq and therefore divisible by p and by q; that is, $x^2 - a \equiv 0 \pmod{p}$ and $x^2 - a \equiv 0 \pmod{q}$. So, a is a quadratic residue of p and a quadratic residue of q.

Conversely, suppose a is a quadratic residue of p and a quadratic residue of q. Then for some integers x_p and x_q:

$$x_p^2 \equiv a \pmod{p}$$
$$x_q^2 \equiv a \pmod{q}$$

By the Chinese Remainder Theorem—Sect. 2.3, with proof below—there is an x such that

$$x \equiv x_p \pmod{p}$$
$$x \equiv x_q \pmod{q}$$

So:

$$x^2 \equiv a \pmod{p}$$
$$x^2 \equiv a \pmod{q}$$

and by Lemma A.2

$$x^2 \equiv a \ (\text{mod} \ pq)$$

Therefore a is a quadratic residue of pq.

The proof of the Theorem 2.5 makes use of Lemma A.3.

Lemma A.3 *Let p be an odd prime and g be a primitive root of p. Then a quadratic residue of p is congruent modulo p to an even power of g, and a quadratic nonresidue of p is congruent modulo p to an odd power of g.*

Proof We show that a is a quadratic residue of p if and only if a is congruent to an even power of g.

If a is a quadratic residue of p, then for some x_a such that $1 \leq x_a < p$,

$$x_a^2 \equiv a \ (\text{mod} \ p)$$

And for some positive integer k, $x_a = g^k$. So,

$$x_a^2 \equiv g^{2k} \ (\text{mod} \ p)$$

Conversely, $g^2, g^4, g^6, \ldots, g^{p-1}$ are quadratic residues of p: g^j is a solution of $x^2 \equiv g^{2j} \ (\text{mod} \ p)$, for $k = 1, 2, \ldots, p - 1$.

Corollary A.4 *For an odd prime p, half of the elements in $\{1, 2, \ldots, p - 1\}$ are quadratic residues, and half are quadratic non-residues.*

◊

Theorem 2.5 *With respect to a given modulus that is an odd prime:*

(i) *The product of two quadratic residues or two quadratic nonresidues is a quadratic residue.*
(ii) *The product of a quadratic residue and a quadratic nonresidue is a quadratic nonresidue.*

Proof Let g be a primitive root of an odd prime p. By Lemma A.3, if a and b are quadratic residues of p, then

$$\begin{aligned} ab &\equiv g^{2k} g^{2j} \ (\text{mod} \ p) \\ &\equiv g^{2(k+j)} \ (\text{mod} \ p) \end{aligned}$$

for some k and j. And if they are quadraticnonresidues, then

$$ab \equiv g^{2k+1}g^{2j+1} \pmod{p}$$
$$\equiv g^{2(k+j+1)} \pmod{p}$$

for some k and j. The result is an even power of g in both cases.

For a quadratic residue and a quadratic nonresidue we get an odd power of g:

$$ab \equiv g^{2k}g^{2j+1} \pmod{p}$$
$$\equiv g^{2(k+j)+1} \pmod{p}$$

◊

Theorem 2.6 *For an odd prime p:*

(i) $\left(\dfrac{ab}{p}\right) = \left(\dfrac{a}{p}\right)\left(\dfrac{b}{p}\right)$

(ii) $\left(\dfrac{a}{p}\right) = \left(\dfrac{b}{p}\right)$ if $a \equiv b \pmod{p}$

Proof For (i), by Theorem 2.3 and Corollary 2.3:

$$\left(\frac{ab}{p}\right) \equiv (ab)^{(p-1)/2} \pmod{p} \qquad \text{See Eq. A.1}$$

$$\equiv a^{(p-1)/2}b^{(p-1)/2} \pmod{p}$$

$$\equiv \left(\frac{a}{p}\right)\left(\frac{b}{p}\right) \pmod{p}$$

The value of the Legendre symbol is $+1$ or -1. So, were it that

$$\left(\frac{ab}{p}\right) \neq \left(\frac{a}{p}\right)\left(\frac{b}{p}\right)$$

we would have $-1 \equiv 1 \pmod{p}$; i.e., $2 \equiv 0 \pmod{p}$. That is impossible, since p is an odd prime. So it must be that

$$\left(\frac{ab}{p}\right) = \left(\frac{a}{p}\right)\left(\frac{b}{p}\right)$$

For (ii): if $a \equiv b \pmod{p}$, then

$$x^2 \equiv a \pmod{p}$$
$$x^2 \equiv b \pmod{p}$$

either have the same solutions or no solutions at all, whence

$$\left(\frac{a}{p}\right) = \left(\frac{b}{p}\right)$$

◊

The proof of the Chinese Remainder Theorem uses the following two lemmas.

Lemma A.4 (Bezout's Lemma) *If a and b are nonzero, then there exist x and y such that*

$$\gcd(a, b) = ax + by$$

Proof Let d be the smallest integer such that for some integers x and y

$$d = ax + by$$

Now, there are q and r such that

$$a = qd + r \quad 0 \leq r < d$$

Therefore:

$$r = a - qd$$
$$= a - q(ax + by)$$
$$= a(1 - qx) + b(-qy)$$

That is, r is a linear combination of a and b. Since d is the smallest integer that can be expressed in that form, with $0 \leq r < d$, it must be that $r = 0$ and $a = qd$, which means that a is divisible by d. In a similar manner, we can show that b is divisible by d and thus conclude that d is a common divisor of a and b.

To show that d is the greatest common divisor of a and b, suppose there is a greater common divisor, d'. Then for some integers j and k, $a = jd'$ and $b = kd'$, and so

$$d = ax + by$$
$$= (jd')x + (kd')y$$
$$= d'(jx + ky)$$

That is, d is divisible by d', which means that $d' \leq d$. That contradicts the assumption that $d' > d$.

Lemma A.5 *If c is divisible by a and by b and* $\gcd(a, b) = 1$, *then c is divisible by ab.*

Proof If c is divisible by a and b, then there are some j and k such that $c = ja$ and $c = kb$. By Lemma A.4, for some integers x and y:

$$1 = ax + by$$

Multiplying both sides of the last equation by c:

$$
\begin{aligned}
c &= c(ax + by) \\
  &= acx + bcy \\
  &= a(kb)x + b(ja)y \\
  &= ab(kx + jy)
\end{aligned}
$$

Therefore, c is divisible by ab.

◊

Chinese Remainder Theorem *Let* $m_1, m_2, \ldots, m_k$ *and* $a_1, a_2, \ldots, a_n$ *be such that* $0 \le a_i < m_i$ *and* $\gcd(m_i, m_j) = 1$, $i = 1, 2, \ldots, n$, $j = 1, 2, \ldots, n$. *Then the set of equations*

$$x \equiv a_1 \pmod{m_1}$$
$$x \equiv a_2 \pmod{m_2}$$
$$\cdots$$
$$x \equiv a_n \pmod{m_n}$$

has the unique solution

$$x \equiv a_1 \left| M_1^{-1} \right|_{m_1} M_1 + a_2 \left| M_2^{-1} \right|_{m_2} M_2 + \cdots + a_n \left| M_n^{-1} \right|_{m_n} M_n \pmod{M}$$

where

$$M = \prod_{i=1}^{n} m_i$$

$$M_i = \frac{M}{m_i} \quad i = 1, 2, \ldots, n$$

$$\left| M_i^{-1} \right|_{m_i} = \text{ the multiplicative inverse of } M_i \text{ with respect to } m_i$$

Proof Since $\gcd(M_i, m_i) = 1$, the inverses $\left|M_i^{-1}\right|_{m_i}$ exist. If $i = j$, then

$$a_i \left|M_i^{-1}\right|_{m_i} M_i \equiv a_i \pmod{m_i} \quad i = 1, 2, \cdots n$$

And if $i \neq j$, then M_i is divisible by m_j; so,

$$a_i \left|M_i^{-1}\right|_{m_i} M_i \equiv 0 \pmod{m_j} \quad j = 1, 2, \cdots n$$

From these two equations we may conclude that

$$x \equiv a_i \pmod{m_i}$$

To confirm uniqueness, suppose x' is another solution to the set of equations. Then

$$x' \equiv a_i \pmod{m_i} \quad i = 1, 2, \cdots n$$
$$\equiv x \pmod{m_i}$$

Since $x' - x$ is divisible by m_i for each i, by Lemma A.6 it is divisible by M. Therefore

$$x' \equiv x \pmod{M}$$

$\Diamond$

Theorem 3.1 $(a \bmod km) \bmod k = a \bmod k$

Proof Suppose $a \bmod km = b$. Then $b = a - jkm$ for some j, and

$$(a \bmod km) \bmod k = b \bmod k$$
$$= (a - jkm) \bmod k$$
$$= [(a \bmod k) - (jkm \bmod k)] \bmod k$$
$$= a \bmod k$$

$\Diamond$

Theorem 3.2 *If* $\gcd(k, m) = 1$, *then* $a \equiv b \pmod{km}$ *if and only if* $a \equiv b \pmod{k}$ *and* $a \equiv b \pmod{m}$.

Proof See Lemma A.2.

Theorem 3.3 *If* $0 < x < p$, *with* p *prime, and* $a \equiv b \pmod{p-1}$, *then* $x^a \equiv x^b \pmod{p}$.

Proof If $a \equiv b \pmod{p-1}$, then $a = b + k(p-1)$ for some k. So:

$$x^a \bmod p = \left[x^b \left(x^k \right)^{p-1} \right] \bmod p$$

$$= \left[\left(x^b \bmod p \right) \left(\left(x^{p-1} \right)^k \bmod p \right) \right] \bmod p$$

$$= x^b \bmod p \qquad \text{by Fermat's Little Theorem, SINCE } \gcd(x, p) = 1$$

◇

Theorem 3.4 *Let* p *and* q *be primes, with* q *a factor of* $p-1$. *And let* g *be a generator such that* $g^q \bmod p = 1$. *If* $a \equiv b \pmod{q}$, *then* $g^a \bmod p = g^b \bmod p$.

Proof If $a \equiv b \pmod{q}$, then $a = b + kq$ for some k. Therefore

$$g^a \bmod p = \left[g^b \left(g^k \right)^q \right] \bmod p$$

$$= \left[\left(g^b \bmod p \right) \left(g^q \bmod p \right)^k \right] \bmod p$$

$$= g^b \bmod p \qquad \text{since } g^q \bmod p = 1$$

◇

Theorem 6.3 (Bezout's Lemma) *If* a *and* b *are nonzero, then there exist* x *and* y *such that*

$$\gcd(a, b) = ax + by$$

Proof See proof of Lemma A.4.

◇

Corollary 6.3 *If* $\gcd(a, m) = 1$, *then there exists a unique* x—*the* **inverse** *of* a—*such that*

$$ax \equiv 1 \pmod{m}$$

Proof By Lemma A.4, for some x and y,

$$1 = ax + my$$

Therefore,

$$ax - 1 = m(-y)$$

which means that $ax - 1$ is divisible by m, and so $ax \equiv 1 \pmod{m}$.

To show that the inverse is unique, consider any x' such that $ax' \equiv 1 \pmod{m}$: $ax \equiv ax' \pmod{m}$

> iff $(ax - ax')$ is divisible by m
>
> iff $a(x - x')$ is divisible by m
>
> iff $x - x'$ is divisible by m since $\gcd(a, m) = 1$

Therefore, $x \equiv x' \pmod{m}$.

$\Diamond$

The proof of Theorem 6.4 uses Lemma A.6.

Lemma A.6 *If $a = qb + r$, then $\gcd(a, b) = \gcd(b, r)$.*

Proof Suppose $d = \gcd(a, b)$. Then d is a divisor of a and b and, therefore, of $a - qb$; i.e., d is a divisor of r. So, d is a common divisor of b and r. To show that it is the greatest, suppose d' is any divisor of b and r. Then d' is a divisor of $qb + r$. That is, d' is a divisor of a and, therefore, a common divisor of a and b; so $d' \leq d$.

$\Diamond$

Theorem 6.4 *The Euclidean Algorithm (Eqs. 6.44–6.47) computes $R_n = \gcd(a, b)$.*

Proof Note that $R_i \geq R_{i+1} \geq 0$. So, at some point the computation must end with a remainder of zero. And by Lemma A.6:

$$\gcd(a, b) = \gcd(b, R_0) = \gcd(R_1, R_2) = \cdots = \gcd(R_n, R_{n+1}) = \gcd(R_n, 0) = R_n$$

A.2   Part III of Main Text

The proof of Theorem 10.1 uses some preliminary results that we get to via the concept of *coset decomposition*.

Let H, with the set of elements $\{h_1 = 1, h_2, h_3, \ldots, h_m\}$, be a subgroup of a finite group G under the operation $\circ$. One will denote the identity element, and x^{-1} will denote the inverse of x.

The *coset decomposition*[1] of G with respect to H is the array

$$
\begin{array}{ccccc}
h_1 = 1 & h_2 & h_3 & \cdots & h_m \\[1ex]
g_2 \circ h_1 = 1 & g_2 \circ h_2 & g_2 \circ h_3 & \cdots & g_2 \circ h_m \\[1ex]
g_3 \circ h_1 = 1 & g_3 \circ h_2 & g_3 \circ h_3 & \cdots & g_3 \circ h_m \\[1ex]
\vdots & \vdots & \vdots & \vdots & \vdots \\[1ex]
g_n \circ h_1 = 1 & g_n \circ h_2 & g_n \circ h_3 & \cdots & g_n \circ h_m
\end{array}
$$

where the first row consists of the elements of H, with each element appearing exactly once; in the second row, g_2 is an element of G that does not appear in the first row; in the third row, g_3 is an element of G that does not appear in the first two rows; in the fourth row, g_4 is an element of G that does not appear in the first three rows; and so on. The array is finite because G is finite. Each row of the array is a *coset*.

Theorem A.7 *Every element of G appears exactly once in a coset decomposition of G.*

Proof Every element of G appears at least once. Suppose some element appeared twice in the same row, i. Then for some j and k:

$$g_i \circ h_j = g_i \circ h_k$$
$$g_i^{-1} \circ g_i \circ h_j = g_i^{-1} \circ g_i \circ h_k$$
$$h_j = h_k$$

which is not possible if each element of H appears exactly once in the first row.

And suppose that some element appeared in two different rows, k and i, with $k < i$. That is, there are j and l such that

$$g_i \circ h_j = g_k \circ h_l$$

Then:

$$g_i \circ h_j \circ h_j^{-1} = g_k \circ h_l \circ h_j^{-1}$$
$$g_i = g_k \circ \left(h_l \circ h_j^{-1} \right)$$

---

[1] Strictly, the following is a *left-coset* decomposition; a *right-coset* decomposition would be formed with row entries of the form $h_i \circ g_j$.

Since $h_l \circ h_j^{-1}$ is in H, this places g_i in the kth coset. That is not possible, because row i is started by selecting g_i on the basis that it is not in any of the preceding rows.

$\Diamond$

Corollary A.8 *Let H be a subgroup of a finite group G. And let m be the order of H and n be the order of G. Then n is divisible by m.*

Proof From the construction above, it is evident that if the number of cosets is c, then $mc = n$.

$\Diamond$

Lemma A.9 *Let G be a finite group of order n and a be an element of G. If the order of a is m, then n is divisible by m.*

Proof G contains the cyclic subgroup generated by a, and the result follows from Corollary A.8.

$\Diamond$

Theorem 7.1 *For every element a in GF(q):*

$$a^q = a$$

Proof This is evident if $a = 0$. Suppose $a \neq 0$. The nonzero elements of GF(q) form a multiplicative group of order $q - 1$. If k is the order of a in that group, then, by Lemma A.9, $q - 1$ is divisible by k. So:

$$a^q = a^{q-1}a = \left(a^k\right)^{(q-1)/k} a$$
$$= 1^{(q-1)/k} a$$
$$= a$$

$\Diamond$

Corollary 7.1 *If α is a nonzero element of GF(q), then it is a root of $x^{q-1} - 1$.*

$\Diamond$

The proof of Theorem 7.2 uses the following result.

Lemma A.10 *Let k be the order of a in a field. Then $a^n = 1$ if and only if n is divisible by k.*

Proof Suppose $a^n = 1$. Since k is the smallest m such that $a^m = 1$, we have $n \geq k$. Therefore, there are some q and r such that $n = qk + r$, with $0 \leq r < k$. So:

$$\alpha^n = \left(a^k\right)^q a^r$$
$$= 1^q a^r$$
$$= 1 \qquad \text{by assumption}$$

which implies that $r = 0$; i.e., $n = qk$.

Conversely, suppose n is divisible by k; i.e., $n = jk$ for some positive j. Then

$$a^n = \left(a^k\right)^j$$
$$= 1$$

◊

Theorem 7.2 *The nonzero elements of GF(q) form a cyclic group under multiplication.*

Proof We may assume that $q > 2$. Let $p_1^{r_1} p_2^{r_2} \cdots p_m^{r_m}$ be the prime factorization of $h = q - 1$. For every i, $1 \leq i \leq m$, the polynomial $f_i(x) = x^{h/p_i} - 1$ has at most h/p_i roots in GF(q), which means that there is at least one nonzero element of GF(q) that is not a root of $f_i(x)$. Let a_i be such an element and set

$$b_i = a_i^{h/p_i^{r_i}}$$

Then $b_i^{p_i^{r_i}} = a^{q-1} = 1$ and, by Lemma A.10, $p_i^{r_i}$ is divisible by the order of b_i, which order therefore has the form $p_i^{s_i}$ for some s_i such that $0 \leq s_i \leq r_i$.

On the other hand

$$b_i^{p_i^{r_i}-1} = a^{h/p_i} \neq 1$$

Therefore, the order of b_i is exactly $p_i^{r_i}$. We next show that $b = b_1 b_2 \cdots b_m$ is of order $h = q - 1$, and, therefore, the group is cyclic, generated by b.

Suppose, on the contrary, that the order of b is a proper divisor of h and, therefore, a divisor of at least one of the m integers h/p_i, $1 \leq i \leq m$—say, without loss of generality, of h/p_1. Then

$$1 = b^{h/p_1}$$
$$= b_1^{h/p_1} b_2^{h/p_1} \cdots b_m^{h/p_1}$$

For $1 < i \leq m$, h/p_1 is divisible by $p_i^{r_i}$. This means that $b_i^{h/p_1} = 1$, which forces $b_1^{h/p_1} = 1$. This implies that h/p_1 is divisible by the order of b_1, which is impossible, since the order of b_1 is $p_1^{r_1}$. Therefore the group is cyclic, with generator b.

◊

The proof of Theorem 7.3 uses the following result.

Lemma A.11 *Let $f(x)$ be an irreducible polynomial of degree m over GF(p) with roots in GF($p^m - 1$). Then all the roots of $f(x)$ have the same order.*

Proof Let α be a root of $f(x)$ and k be its order. By Corollary 7.1, $a^{p^m-1} = 1$. So, by Lemma A.10, $p^m - 1$ is divisible by k.

If p is even, then k is odd. Also

$$\left(\alpha^k\right)^{2^j} = 1^{2^j}$$
$$= 1$$
$$= \left(\alpha^{2^j}\right)^k$$

Therefore, if l is the order of α^{2^j}, then, by Lemma A.10, k is divisible by l.
And

$$\alpha^{2^j l} = 1$$

which means that, by Lemma A.10, $2^j l$ is divisible by k. Since k is odd, l is divisible by k

So $l = k$.

A similar argument, with 3^j and "even" for "odd," shows that $l = k$ if p is odd.

Theorem 7.3 *A root α in $GF(p^m)$ of a primitive polynomial of degree m over $GF(p)$ is of order $p^m - 1$ and is therefore a primitive element of $GF(p^m)$.*

Proof Let $f(x)$ be a primitive polynomial of degree m over $GF(p)$ and α be a root of $f(x)$. Then,[2] $x^{p^m-1} - 1$ is divisible by $f(x)$. Therefore, α is a root of $x^{p^m-1} - 1$ as well. That is, $\alpha^{p^m-1} = 1$. So, if k is the order of α, then, by Lemma A.10, $p^m - 1$ is divisible by k.

Now, let β be any root of $x^k - 1$. Since $p^m - 1 = jk$ for some positive integer j,

$$\beta^{p^m-1} = \left(\beta^k\right)^j = 1$$

So, β is a root of $x^{p^m-1} - 1$ as well. That is, all the roots of $x^k - 1$ are also roots of $x^{p^m-1} - 1$. Therefore, $x^{p^m-1} - 1$ is divisible by $x^k - 1$.

$f(x)$ is irreducible; so, by Lemma A.11, all its roots have the same order. Therefore, all the roots of $f(x)$ are roots of $x^k - 1$, which means that $x^k - 1$ is divisible by $f(x)$. So, $k \geq p^m - 1$. But $p^m - 1$ is divisible by k; so we must have $k = p^m - 1$.

◊

Lemma A.12 *If a and b are elements of $GF(p^m)$, then for all positive n:*

$$(a + b)^{p^n} = a^{p^n} + b^{p^n}$$

---

[2]Recall the definition of a *primitive polynomial*.

Proof The proof is by induction on n. For $n = 1$:

$$(a + b)^p = \sum_{k=0}^{p} \binom{p}{k} a^k b^{p-k}$$

$$= \sum_{k=0}^{p} \frac{p!}{k!(p-k)!} a^k b^{p-k}$$

$$= a^p + \sum_{k=1}^{p-1} \frac{p!}{k!(p-k)!} a^k b^{p-k} + b^p$$

$p!$ contains a factor of p, but for $1 \le k \le p - 1$ neither $k!$ nor $(p - k)!$ contains a factor of p. Therefore,

$$\binom{p}{k} \equiv 0 \pmod{p} \qquad 1 \le k \le p - 1$$

Therefore, $(a + b)^p = a^p + b^p$.

Now, assume the claim holds for all j such that $1 \le j \le n$. Then it holds for $j + 1$:

$$(a + b)^{p^{j+1}} = \left[(a + b)^p\right]^{p^j}$$

$$= \left(a^p + b^p\right)^{p^j}$$

$$= a^{p^{j+1}} + b^{p^{j+1}}$$

$\Diamond$

Theorem 7.4 *Let β be a root in $GF(p^m)$ of an irreducible polynomial $f(x)$ of degree m over $GF(p)$. Then all the roots of $f(x)$ are*

$$\beta, \beta^p, \beta^{p^2}, \ldots, \beta^{p^{m-1}}$$

Proof Suppose β is a root of $f(x) \overset{\triangle}{=} a_0 + a_1 x + a_2 x^2 + \cdots + a_m x^m$, with $a_i \in GF(p)$, $1 \le i \le m$. Then:

$$g\left(\beta^p\right) = a_0 + a_1 \beta^p + a_2 \beta^{2p} + \cdots + a_m \beta^{mp}$$

$$= a_0^p + a_1^p \beta^p + a_2^p \beta^{2p} + \cdots + a_m^p \beta^{mp} \qquad \text{by Theorem 7.1}$$

$$= \left(a_0 + a_1 \beta^p + a_2 \beta^2 + \cdots + a_m \beta^m \right)^p \qquad \text{by Lemma A.12}$$

$$= [f(\beta)]^p$$

$$= 0$$

And by iterating, we determine that $\beta^{p^2}, \beta^{p^3}, \ldots, \beta^{p^m-1}$ are also roots of $f(x)$.

(Strictly, it is also necessary to show that all the elements of $\{\beta, \beta^p, \beta^{p^2}, \beta^{p^3}, \ldots, \beta^{p^{m-1}}\}$ are distinct, which they are. We omit the details and refer the reader to standard texts—e.g., [4, 6].)

Theorem 10.1 *See Theorem 7.1.*

Theorem 10.2 *See Theorem Lemma A.4.*

Theorem 11.1 *See Theorem 7.1.*

References

1. A. Burton. 2010. *Elementary Number Theory*. McGraw-Hill Education, New York.
2. G. H. Hardy and E. M. Wright. 2008. *An Introduction to the Theory of Numbers*. Oxford University Press, Oxford, U.K.
3. J. B. Fraleigh. 2002. *A First Course in Abstract Algebra.*. Addison-Wesley, Boston, USA.
4. R. Lidl and H. Niederreiter. 1994. *Introduction to Finite Fields and their Applications*. Cambridge University Press, Cambridge, UK.
5. R. E. Blahut. 1983. *Theory and Practice of Error Control Codes*. Addison-Wesley, Reading, Massachusetts, USA.
6. S.B. Wicker. 1995. *Error Control Systems for Digital Communications and Storage*. Prentice Hall, Upper Saddle River, USA.

Index

© Springer Nature Switzerland AG 2020
A. R. Omondi, *Cryptography Arithmetic*, Advances in Information Security 77,
https://doi.org/10.1007/978-3-030-34142-8

Printed in the United States
By Bookmasters